Viruses: Intimate Invaders

Van G. Wilson

Viruses: Intimate Invaders

palgrave
macmillan

Van G. Wilson
Microbial Pathogenesis & Immunology
Texas A&M University College of Medicine
Bryan, TX, USA

ISBN 978-3-030-85486-7 ISBN 978-3-030-85487-4 (eBook)
https://doi.org/10.1007/978-3-030-85487-4

This Springer imprint is published by the registered company Springer Nature Switzerland AG.
The registered company address is: Gewerbestrasse 11, 6330 Cham, Switzerland

This book is dedicated to everyone who ever thought about viruses and wondered how or why?

Preface

This is a viral world: collectively, there are more viruses on Earth than any other life form. The original viruses were evolutionarily ancient and arose at the dawn of life forms on Earth. Now every single species of plants, animals, and bacteria has its cadre of viruses that can attack and infect it. These infections can range from benign, inapparent interactions to debilitating disease or death. During infection, viruses reproduce very quickly and can go through hundreds of generations in the course of a single infection, often acquiring mutations that can change their properties. Many of these viruses can also jump across species to gain new hosts, start new infections, and increase their diversity. Simultaneous co-infection of a host with different viruses can generate new hybrid viruses with unique and sometimes dangerous new characteristics. As a result of these properties, viruses are not static and instead are highly protean entities that are constantly changing and adapting. As viruses evolved, the incessant infections of their host species exerted evolutionary pressure on the hosts to develop anti-viral defenses. This endless conflict of attack and counterattack has driven the evolution of all life on Earth and continues to this day.

Like every other life form, humans are enmeshed in this dynamic and chaotic viral world. We are exposed daily to a continuous onslaught of these invisible marauders. Some infections go unnoticed, some cause disease, and some can even be fatal. Did you ever wonder why the outcomes of infections can be so different, what viruses look like, where they live, how they infect us, how they reproduce, or why antibiotics don't work on viruses? This book will address these questions and many more to examine the nature of viruses and their complex interactions with humans. Any discussion of viruses must start with basic science to explain their structure, function, and origins. Some would call this the "hard" science, but I've tried to make this information accessible to the general reader. With this foundation established, we can explore disease mechanisms and the host immune response to these organisms, as this remarkably complex interplay determines the outcome of each human-virus encounter. While our individual outcomes are critical for each of us, viruses have exerted much larger impacts by affecting whole populations and cultures. Subsequent chapters will present the broader social and historical significance of viruses, followed by a look to the future of how we can use viruses to our benefit in treating other diseases.

I've spent over 40 years of my life studying viruses, and my goal is to convey the wonder, beauty, and power (both good and bad) of these minute creatures. This is their world, and as a species, humans have been lucky to survive through biological good fortune. Beginning in the twentieth century, the development of modern science has finally given us the tools to study, understand, and defend ourselves against many viruses. No longer are we completely at the mercy of chance and our personal genetics. As the SARS-CoV-2 pandemic has shown us, our survival going forward depends now on our intellectual abilities to develop novel strategies to prevent and/or treat viral infections. The old adage to know thy enemy is especially critical for these infectious agents as they continue to change and evolve all around us. I hope this book can answer those questions and along the way educate, inspire, and help prepare the readers to deal with our dangerous and fascinating viral world.

Bryan, TX, USA   Van G. Wilson

Acknowledgment

Any scholarly creation owes a great debt to the multitude of people whose contributions to the author's life, training, and work helped shape the final product. First, I need to thank my mentors: Dr. Edward Fincher at the Georgia Institute of Technology who introduced me to microbiology, Dr. Robert Hogg at Case Western Reserve University who taught me how to do science, and Dr. Peter Tegtmeyer at SUNY Stony Brook who made me a virologist. Next, I am grateful to all the students, trainees, and colleagues who graced my life and my career. These individuals challenged me, inspired me, and taught me so much more than I taught them. Collectively, there is a little bit of every one of these people on each page of this book. Lastly, neither my career nor this book would have occurred without the support, encouragement, and patience of my wonderful, non-scientist wife, Patricia. She read every page of this book several times and never failed to spot the awkward phrase, the discordant or repetitive word, and the poorly explained concept. I know this book is much improved by her helpful and thoughtful comments. My heartfelt thanks to her and all the others who made this book a reality.

Contents

List of Figures

also taken up by endocytosis while others enter via a membrane fusion process. In the latter process, the virion membrane fuses with the cell membrane which releases the nucleocapsid into the cytoplasm. (Created with BioRender.com)   32 and red RNA segments for virion B). The surface proteins of these two parental viruses are antigenically different as represented by the different colors for these proteins. During a mixed infection, the 8 genome segments from each parental virus will be replicated, and the cells will fill with a mixture of many red and blue copies of each segment. As new virions assemble, each progeny virion must package one copy of RNA segments 1–8. For each of these 8 RNA segments packaged there is a random choice between a blue copy and a red copy. Some progeny virions may contain all blue copies or all red copies identical to the parental viruses. However, most virions likely possess a mixture of some red and some blue copies and these are the reassortants. The reassortant shown has 6 blue RNA segments derived from Parental Virus A and 2 red segments derived from Parental Virus B, but every combination of red and blue mixed reassortants is possible. In the example shown, the mixed RNA segments in the reassortant encode surface proteins from each parent virus resulting in a reassortant that is antigenically different from either parent virus. Such changes in reassortants can lead to novel viruses with properties very different from the parental viruses. (Created with BioRender.com)

which generally induces a robust and long-lived B and T cell immune response. However, attenuated vaccines may still cause mild disease in normal recipients and should not be used in immunocompromised patients or pregnant women in whom they might cause serious disease. Subunit vaccines use only a portion of the virion, typically one or more surface proteins that will invoke neutralizing antibodies against the whole virion. The vaccine can consist of purified viral protein or can be the gene (DNA) or mRNA that encodes the target protein. The pathogen gene is often administered by cloning it into a harmless virus to make a chimeric delivery virus while mRNA is more commonly delivered via lipid nanoparticles. Subunit vaccines contain neither the whole virion nor the complete viral genome, therefore, they cannot cause the viral disease. (Created with BioRender.com)

1

The Question of Life

Keywords Virus • DNA • RNA • Protein • Virion • Capsid • Obligate intracellular parasite

> *It's alive! It's alive!*
> *Mel Brook's Young Frankenstein*

Weird Facts and Big Numbers

For all our hubris as the dominant species on the planet, we humans aren't entirely human as we are genetically 5–8% viruses. One of the stunning and remarkable findings from human genome sequencing is that our DNA is not completely our own. Over the evolutionary eons, numerous viruses have invaded our genomes and inserted viral genetic information permanently into our DNA. These accumulated viral DNA sequences, known as endogenous viral elements (EVEs), are scattered throughout our 23 chromosome pairs like weeds in a field of corn. The precise effects and consequences of EVEs in our genomes are still being investigated, but this non-human genetic contribution has impacted our evolutionary history and our very biology (Chap. 9). In addition to EVEs, billions of viruses are everywhere in our world, and because of their prevalence, both our personal genetics and our daily health can be at risk from these invisible marauders. The recent COVID-19 coronavirus outbreak is just the latest example of a novel virus invading the human population, but similar attacks occurred repeatedly during human evolution. Think about that again: We are part virus and we are constantly under attack from

V. G. Wilson, *Viruses: Intimate Invaders*, https://doi.org/10.1007/978-3-030-85487-4_1

new viruses. Given this perpetual onslaught by viruses, we need to know this invader yet how many people really know what a virus is? Everyone knows the word, but beyond those five letters what are viruses really? Where do they come from, what are their properties, and how do they cause disease? These are all questions that modern science can address now, and the answers, even when still incomplete, provide a fascinating insight into our complex relationship with these microscopic agents of disease.

The word virus comes from Latin and means poison, venom, or slimy liquid. Certainly, this is an apt description for a group of infectious agents that have killed, injured, and sickened so many of us over thousands of generations. In addition to invading our genomes, the list of infectious viral diseases is long and has spurred intensive research efforts over the last century to control or eradicate what were once devastating scourges for both individuals and whole populations. While dramatic and common viral diseases such as smallpox, poliomyelitis, and HIV (human immunodeficiency virus) have killed hundreds of millions of us and have helped shape human history (Chap. 6), many other viruses have had much less obvious though incredibly important impacts on humans. Viruses can subtly influence our daily health, can weaken our immune system, can alter our genetics, and can even predispose us to the development of certain cancers. In subsequent chapters we will explore many facets of what viruses are, how they originated, how they reproduce, how they interact with humans, and how they are inexorably intertwined with human biology. But first, we need to know some basic information about these incredible disease agents with which we share our planet.

10^{31}. That's 10 followed by 30 zeros, a number so large that it has no real concept in the human brain. This impossibly large number is one estimate of the number of individual virus particles, called virions, present at any time on the planet earth. While these particles are so small that they are invisible to the human eye, their sheer aggregate number is staggering. To try to grasp and appreciate what 10^{31} denotes, let's examine it in the context of three more visible and meaningful examples of hugely large numbers. First, sand is one of those prodigious substances that are ubiquitous around the world. Miles and miles of beaches and deserts are composed of tiny sand granules made mostly of quartz or calcium carbonate. Estimates of the number of granules of sand comprising all the beaches and deserts on earth vary widely but range only from 10^{19} to 10^{23}. In comparison, this is at least 100 million times less than the estimated number of viruses, so the total of this seemingly endless supply of sand doesn't even come close to the number of virus particles on earth. The number of stars in the universe is another example of something that we often consider so large a number as to seem infinite. Like sand, an estimate of the number of existing stars varies greatly among arguing astrophysicists, from a low of 10^{19} to a high of 10^{24}. Nonetheless, since this range for stars is similar to the range of sand granules, the number of virus particles on earth also greatly exceeds the total number of stars in the entire known universe. Lastly, let's consider the human body composition. Our bodies consist of multiple organs such as the heart, kidneys, and lungs, each composed of millions of microscopic cells that aggregate to form these magnificent, functional organs. There is much tighter agreement on how many cells are present in the average adult human body with a range of only 10^{13}–10^{14}, but this is a paltry and insignificant number compared to the aforementioned examples. With all the complexity and functional variation that our different organs exhibit, our individual cell numbers don't come close to matching the diversity present in even a small fraction of the viral population on our planet. And when you think of the number of individual virus particles compared to the roughly 7.8×10^9 humans alive in 2021, this panoply of viruses dwarfs the number of our meager species! Even considering the totality of cells in all the humans alive today the number of cells still only reaches around 10^{24}, again an amount vastly smaller than the total number of viruses on our planet.

Although the number of virus particles is immense beyond real comprehension, humans only began to recognize and understand their existence in modern times. Because individual virions are so small that they cannot be seen nor felt by humans, their very existence was not appreciated until the late 1800s. The majority of viruses have a roughly spherical shape, and among human viruses, they range in size from parvoviruses with a diameter of 20 nanometers (a nanometer is one-billionth of a meter or about 1 billionth of a yard) to the 200-nanometer poxviruses. If we consider non-human viruses also, this range in size is even greater, going from the 17 nanometer porcine (pig) circovirus to the 1000 nanometer pandoravirus, a recently discovered member of the group of so-called giant viruses. To put these small sizes in perspective, let's consider a typical spherical virion like HIV. With a diameter of about 150 nanometers, the HIV virion is approximately 10 times smaller than the microscopic *E. coli* bacteria, 50 times smaller than a human red blood cell, 500 times smaller than the diameter of an average human hair, and nearly 500,000 times smaller than a tennis ball; no wonder it took humans so long to finally detect these minute particles. Since it's often hard to conceptualize these extraordinarily small sizes, we can try to imagine the converse situation. If every virion on earth was the size of a grain of sand, their aggregate volume would be around 10^{22} cubic feet. By comparison, the volume of the 1454-foot tall Empire State building is only about 40 million (4×10^7) cubic feet. To accommodate the entire 10^{22} cubic feet of viruses the Empire State building would need to be 70 trillion miles tall. This height is nearly three times longer than the distance from Earth to the nearest star outside our solar system, Alpha Centauri! Thank goodness virions are so small or there would be no space left on earth for the rest of us! Still, it is clear that humans share our planet with a huge and diverse community of viruses, and this book will explore ways that this multitude of viruses impact human activity on many levels.

From Ignorance to Fascination

Viral diseases have been with humans as long as *Homo sapiens* have existed, but the written record of viral disease goes back only 3500 years with the depiction on an Egyptian carving of an individual with a foot drop deformity often seen after polio infection. Subsequent writings from ancient cultures in the Middle East, India, and Asia described diseases that we now believe were smallpox and measles, as well as polio. However, all these historical accounts merely described the disease symptoms and effects with no insight into the causative agent. Diseases in those times were viewed more as a curse from the gods or a chance encounter with evil spirits. It wasn't until the late nineteenth century that we developed an awareness of viruses as a distinct class of infectious agents. By that time bacteria were well acknowledged as they had become visible through the invention of light microscopy (first demonstrated by Antonie van Leeuwenhoek in 1676). There was now a burgeoning recognition of the causal association of bacteria with diseases, pioneered by Louis Pasteur in Paris and Robert Koch in Germany. Transmission studies showed that you could take the blood or bodily fluid from animals with certain diseases and cause the same disease in healthy animals simply by injecting these samples. This quickly led to the concept that these samples contained the disease-causing agent that could be physically transmitted to a new host to initiate disease all over again. Coupling transmission studies with visualization of bacteria in the samples helped establish the disease-causing capabilities of bacteria.

One of the significant principles that developed secondary to the transmission studies was the concept of filter sterilization. An important discovery was that passing the infected fluid samples through filters with very small pores eliminated the capacity of the samples to cause disease in the recipient animals. Functionally it became clear that the pores were too small to allow the causative bacteria to pass through, thus the liquid that did pass through the filter was now bacteria-free and sterile for disease transmission. Disconcertingly, filter sterilization only worked for a subset of infectious diseases. Testing more and more diseases by this method revealed that there were two classes of agents, those that could be filtered out and those that could not. Those agents that could be filtered out could be seen by microscopy and were visibly bacteria. The second class of agents was invisible by microscopy and must be something much smaller than bacteria that could still pass through the tiny pores. Consequently, the first working definition of what was to become known as viruses was that they were an infectious agent smaller than typical bacteria that could not be removed by filter sterilization. Soon afterward, in 1898, the Dutch botanist, Martinus Beijerinck, coined the term virus to describe an infectious agent of the tobacco plant that passed through the typical filters used in these studies. However, it would take another 40 years to actually "see" the first viruses by electron microscopy and begin to understand their true forms and structures.

Since those early years, the concept of viruses has become firmly entrenched in biology and medicine, and the word virus is commonly used by the general populace: "I have a nasty virus", "I'm going to the doctor to get something for this virus", or "I just got over a bad virus". This is often how we write and speak about viruses, and all these phrases sound like we are dealing with typical living organisms, but are we? Though much of what we hear or read about human infections concerns viruses, it is often without any comprehension of their biological and physical properties, and without much thought about how these properties are important for our interactions with these agents. But viruses are so different and unique from all other living organisms, that it is debatable if viruses can even be considered alive! Understanding their novel features and properties is crucial to fathom their role in our lives and our biosphere.

It is also important to note that we all encounter viruses frequently in our daily lives. When we contract certain illnesses, such as colds or the flu, the encounter becomes obvious. Fortunately, most encounters do not produce frank illness so we are unaware of the myriad of viral exposures we have daily from other humans, animals, and the environment. By some estimates, every species on earth has more than 50 different types of viruses that can infect it, and some of these types have dozens to hundreds of different subtypes known as strains. Collectively, this yields a total of tens of millions of different types of viruses inhabiting our planet, the vast and overwhelming majority of which have not yet been isolated or identified. Each of these millions of different virus types exists in huge numbers of copies, so collectively they give rise to the 10^{31} total virions we've already considered. We also know that many viruses can sometimes jump from their normal host species to other species, so our repertoire of exposures and potential risk is much greater than the hundreds of known human virus types and subtypes. These constant and life-long exposures influence our personal health and can exert massive social effects on human cultures. The enormous importance of viruses at both the personal and societal levels has spurred decades of intense research aiming at understanding their biology, attempting to thwart their negative effects on humans, and more recently trying to adapt them for ingenious therapeutic uses (Chaps. 12 and 13). But before we delve into these facets of viruses themselves, we need to understand the larger context of life on earth so that we can effectively appreciate the uniqueness of viruses compared to other life forms.

A Biological Primer – Multicellular Organisms

To begin our journey, let's start with the question of life. Before we can address how viruses differ from other life forms, we have to identify and agree on the general characteristics that define life. I'm pretty sure most of us think we know what life is and can identify it when we see it, at least on the macroscopic scale of the things we can easily observe. Of course we humans are alive, as are all the typical animals we see and know such as cats, dogs, cows, horses, sheep, and so on. However, these examples are all one subclass of animals known as mammals because they have mammary glands for producing milk to nurse their young. What scientists call the biological kingdom of Animalia is much larger and includes many other types of animals such as birds, fish, and insects, just to name a few. This myriad of animals can run, or swim, or climb, or fly across our globe with incredible variation in size and color. But even if it's a mammal, insect, or bird that we've never seen or heard of before, when we view it in a zoo, in the media, or on our travels we immediately can say "that's alive". So what are some of the common characteristics that we associate with being alive and which make it easy for us to recognize "life"? Certainly, for animals we think of some typical physical attributes like eyes, ears, and mouth; a body structure with some sort of head, trunk, and limbs (the so-called body plan); and the ability to move. Additionally, at the microscopic level, animal bodies consist of millions of cells with specialized functions. Each individual cell is a living entity, and they combine collectively and collaboratively to form our different organs like the lungs or kidneys. Though not all animals have every one of these features, humans are pretty good at generalizing, and just by looking at a new creature, we can readily discern that it is alive and not just some inanimate material in nature.

In addition to movement and appearance, there are other obvious criteria associated with animals such as consuming organic materials (eating) and oxygen (breathing). Biologically, eating and breathing provide the raw materials and energy source for the growth and maintenance of the animal. After eating, nutrients eventually pass to the cells. In the cells, through a process known as catabolism, consumed organic materials break down from complex molecules into simpler compounds. Catabolism works in conjunction with oxygen-based respiration to generate a universal form of energy currency in the form of the compound called ATP (**a**denosine **trip**hosphate). Much of this ATP formation occurs in mitochondria which are a smaller organelle within animal cells. Subsequently, this energy drives the assembly of simple organic molecules into the complex components of the particular animal in a process called anabolism. Collectively, catabolism and anabolism comprise metabolism which is inherent in all animals as the mechanism to survive, grow, and ultimately reproduce.

Reproduction is another less often observed but equally critical biological criterion for animals. Animals reproduce by some form of a sexual mating process where male and female partners contribute genetic material to the offspring. All animals on earth use DNA, **d**eoxyribo**n**ucleic **a**cid, as their genetic material. Each cell of an animal contains a complete copy of the total of their DNA referred to as their genome. DNA is a wonderfully simple, yet infinitely complex molecule that is relatively stable and thus an effective long-term repository of genetic information. Chemically, DNA is composed of only four subcomponents, the nucleotides designated A (adenine), C (cytosine), G (guanine), and T (thymine) (Fig. 1.1). DNA itself is just a long strand of these four nucleotides repeated over and over. However, animal DNA is double-stranded and normally exists as two paired strands twisted into the canonical double helix. Importantly, the strands in the double helix are complementary which simply means that an A on one strand is always paired with a T on the other strand, and likewise, G is always paired with C. Functionally, the nucleotides are analogous to an alphabet of four letters. Just as the 26 letters of our Standard English alphabet combine in virtually infinite arrays to form words and sentences, the four nucleotides can form unlimited combinations along the length of DNA to create the various genes that define each organism. In the reproductive cells, the maternal DNA is carried in the egg and the paternal DNA in the sperm. I think most of us are familiar with how we humans and other mammals perform our mating via direct male and female contact. In contrast, there are many animals such as fish and amphibians where the parents do not have direct contact during reproduction, and the egg and sperm meet out in the environment. Regardless of whether or not there is direct parental contact, the productive result of mating is fertilization, the joining of an egg and sperm to start a new individual with a genetic composition contributed by both parents. The scrambling of parental genetic information into different eggs and sperm, coupled with the relatively

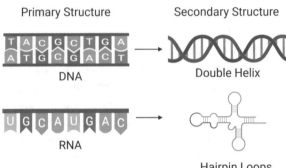

Fig. 1.1 The composition and structure of DNA and RNA. DNA is composed of four deoxyribonucleotides designated A, C, G, and T that are linked together in a linear strand. This linear sequence is called the primary structure. Our genomic DNA is double-stranded so two parallel chains bind to each other through the nucleotide pairs; A always pairs with T and C always pairs with G. Normally the double-stranded DNA is coiled into a helical form and this shape is referred to as the secondary structure. RNA is composed of similar nucleotides called ribonucleotides. For RNA there is no T ribo-nucleotide and instead, there is a ribonucleotide called U that pairs with A. RNA is typi-cally single-stranded but can form short double-stranded regions through self-base pairing that creates structures called hairpin loops. These hairpin loops constitute the secondary structures of RNA. (Created with BioRender.com)

random nature of a sperm finding and penetrating an egg, results in all the wonderful genetic diversity that makes each animal a unique individual differ-ent from either parent, siblings, or other members of its species.

We've defined several characteristics of animals that include movement, physical appearance, cellular composition, energy production, and reproduc-tive strategy, but you may wonder about plants; don't we consider members of the kingdom of Plantae alive? An enormous diversity of plants populate the world from tiny watermeal to towering sequoias. We watch plants germinate, grow, and die, and we've learned to cultivate them for food and aesthetics, so we intuit that they are living organisms. However, most plants don't move, at least not in the fashion of animals, they don't share the same physical appear-ance as animals, and they don't eat like animals, so what are the key attributes shared by plants and animals that invoke us to consider them both living organisms? If movement and body organization aren't important then other common features must denote life. In studying plants it is apparent that they do consume nutrients to produce energy and grow (metabolism). While lack-ing mouths to eat, they easily take in raw materials through their root struc-tures (water and minerals) and leaves (carbon dioxide), and they utilize photosynthesis to convert these nutrients to new plant material. Additionally, plants also have DNA genomes and do reproduce with male (stamens) and female (carpels) structures. These plant reproductive organs produce the egg and sperm equivalents that mix their parental DNAs (again, sexual reproduction). Lastly, plants are also composed of millions of individual cells that combine to form the different types of plant structures, such as stems and flowers. So even for all their differences, plants and animals share these four qualities: metabolic capability, DNA genome, sexual reproduction, and multicellular composition. Therefore, these four features constitute some of the definition of life. But can we parse this any finer to identify the absolute minimum requirements for something to be alive?

Smaller, But Still Alive – Unicellular Organisms

Fortunately, several other kingdoms of life exist that can provide further refinement to a definition of a living organism. In particular, the kingdoms of Protozoa and Bacteria represent some of the simplest life forms on this planet. Both of these kingdoms are comprised of single-cell organisms that lack all the sophisticated organs and assemblies seen in multicellular life forms. Consequently, those complex structures such as hearts and lungs and leaves and roots that are typical of animals and plants clearly aren't necessary to define life. One thing the unicellular organisms do share with their multicellular brethren is the presence of a genome. Like plants and animals, protozoa and bacteria use DNA as their genetic material. In protozoa, the DNA localizes within an internal structure called the nucleus. The nucleus encloses the DNA genome and physically separates it from the aqueous interior of the cell known as the cytoplasm. Plants and animals similarly package their DNA in a nucleus, and all organisms with a nucleus are designated eukaryotes. In contrast, bacteria have no nucleus and their DNA simply subsists within the cytoplasm; organisms lacking a nucleus are called prokaryotes. Regardless of how they store their DNA, protozoa and bacteria both normally exist as solitary living cells, unlike the multicellular plants and animals which we have already considered. Each of these single cells is a single entity that is capable of surviving, growing, and reproducing without help from any other cells of its same species. While a free-living cell may at times functionally interact with other cells, such interactions are not necessary or essential to their individual lives, so multicellularity is not a requirement for life. Instead, each bacterial or protozoan cell is a completely self-sufficient life form that can thrive and propagate in its normal biological niche. Then what common properties are inherent in animals, plants, and single-cell organisms that endow them all with life?

Let us focus on bacteria as simple models of living organisms from which we can extract some fundamental requisites for life. Most bacterial species fall into one of two common shapes, either a rod-shaped form (bacillus) or a spherical form (coccus), though some species do have more complex shapes such as spirals or filamentous forms. The shape and rigidity of bacteria are determined by a complex, multi-layered lattice structure called the cell wall – think of a piece of chain link fence fashioned into a sphere or rod shape and then repeated in multiple layers. Fortunately, as with layers of chain link, the cell wall remains relatively porous so that water and small nutrients can easily diffuse through. Inside the cell wall is a membrane structure that provides the barrier function that separates the cellular interior (cytoplasm) from the external environment. Membranes are composed of lipids that are not porous to aqueous materials, but membranes typically have pores and other structures that allow certain materials to pass across either passively or by active transport mechanisms. In function, the membrane serves as a barricade that controls the influx of nutrients and the export of waste products and other secreted materials. Thus, bacteria eat without a mouth by controlling the passage of nutrients through their membranes. Individual cells of plants and animals also have membranes that likewise control nutrient entry at the cellular level.

Once inside a cell, the conceptual fate of nutrients is consistent across unicellular and multicellular organisms. In every case, environmental nutrients break down (catabolism) within the cell for ATP energy production and to form the building blocks for the synthesis (anabolism) of new cell components. Under favorable conditions, cells from bacteria to us humans use the nutrients to build more and more of their constituent parts, and the overall size of the cell increases until it is roughly 1.5–2 times the size of the starting cell. The specific nutrients required, the methods for bringing nutrients into the cells, and the biochemical mechanisms by which nutrients are metabolized are distinct and unique for different organisms, but the fundamental concept of needing external food for cell growth is universal for all living organisms. One of the cellular components that increases during bacterial growth is their genome which again is always composed of DNA. During the growth phase, the cell uses newly made nucleotides to synthesize an identical copy of its existing genome. This is a very precise and regulated process so that the new copy is as faithful as possible to the original genome in length and sequence. The goal is to create a perfect replica of the original genome so that at the end of the synthesis phase the cell has exactly two identical copies of its entire genome. Once the cell has roughly doubled in size and has replicated its genome it can undergo a cell division process. The two genome copies physically separate and the parental cell splits into two halves (a process called binary fission) with one genome copy going into each daughter cell. This is an asexual reproduction event yielding two new cells that are each identical to the original parental cell in terms of genetic content, so even sexual reproduction isn't a requirement for a living organism. However, the concept that cells reproduce by increasing in size and doubling genome content followed by fission to form two daughter cells is a biological constant across unicellular and multicellular creatures, and the individual cell is considered the basic building block of life. Based on their commonality among both prokaryotes and eukaryotes, we can summarize the inherent properties of living cells to include: (1) possessing a membrane barrier that physically and functionally separates the cell interior from the environment, (2) possessing a DNA genome, (3) having ATP-generating metabolic machinery that provides energy and new raw materials for growth, and (4) using a binary fission-based reproductive scheme. While it is possible to find organisms that partially lack some of these features (e.g. some bacteria of the *Rickettsia* genus cannot produce ATP), these limited exceptions do not invalidate these four characteristics as being central tenets of the biology of organic life forms. Thus, our brief interrogation of multicellular and unicellular life forms has informed us of some very basic and fundamental principles of living organisms.

Biochemistry – You Can't Escape It

Our exploration of different life forms in the previous sections focused mostly on properties at the cellular level: having a membrane, having a genome, having metabolism, and reproducing by binary fission. However, it would be remiss to neglect some important features of living organisms at the molecular level, what we term the biochemistry of the cell. There is a bewildering array of biomolecules produced during growth and reproduction, and any comprehensive review is way beyond the scope of this book. Fortunately, among this myriad of biological compounds, there is only a small subset that we need to consider due to their universal and critical roles in living cells. Having already introduced two important biomolecules, DNA and ATP, there are two other types of molecules warranting our special attention, protein and RNA.

Proteins are fairly large molecules made up of linear chains of individual subunits known as amino acids. As with the previous DNA nucleotide example, you can think of amino acids as the letters of the protein alphabet. While our written alphabet has 26 characters, there are 20 primary amino acids that link together to form protein chains, and this linear sequence is called the primary structure (Fig. 1.2). These 20 amino acids are conserved across most life forms and are used to build proteins in all organisms from bacteria to humans. Each protein is like a unique "word" composed of a different number and combination of amino acids. Just as the length of words varies, the lengths of different proteins vary, ranging from small proteins, such as the 20 amino acid long TRP-Cage protein from the saliva of Gila monsters, to the giant 34,350 amino acid long Titin protein found in muscles. In the repertoire of all possible protein sequences, each position in the chain could be occupied by any amino acid, so there are 20 possible amino acids at position number 1, 20 possible amino acids at position 2, and so on for every position in the protein chain. Mathematically the number of possible combinations of amino acid sequences is calculated as 20^n where n is the total number of amino acids in the protein chain. Even for the short TRP-Cage protein of 20 amino acids, there are 20^{20} possible combinations that evolution could have chosen for this protein. Since the average length of a human protein is typically around 400 amino acids, the potential number of different possible sequences for a protein of this length is an incomprehensible 20^{400} (and you thought the number of virions on earth was big)! This enormous combinatorial diversity means that there is essentially no limit to what proteins can have

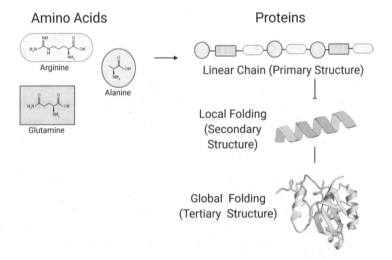

Fig. 1.2 Protein composition and structure. Proteins are comprised of 20 individual subunits called amino acids (3 are depicted). Amino acids are linked together into long chains that constitute the primary structure of proteins. Localized regions of these chains can fold in specific ways that are designated secondary structures. Finally, the entire chain will assume a specific three-dimensional shape that is referred to as the tertiary structure. Evolutionarily related proteins will exhibit similarities in the sequences and structure. (Created with BioRender.com)

evolved to accomplish. Structurally, these protein chains fold up into localized regions with a specific shape called the secondary structure (Fig. 1.2). Finally, the entire assembly of amino acids will fold into a specific 3-dimensional shape known as the tertiary structure (Fig. 1.2). These unique tertiary structures are often conserved for related proteins across widely disparate organisms – we'll revisit this in Chap. 3. Functionally, proteins are involved in nearly everything a cell does, and they can serve as anything from relatively inert scaffolding material, to transport molecules, or to highly active enzymes that catalyze all of our critical biochemical reactions in the cell.

The reason proteins occupy a unique niche in biology is not only because of their functional importance but also because they serve as the direct incarnation of our genetic information. The DNA of our genomes contains many genes, each consisting of a specific string of nucleotides. The one gene equals one protein hypothesis predicted that each gene encodes a single protein – this is now known to be oversimplified as a single gene can express multiple related versions of a protein, but for our purposes, the one gene equals one protein model is sufficient. To make a protein, the string of nucleotides in a DNA gene first has to be copied into a slightly different type of molecule, a single-stranded nucleic acid called messenger RNA (mRNA), by a process known as transcription. RNA (**ribo**nucleic **a**cid) is chemically the same as DNA except that it contains an additional oxygen-hydrogen group (called a hydroxyl moiety) at a single position in each nucleotide (Fig. 1.1). Also, while DNA uses the A, T C, and G nucleotides, RNA replaces the T with a related nucleotide called uracil, U for short. While DNA is fairly stable, making it an excellent long-term storage molecule for genetic information, the presence of the hydroxyl group makes the RNA molecule much less stable under aqueous cellular conditions. Thus RNA is less suitable as a medium for the storage of information in long-lived cells and organisms but is acceptable as a transient, messenger molecule.

After mRNAs are produced from our genes, they are transported to cellular structures we call ribosomes (Fig. 1.3). Ribosomes are biochemical machines that read the nucleotide sequence in multiples of three nucleotides, called a codon, and translate the codon information into protein sequences. Since a codon is three nucleotides long there are only 64 possible combinations that can be made from the four nucleotides. (Any of the four possible nucleotides could occupy each of the 3 positions of a codon, so the number of combinations is $4 \times 4 \times 4 = 64$.) Three of these 64 combinations act as "punctuation" that tells the ribosome to stop making protein. The other 61 codons each encode a specific amino acid. As there are only 20 amino acids and 61 codons it turns out that there is redundancy in the code, i.e. some amino acids have

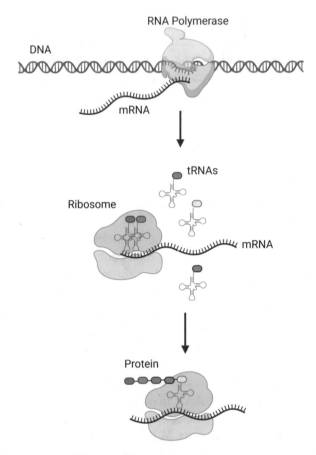

Fig. 1.3 DNA to Protein. The so-called "central dogma" of biology is that genetic information flows from DNA to mRNA to proteins. DNA contains an organism's genes that are passed from generation to generation. To express a gene it is transcribed into a complementary messenger RNA (mRNA) by the enzyme RNA polymerase. After its synthesis, the nucleotide sequence of the mRNA is read by ribosomes in increments of 3 nucleotides called codons. Transfer RNAs (tRNAs) carry amino acids to the ribosome and each tRNA recognizes a specific codon. The ribosome helps position the tRNAs on the mRNA and then catalyzes the joining of adjacent amino acids to form the protein chain. The finished protein is released from the ribosome to perform its function in the cell. (Created with BioRender.com)

more than one codon. For example, the amino acid alanine is specified by four codons: GCT, GCC, GCA, and GCG. As the ribosome scans the mRNA and reads the codons one after another it decodes the information that was originally contained in the linear DNA sequence and translates it into consecutive amino acids that are permanently linked together to form the new protein chain. The amino acids are carried to the ribosome by another type of

RNA known as the transfer RNA (tRNA). If this biochemical explanation is too abstract, think instead of those increasingly popular 3-D printers. The printer device itself would be the ribosome equivalent, i.e. a machine for turning code into the product. The printer can be given a set of instructions in computer code (the mRNA equivalent) consisting of ones and zeros. (Note that while computer code only uses 2 numbers making it a binary code, the DNA/mRNA code is quaternary as it uses the 4 nucleotides, A, C, T, and G.) This computer code contains the information to specify some final object (the protein equivalent). Like the ribosome, the printer is the decoder/synthesizer that reads the inherent information in the computer code and then translates it into a real-world object. Parenthetically, just as the printer needs electrical energy to function, so too the ribosome needs ATP energy to perform its task. Ultimately then, the protein is the direct functional readout of our DNA, and without mRNA and ribosomes, proteins could never be produced. As you might expect from this explanation, all living organisms must have RNA and ribosomes to convert their static genomic DNA information into their dynamic protein products thereby adding RNA and ribosomes to our list of essentials for life along with DNA and ATP.

Viruses At Last

We've briefly covered information about virus quantity, size, and historical discovery, which I hope piqued your curiosity to learn more about these tiny but powerful infectious agents. We then ventured far from viruses and explored some basic biology of multi- and unicellular organisms, as well as some elementary biochemistry, to illustrate the basic principles of living creatures. All of this digression provides a suitable framework, vocabulary, and context for our discussion of viruses in this and the subsequent chapters. From that overview we've seen that cells are the fundamental unit of living organisms and that cells have certain common features, including a membrane to separate their cytoplasm from the environment; metabolic capacity to generate ATP and raw materials for growth; a double-stranded DNA genome; ribosomes and mRNA to translate their genetic information into functional proteins; and a reproductive scheme where the cell grows in size, double their DNA content, and divide by fission to produce two daughter cells. Now, to return to the original question, what is unique and different about viruses compared to plants, animals, bacteria, and other life forms? Stated another way, which of the cellular properties, if any, are manifest in viruses?

Let us start with the basic structure and composition of a virus and see how that compares to the properties of a living cell. For illustrative purposes, we'll consider one of the smallest and simplest viruses, a pig virus called porcine circovirus (PCV). The PCV virion consists of two components, a genome comprised of DNA and a capsid made of a viral protein. Functionally, the capsid is a protective coat that forms a roughly spherical structure around the DNA, thus containing and sequestering the DNA from environmental exposures which could damage or destroy the sensitive genetic information. Capsids also serve as the interface between a virus and the host cell that it will infect, and to enter a cell there must be a "receptor" on the cell that specifically recognizes and binds the capsid (we'll delve into this in much more detail in subsequent chapters). While most viruses are larger and more complex in their components than PCV (see Chap. 2), they nonetheless all share these two fundamental features possessed by PCV: a genome and a capsid to protect it. Since the presence of a genome is one key element of cells, this is a feature that viruses do share with other life forms. However, as we'll see in Chap. 2, there is some interesting and unexpected diversity among viral genomes that is not present in any other living organism.

Beyond having a genome, it should be obvious from the above description that PCV is missing many of the other standard features that we've defined as comprising a living cell. First, it has no membrane structure; while many viruses do have a lipid layer outside of and surrounding their capsids (see Chap. 2), many do not, so membranes are not a required feature for all viruses. Second, there is no metabolic machinery or ATP in the PCV virion, only the structural capsid and the nucleic acid genome. By itself, the virion has no mechanism to produce energy or to create new raw materials to reproduce its capsid or its DNA. Furthermore, there are no ribosomes or mRNAs in the virion so by itself it cannot even translate its DNA into new capsid proteins. Functionally, the virion on its own is essentially inert, lacks most of the critical components of cells, and cannot perform any biochemical activity typical of cells. Simply put, if you had a bottle with 1000 virions sitting on your desk it would just be static with no change in size or number of virions over time. Even if you gave it ATP and some basic compounds, like nucleotides and amino acids, nothing would happen. This is a serious quandary for viruses as they must reproduce to survive and propagate, yet they cannot intrinsically do any of the basic cellular activities needed for reproduction.

If virions can't metabolize, and can't make new proteins or other essential biomolecules, then how do they grow and divide? The short answer is that they do neither. Once a virion is formed it never enlarges in size, never duplicates its internal genome, and never undergoes binary fission. In contrast to living cells, viruses reproduce in a completely different fashion which does not involve a physical division of the parental virion into two daughter virions. Key to the whole process of viral reproduction is the concept that viruses are all obligate intracellular parasites, which simply means that they can only reproduce when inside a true cell. Since viruses lack the cellular components needed for reproduction they must "borrow" these components from the cells that they infect. For all viruses, the process of reproduction starts with the attachment of the virion to a susceptible cell followed by penetration of the cell membrane to deposit the virion within the cell. Once inside the cell, the virion particle undergoes a dissociation event known as uncoating, where the capsid is completely or partially removed to liberate the viral genome. Freeing the viral genome from the protective capsid gives the nucleic acid access to everything within the cell, including the cell's ATP, enzymes, nucleotides, amino acids, and ribosomes. The virus can now use cellular materials to transcribe its genes into mRNAs and will then use the cell's amino acids and ribosomes to translate those viral mRNAs into thousands of copies of new viral proteins. Some of these new viral proteins will use the cell's ATP and nucleotides to synthesize new copies of the viral genome, so a single genome brought in by the infecting virion can reproduce to make hundreds to thousands of identical progeny copies within the cell. Ultimately, as the new viral proteins and new viral genomes accumulate, they come together to self-assemble into new virions, so a single cell infected with one virion can produce thousands of new copies of the parental virion. Conceptually, this scheme for viral reproduction is nothing like cellular reproduction and is instead more akin to how factories mass-produce cars. To create a new car the individual components are manufactured separately and are then brought together in an ordered assembly process to construct the final product. Once made neither cars nor virion increase in size, and they certainly don't divide in half to form two new copies! In this analogy, the cell is a factory that is normally producing new copies of itself. However, the viral infection repurposes the factory now to make virions instead of cellular components. So in essence, a virus is a self-replicating nano-machine that usurps the host cell and uses all the host resources to make new copies of the virions. This hijacking of the cell by the virus usually has adverse consequences for normal cellular function, and viral-induced disruption of cell function contributes to disease expression as we'll see in subsequent chapters. Once assembled, the newly released virions go out into the host body and/or the environment where they remain inert until they infect a new cell to start the process all over again. Since one infected cell can produce thousands of progeny virions that can each infect another cell, the total number of infected cells in a plant or animal can increase quickly to the millions after a few cycles of infection, reproduction, and release. Millions of cells dying and/or killed cause dysfunction of the infected tissue or organ leading to a characteristic disease presentation.

To summarize, the basic reproductive scheme of all viruses is to infect a susceptible host cell, reprogram the host cell machinery to make viral components (e.g. proteins and nucleic acid), assemble the components into new virions, then release hundreds to thousands of new virions from the infected cell to spread the infection and amplify it rapidly. The virion itself doesn't eat, drink, or breathe; it doesn't metabolize to make raw materials or ATP; it lacks ribosomes and can't make proteins; and it doesn't grow or divide, so viruses lack almost all of the characteristics that we attribute to living organisms. Compared to cells, virions are mostly much simpler static structures designed primarily to house and protect the viral genetic information and to facilitate its delivery to host cells. Outside the cell, virions are essentially inert with no movement or biochemical activity. This list of viral properties is so different from cell-based life forms that viruses confound the very definition of being alive. Yet viruses do reproduce by taking over living cells, they can spread within and between individuals, they can cause disease, and their genetic material undergoes the same types of mutations as do our genomes. So are they actually alive or are they just sophisticated molecular machines? Even among scientists opinions differ, so I'll let the readers decide for themselves. Alive or not, I hope you will agree that viruses are unique parasites that drastically differ from all other life forms on earth.

Additional Reading

1. "A Strategy to Estimate Unknown Viral Diversity in Mammals". mBio 4:1–15, 2013.
2. "Are Viruses Alive?" Microbiology Today, May, 2016.
3. "To be or not to be alive: How recent discoveries challenge the traditional definitions of viruses and life." Studies in History and Philosophy of Biological and Biomedical Science 59: 100–108. 2016

Definitions

Adenine – A biomolecule found in nucleotides and used in both DNA and RNA.

Adenosine triphosphate (ATP) – This is a molecule used to store energy in the cell for use in biochemical reactions.

Amino acids – Small organic compounds containing both an acidic group and a nitrogen-containing amino group. There are 20 common amino acids that are the building blocks of proteins.

Anabolism – The biochemical process whereby larger biomolecules are synthesized from smaller precursor molecules. Typically energy in the form of ATP is required to join the smaller molecules together.

Bacillus – Any bacterium that is cylindrical or rod-shaped.

Bacteria – Microscopic, single-celled organisms that lack a nucleus. Bacteria are prokaryotes.

Binary fission – The process of asexual reproduction where a single cell divides in half to create two daughters cells that each contain the complete parental genome.

Biomolecule – Any type of chemical compound produced in living cells or organisms.

Biosphere – All the ecosystems on Earth where life exists.

Capsid – A protein shell that contains and protects viral nucleic acids.

Catabolism – The biochemical process whereby large biomolecules are broken down into small molecules. Typically this releases energy which is stored as ATP.

Cell wall – A rigid, permeable structure located outside of the cell membrane of plants, bacteria, fungi, and algae. Cell walls provide shape and protection to the cell.

Chromosome – A structure composed of protein and nucleic acid (DNA or RNA). The proteins wrap and condense the nucleic acid. Each organism or virus has a characteristic number of chromosomes.

Circovirus – A family of viruses with small, circular, single-stranded DNA genomes.

Coccus – Any bacteria with a spherical shape.

Codon – A series of three consecutive nucleotides on messenger RNA (mRNA) that are read by ribosomes in the translation process. There are 64 possible codons and each codon specifies either an amino acid or a stop signal.

Cytoplasm – The aqueous environment enclosed by a cell membrane, i.e. the interior of the cell. In eukaryotic cells that contain a nucleus, the cytoplasm is the region inside the cell membrane and outside the nucleus.

Cytosine – A biomolecule found in nucleotides and used in both DNA and RNA.

Deoxyribonucleic acid (DNA) – The genetic material for all cells and some viruses. DNA is a long chain composed of four deoxyribonucleotides: A, C, T, and G.

Double helix – The molecular shape of double-stranded DNA. Two parallel DNA molecules adhere together through the complementary pairing of their nucleotide sequences and then twist around each other to form a spiral helix.

Endogenous viral elements (EVEs) – Viral nucleic acid sequences that invaded cellular genomes in the evolutionary past and became a permanent part of the cellular DNA.

Enzyme – A biomolecule that can catalyze a biochemical reaction. Enzymes are usually proteins, but some RNA molecules also have enzymatic activity.

(continued)

20   V. G. Wilson

(continued)

Escherichia coli (E. coli) – A common, rod-shaped bacteria found in the digestive tract of humans and other warm-blooded organisms. E. coli is widely used as a model to study bacterial biology and for cloning genes.

Eukaryotes – Organisms whose cells possess a nucleus, e.g. plants and animals.

Filter sterilization – The process by which bacteria are removed from a liquid by passing the liquid through a filter that retains the bacteria while allowing the liquid to flow through.

Gene – The fundamental unit of heredity. A gene is a specific sequence of nucleotides on the genome that encodes either a protein or an RNA such a transfer RNAs (tRNAs)

Genome – The complete genetic information of an organism. Genomes are DNA for all cellular organisms and some viruses. For other viruses the genomes are RNA.

Guanine – A biomolecule found in nucleotides and used in both DNA and RNA.

Hairpin loop – A type of secondary structure that can form in RNA or DNA. The structure will have a double-strand stem region supporting a single-strand loop.

Human immunodeficiency virus (HIV) – An RNA virus of the retrovirus family that is the causative agent of acquired immune deficiency syndrome (AIDS).

Hydroxyl moiety – A chemical group consisting of one hydrogen atom and one oxygen atom.

Immunodeficient – The condition of having an impaired or absent immune response.

Lipid –Organic compounds that are insoluble in water, including fats and oils. Lipids are a primary component of cell membranes.

Membrane – The structure that forms the boundary of living cells and some internal cellular compartments. Membranes, which are composed primarily of lipids and proteins, separate the cytoplasm from the environment.

Messenger RNA (mRNA) – A single-stranded molecule transcribed from DNA by an RNA polymerase enzyme. Messenger RNA is translated by ribosomes to produce proteins.

Metabolism – The combined processes of catabolism and anabolism that generate energy and synthesize new materials for cell growth.

Mitochondria – A membrane-bound structure located in the cytoplasm of cells. Mitochondria produce energy for the cell and store it as ATP.

Mutation – A change in the nucleotide sequence of a gene.

Nucleic acid – The general term for the type of organic molecules that include DNA and RNA.

Nucleotides – The building blocks of DNA and RNA consisting of a base [adenine (A), cytosine (C), guanine (G), thymine (T), or uracil U)], a phosphate group, and a sugar. DNA uses deoxyribose as its sugar along with the bases A, C, G, and T. RNA uses ribose with the bases A, C, G, and U.

Nucleus – The membrane-bound structure in eukaryotic cells that contains the chromosomes.

Obligate intracellular parasite – An organism, such as a virus, that can only reproduce inside of a host cell.

(continued)

(continued)

Pandoravirus – A genus of giant viruses.

Parvovirus – Nonenveloped viruses with small (~ 5000 nucleotides), single-stranded, linear DNA genomes.

Poxvirus – Double-stranded DNA virus with a complex envelope and a large (>130,000 nucleotide pairs) genome.

Primary structure – The linear amino acid sequence for proteins or the linear nucleotide sequence for nucleic acids.

Prokaryotes – Unicellular organisms such as bacteria that lack a nucleus.

Protein – A biomolecule comprised of a linear chain of amino acids. Proteins are encoded by genes and have many functions in the cell, including structural, regulatory, and enzymatic.

Protozoa – Unicellular organisms such as amoeba that possess a nucleus.

Receptor – A molecule that binds a specific partner molecule with the binding triggering a signal or other response in the cell. Cells have numerous receptors that are involved in growth, cell division, immune response, cell-to-cell communication, and other biological processes. Some molecules on the cell surface serve as receptors for virus attachment.

Ribonucleic acid (RNA) – RNA is a long chain composed of four ribonucleotides: A, C, T, and U. In cells there are several forms of RNA including mRNA and tRNA. For some viruses, RNA serves as the genome.

Ribosome – A complex biomolecule in the cytoplasm that synthesizes proteins. The ribosomes read mRNA codons and convert the information into the amino acid sequence of the new protein.

Secondary structure – Localized folding of amino acids in a protein into specific spatial arrangements such as a helix. Individual proteins can have multiple regions of secondary structure along the primary amino acid sequence.

Tertiary structure – The overall 3-dimensional shape into which a biomolecule folds. Proteins generally have specific tertiary structures that are critical to their function. Disruption of the tertiary structure by mutation or environmental conditions can reduce or eliminate a protein's normal activity.

Thymine – A biomolecule found in nucleotides and used in both DNA and RNA.

Transcription – The process of synthesizing an RNA sequence by an RNA polymerase enzyme. Cells and DNA viruses use DNA as the template for RNA polymerase while RNA viruses use RNA as the template.

Transfer RNA (tRNA) – The type of RNA that carries amino acids to the ribosome for protein synthesis.

Uracil – A biomolecule found in nucleotides and used only in RNA.

Virions – The entire virus particle consisting of a nucleic acid genome enclosed in a capsid and for some viruses also surrounded by a membrane envelope.

Abbreviations

A – adenine
ATP – adenosine triphosphate
C – cytosine
DNA – deoxyribonucleic acid
EVEs – endogenous viral elements
G – guanine
HIV – human immunodeficiency virus
mRNA – messenger RNA
PCV – porcine circovirus
RNA – ribonucleic acid
T – thymine
tRNA – transfer RNA
TRP – transient receptor potential
U – uracil

2

Families, Forms, and Functions

Keywords Taxonomy • Viral life cycle • Viral pathogenesis

A virus is a piece of bad news wrapped in protein.
Sir Peter Medawar

Taxonomy and Other Geeky Things

In 1735 Carl Linnaeus published his *Systema Naturae* which defined an organizational scheme for classifying life forms into kingdoms, phyla (singular: phylum), classes, orders, families, genera (singular: genus), and species; a system that became widely adopted in biology. While viruses may or may not be living organisms, they still constitute a diverse group of infectious entities that needs classification to bring order and understanding to their functional and evolutionary relationships. However, initially there was no systematic classification or naming of viruses. Many viruses were simply named after diseases they caused, such as the smallpox virus or the chickenpox virus, since disease manifestation was their major discernible property prior to modern biological analysis. Some viruses were named after the anatomical site from which they were first isolated such as the respiratory rhinoviruses from rhino meaning "nose" in Greek. Other viruses were named after the geographic site where they were first identified. For example, the Ebola virus was named after a river in the Democratic Republic of Congo, and the Zika virus was named after a forest in Uganda. Alternatively, some viruses were named after their discoverers, such as the eponymous Epstein-Barr virus (which causes mononucleosis),

V. G. Wilson, *Viruses: Intimate Invaders*, https://doi.org/10.1007/978-3-030-85487-4_2

named after doctors Epstein and Barr. While these so-called "common names" are still in wide use, such names do little to denote shared features or clarify genetic relatedness between viruses. To build a scientifically valid classification system the actual physical properties of viruses needed identification and categorization across large numbers of different viruses.

During the first half of the twentieth century, developments in technology allowed significant advances in the kinds of information that could be collected about individual types of viruses. In particular, electron microscopy allowed visualization of individual virions for the first time, providing enormous new insights into the sizes and shapes of different types of viruses. Further refinements in visualization were obtained with X-ray crystallography that allowed highly detailed elucidation of virion structure. In parallel with these imaging techniques, new approaches in biochemical characterization began to define the composition of viruses. One of the most important biochemical discoveries was the chemical nature of viral genomes. As expected, many viruses have DNA genomes that are chemically identical to the DNA of all other organisms. However, unlike any other organism on earth, many viruses do not possess genomes composed of DNA. Instead, a large number of viruses possess RNA genomes, a startling and confounding observation that served to illustrate again how distinct viruses are from other life forms.

In the second half of the twentieth century, with the burgeoning data on virus morphology and composition, it became possible to explore the development of systematic nomenclature and classification. While several classification schemes were proposed by various scientists, the International Committee on Taxonomy of Viruses (ICTV; https://talk.ictvonline.org/) was formed in 1962 to develop an internationally recognized common taxonomy for all viruses. The ICTV still exists and is instrumental in maintaining and expanding the catalog of viruses as new information and new viruses are discovered. Ultimately, a Linnaean type system was adopted that contained orders, families, sub-families, genera, and species. More recently, the ICTV expanded viral classification to include higher levels of Linnaean classification (Fig. 2.1), a move necessary to account for the wealth of new viral data generated by large-scale genomic sequencing. The system for viruses was originally based primarily on their physical characteristics with genome type, either DNA or RNA, being the first major dividing point for all viruses. This is not simply a biochemical distinction but has functional implications as well. DNA viruses have evolved to utilize much of the cellular machinery that functions on our own DNA genomes. This machinery includes the enzymes, components, and complexes that transcribe our genes to messenger RNA (mRNA) and replicate our own genomic DNA. Since this cellular machinery

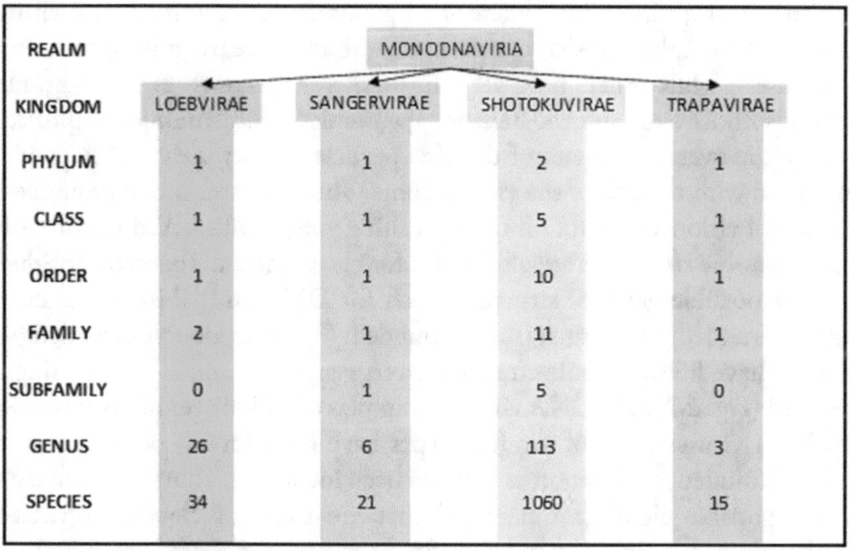

REALM	MONODNAVIRIA			
KINGDOM	LOEBVIRAE	SANGERVIRAE	SHOTOKUVIRAE	TRAPAVIRAE
PHYLUM	1	1	2	1
CLASS	1	1	5	1
ORDER	1	1	10	1
FAMILY	2	1	11	1
SUBFAMILY	0	1	5	0
GENUS	26	6	113	3
SPECIES	34	21	1060	15

Fig. 2.1 Linnaean classification of a representative Virus Realm. The figure depicts the Realm of Monodnaviria and its descending subdivisions from Kingdoms to Species. This Realm has four Kingdoms, and each Kingdom has the subdivisions shown below the Kingdom. The number of further subdivisions within each classification is as indicated. For example, the Kingdom Loebivirae has 1 Phylum, 1 Class, 1 Order, 2 Families, 0 Subfamilies, 26 Genera, and a total of 34 Species

is located in the nucleus of our cells, DNA viruses typically have a major portion of their life cycle in our nucleus. The invasion of our nuclei by DNA viruses imparts many of them with the ability to persist long past the initial infection (Chap. 4) and in some cases to promote cancers (Chap. 8). In contrast, viruses with RNA genomes have unique mechanisms for expressing their genes and replicating their genomes that are not dependent on our nuclear systems, thus most RNA viruses reproduce in our cellular cytoplasm, lack a nuclear phase, and are not oncogenic. Additionally, RNA viruses tend to be less accurate when reproducing their genomes, and they typically acquire mutations at a higher rate than DNA viruses. The high mutation rate makes several RNA viruses particularly difficult for our immune system to deal with since these viruses change faster than our ability to make new antibodies. The influenza virus is the epitome of this problem as the virus changes so rapidly from year to year that we need a new vaccination each year to be protected.

Once the initial separation of viruses into DNA and RNA genome types is made, subsequent stratification within each of these two groups is performed based on other physical characteristics such as virion shape and size, presence or absence of a virion membrane, and genome architecture. Many viruses have a roughly spherical shape, although some are more irregular spheres (e.g. poxviruses), while others have very distinctive shapes such as the long, filamentous Ebola virus and the bullet-shaped rabies virus. Among the spherical viruses, the overall diameter of the viral particle varies widely and is roughly correlated with the size of the viral genome. Thus, the size and appearance of the overall virion are useful for distinguishing viral families. Additionally, different genome structures are also useful for classification. Theoretically, there are four possible genome structures each for DNA and RNA: (1) circular, double-stranded; (2) circular, single-stranded; (3) linear, double-stranded (we humans have linear, double-stranded DNA genomes); and (4) linear, single-stranded (Fig. 2.2). For DNA viruses, examples of all four genome types exist. For RNA viruses three of the four types have been found, but no circular, double-stranded RNA genomes have yet been identified. Thus, there are seven of the possible eight genome types that are used for classifying viruses. Furthermore, for single-stranded RNA genomes, there are two functional types, positive-sense RNAs and negative-sense RNAs. A single-stranded RNA genome that acts as an mRNA and is directly recognized by ribosomes for translation into viral proteins is designated as a positive-sense genome. Conversely, a negative-sense RNA genome cannot be recognized or translated by ribosomes. The negative-sense RNA must first be copied to its complementary positive-strand so that the positive strand can serve as the template for protein production. This distinction between positive-sense and

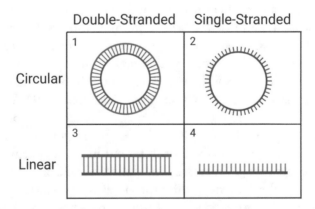

Fig. 2.2 Depiction of the genome types. Shown are the four possible types of DNA or RNA genomes: (1) double-stranded, circular; (2) single-stranded, circular; (3) double-stranded, linear; and (4) single-stranded, linear. Types 1–4 have been found in DNA viruses while only types 2–4 have been observed in RNA viruses. (Created with BioRender.com)

negative-sense RNA viruses is a major classification determinant and reflects an ancient split in the RNA virus lineage (Chap. 3).

Combining all the above features allows the successful distribution of viruses into their unique Orders and Families. More recently, the ability to sequence viral genomes rapidly and completely now allows a much more refined genetic comparison to be added to the previous taxonomic assignment. The incredible advances in nucleic acid sequencing technology over the last 20 years have increased the speed and reliability of the analysis while at the same time drastically reducing costs. Just like human DNA sequencing companies can sequence a person's DNA and identify genetically related individuals and ethnic origins, a similar analysis can now be done for any viral isolate. This level of genetic detail enables a more precise assessment of the evolutionary and biological relationships between viruses from within and between the different families. Ultimately this will improve our understanding of viral evolution, enhance our ability to quickly identify and classify newly discovered viruses, and potentially help with our ability to develop new targeted therapeutics. However, none of this will be quick or easy. As of late 2020, only 6590 species of viruses had formal ICTV recognition. This number of known virus species only represents a small fraction of the viral diversity inhabiting our biosphere which is estimated at over 100,000,000 species! Sorting out relationships among this staggering collection of viruses will take years of intense study by the scientific community.

Virions, the Viral Vehicle

While the number of virus species on Earth is enormous, there is a central problem for all viruses. To reproduce, each viral genome must enter a host cell where it can usurp the cellular functions and reprogram the cellular machinery to produce new virions instead of cellular components. However, the genomes themselves (whether DNA or RNA) are completely inert on their own. They cannot move so how do viral genomes transit from cell to cell and from infected individuals to naïve hosts? Viruses cannot express their genes without the aid of polymerase enzymes that can transcribe the viral genetic information into messenger RNA (mRNA), so where do they get these proteins? Viruses cannot express their mRNA without ribosomes for subsequent translation into new viral proteins, so where do viruses get the ribosomes and the related translational machinery? Viral genetic material is rather fragile and easily broken or damaged, so how do viruses protect their sensitive and critical genomes? And lastly, once viral genomes enter a cell, how do they rapidly appropriate cellular functions and avoid host defenses? Conceptually, this viral problem is similar to what humans face in spaceflight. We can't survive off-world without extensive equipment and supplies, and we need protection from the hostile space environment. The solution for humans is to package ourselves inside a delivery vehicle that protects the cargo and carries all the items needed for the mission, and viruses adopt a similar strategy. The critical viral nucleic acid is packaged inside a structure known as a capsid that fulfills all the virus' needs related to transporting the genome and establishing a successful new infection. However, capsids are not just protective structures harboring the viral genetic material. Capsids can be loaded with different proteins (the viral "equipment and supplies") that enter the host cell upon infection. These transported viral proteins can immediately begin the process of commandeering host cell functions and initiating viral gene expression. A capsid containing a functional viral genome is the basic infectious unit and is called a virion.

While the simplest virions consist of a capsid made up of 1–2 proteins, larger and more complex virions can have significantly more robust capsids composed of dozens of viral proteins fitted together in intricate patterns. Many virions also contain a membrane envelope that surrounds the capsid. An example of complex virions is the poxvirus family (Poxviridae). The most prominent member of this family, smallpox (scientific name – Variola major), was a devastating virus whose impact on world civilization will be explored in Chap. 6. Fortunately, the smallpox virus was eliminated from the natural world by 1980 through a global vaccination program using a close relative known as vaccinia. Vaccinia is an attenuated, or weakened, poxvirus used as the vaccine strain because it induces effective immunity against smallpox without causing significant disease in humans. Even though vaccinia vaccination was discontinued once smallpox was eliminated, vaccinia remains a widely used research tool to explore the biology and general features of this viral family.

The overall vaccinia virion morphology is ellipsoidal or barrel-shaped with the mature virion having an outer membrane layer surrounding the core (the capsid equivalent) (Fig. 2.3). The core itself is also barrel-shaped but with indentations on both sides to give it a figure-eight-shaped appearance. The spaces between the membrane and the core where the indentions are located on each side of the core are known as lateral bodies. Within the core, vaccinia has a large, double-stranded DNA genome that can encode roughly 200 proteins, of which around 90 are found in the virion particle. Approximately 20 of the viral proteins are found in the outer membrane with the remainder

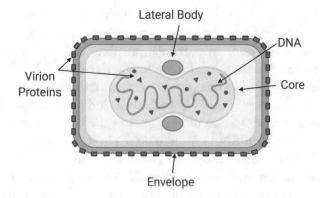

Lateral Body

DNA

Virion Proteins

Core

Envelope

Fig. 2.3 Morphology of Poxviruses. Shown is a poxvirus virion with the structural features labeled. The poxvirus core is the nucleocapsid consisting of a capsid protein structure containing the viral DNA genome and various other viral proteins. The core has a figure-eight shape with protein structures called lateral bodies located in the indentations of the core. The entire core is surrounded by a membrane envelope embedded with viral proteins. (Created with BioRender.com)

located in the core and lateral bodies. This is a large commitment of viral resources to the virion, so what do these proteins do for the virus? Typically we can divide virion proteins into two classes: structural (S) and nonstructural (NS). Structural proteins are the components that form the shape, organization, and appearance of the virion. If we think of the virion as a "vehicle" to carry the viral nucleic acid, then the viral structural proteins are equivalent to things like the frame or chassis that provide the overall shape of the vehicle, and the body panels that provide the exterior façade. In contrast, nonstructural proteins are not typically essential for the structural integrity of the virion. Instead, these NS proteins provide functional activities, often through enzymatic activity, that the virus will need for infection and reproduction. Again by analogy to a vehicle, the NS proteins are equivalent to things like the engine, the air conditioner, and the electronics, i.e. systems that perform functions for the vehicle. Together the structural and nonstructural proteins endow the virion with all the features it needs to initiate the molecular hijacking of the newly infected cells.

For vaccinia, most of the 20 membrane-associated viral proteins are structural and help to provide shape and organization to the membrane layer, although 2–3 have non-structural enzymatic roles that help keep the membrane intact. The remaining roughly 70 viral proteins are located in the core and lateral bodies. At least 19 of these proteins are structural and comprise the core itself, which leaves nearly 50 viral proteins that presumably have functional roles. Many of these proteins have not been well characterized and assigned a specific function, but a number of them do have known activities critical for viral infection. In particular, a subset of the viral NS proteins located in the core is responsible for producing vaccinia mRNA through transcription. Transcription is a complex process that requires enzymes and accessory factors at each step of mRNA synthesis including initiation, elongation, termination, and modification of both ends of the mRNA strand. Normally for DNA viruses their mRNA synthesis occurs in the cell nucleus and uses all the host cell enzymes and factors that generate our mRNAs. However, poxviruses carry all the required proteins for mRNA synthesis within their virions, hence they are not dependent on the host enzymes. Consequently, poxvirus DNA doesn't need to traffic to the cell nucleus to express its genes into mRNA and instead performs this process in the cytoplasm which is a distinctive feature of this DNA virus family. Once inside the cell, the poxvirus NS proteins can utilize the available cellular nucleotides and ATP to synthesize all the viral mRNAs. While the vaccinia example is an extreme version where the virus recreates the entire cellular mRNA synthetic process with its own enzymes, the strategy of using virion proteins to help initiate viral gene expression is a common theme seen with many viral families. Evolution appears to favor this strategy of giving viruses a "head start" by having virions carry in proteins to enhance viral gene transcription. This helps ensure that viruses can immediately express critical replication genes upon entering a new cell and are not dependent on some possibly elusive or limiting cellular factor(s).

While each viral species has its unique virion with a unique set of viral proteins, the common theme is that virions have evolved to carry the specific viral NS proteins as needed to jump-start the viral infection. This may range from one or two proteins in smaller viruses to dozens in larger viruses such as vaccinia. Importantly, besides helping with viral gene expression, virion NS proteins often aid the virus in avoiding host defenses. Humans have a wide array of exquisite defenses against viruses that must be circumvented, at least transiently, if the incoming virus has any chance of reproducing. If viruses had to transcribe their genes into mRNA and then translate their mRNAs into proteins before they could begin to block host defenses, it might give the host defenses such an advantage that viral reproduction would be thwarted. Bringing in premade anti-defense proteins as part of the virion allows immediate action against the host's defense systems which gives the virus a window to begin producing large amounts of viral proteins to overwhelm the cell. The cellular defenses and representative viral countermeasures will be explored in more detail in Chap. 5.

Alive or Not, Viruses Have a Life Cycle

Birth, growth, reproduction, death; the four stages of the animal life cycle are simple in concept though the details for how every species accomplishes each phase are unique and limitless in variation. Similarly, there is a universal set of steps to the life cycle of all viruses though again the details of how each type of virus achieves the steps are remarkable in their diversity. Fortunately, we don't need the intimate details for every virus to understand both the general process and some important overarching principles about how and why all viruses follow the same seven key steps: *attachment, penetration, uncoating, gene expression, genome replication, assembly, and release* (Fig. 2.4). However, before we explore the seven steps there are two important points to keep in mind about the viral life cycle. First, for animal viruses, the generation time (from infection to release of new progeny virions) is on the order of hours to a few days and for bacterial viruses (bacteriophage-Chap. 12) it is on the order of minutes. Contrast that to the human generation time typically considered

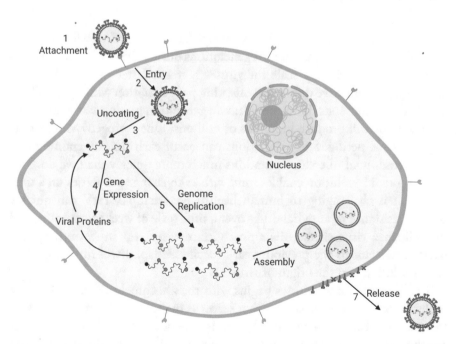

Fig. 2.4 The viral life cycle. The seven steps of the viral life cycle from cell attachment (1) to release (7) are illustrated with the entry step (2) depicted in more detail in Fig. 2.5. The life cycle shown would be typical of an RNA virus since it is occurring in the cytoplasm, while DNA viruses would be performing their gene expression (4) and genome replication (5) steps in the cell nucleus. (Created with BioRender.com)

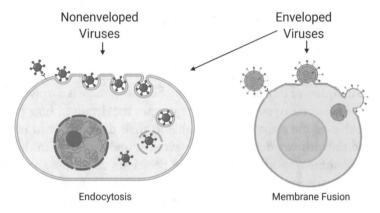

Fig. 2.5 Viral entry mechanisms. Nonenveloped viruses enter cells through the invagination of the cell membrane in a process known as endocytosis. Internalized virions are temporarily contained in intracellular vesicles and must disrupt the vesicle to release the virion into the cytoplasm. Some enveloped viruses are also taken up by endocytosis while others enter via a membrane fusion process. In the latter process, the virion membrane fuses with the cell membrane which releases the nucleocapsid into the cytoplasm. (Created with BioRender.com)

to be 20–30 years. Because of their very rapid life cycle, a virus will have gone through hundreds of thousands of generations during a single human generation, so the evolutionary scale for viruses is much faster than for humans. Secondly, humans have only a few offspring per generation while viruses have hundreds to thousands. If one new human generation produces 2–3 offspring, a single virus will produce hundreds of millions of new progeny virions during that same period. Since mutations can occur each time a genome replicates, viruses tend to collect mutations much more quickly than we do. This relatively rapid evolution enables some viruses to change properties on a time scale that is challenging to human health. For example, HIV can quickly acquire resistance to antiviral drugs over a time scale of weeks to months, and the influenza virus changes its virion surface every year in ways that often make it resistant to the previous year's vaccine. Chap. 3 will further explore virus evolution and the origin of viruses.

The life cycle for all viruses begins with the attachment of the virion to a molecule on the host cell surface that serves as the receptor for the virus. Back to our space vehicle analogy, attachment is like the docking of the space shuttle (the virion) to the space station (the host cell). Successful docking requires a very specific, tight-fitting interaction between a structure on the shuttle surface and a complementary structure on the surface of the station. Without a precise match to these complementary structures, the shuttle would be unable to attach and initiate the transfer of the cargo (the astronauts). Likewise, viral attachment requires a highly specific fit between a virion surface molecule and a receptor molecule on the cell. For viruses, the critical virion molecule is one or more viral proteins exposed on the virion surface. Receptors on the host cell are also commonly proteins, but other types of cell surface molecules such as polysaccharides or certain lipids serve as receptors for some viruses. Importantly, these cell surface receptors are all molecules whose actual function provides some needed activity for the cell. We didn't evolve our cell surface molecules as receptors for viruses; rather viruses evolved their virion surface proteins to bind our cell molecules. For example, rhinoviruses can cause common colds, and they use a cell protein called ICAM-1 as their receptor on cells in the upper respiratory tract. However, ICAM-1's real function is to facilitate cell-to-cell adhesion and the signaling between our cells, and the fact that a virus has evolved to bind ICAM-1 is irrelevant to ICAM-1's biological role. Similarly, every other type of virus has evolved to bind one (or in some cases a few) existing cell surface molecules which they use to latch onto specific target cells as the first step in establishing an infection.

The requirement for the unique virus-receptor interaction to mediate attachment has some important medical implications. First, did you ever wonder why viruses cause different types of disease? Why do some viruses cause respiratory infections (colds or flu), some cause gastrointestinal issues (vomiting and diarrhea), and others cause skin infections such as herpes cold sores? This predilection for infecting certain organ systems or cell types, known as viral tropism, is in part due to the specificity of viral attachment. There are over 200 different cell types in the human body, and any given virus will only infect the subset of these cell types that possess the appropriate receptor. If a virus requires a receptor that is only found on cells in the upper airway (nasal passages, sinuses, and pharynx), then the infection is limited to that site and the result of infection will be respiratory symptoms such as the common cold. Since the requisite receptor molecule isn't found on other cell types the virus won't be able to infect other organs like skin or the gastrointestinal tract, hence there will be no symptoms at those locations. Thus, a virus's receptor molecule distribution among our different organ systems is a prime determinant of where that virus can infect and what kind of clinical illness it can produce. We'll explore this principle of tropism in more detail in Chap. 4 where three distinct models of viral infection will be presented.

A second significant aspect of virus-receptor interactions is that this step is a critical target for our host immune system (Chap. 5). If all viral infections begin with virion attachment to receptor-bearing cells, then blocking this initial step would prevent infection and would help stop the spread of the virus within our bodies. Evolution has taken advantage of this viral vulnerability and has given us antibodies that can act as highly effective blockers. When we are exposed to viruses our immune system detects the surface proteins on the virions, recognizes them as foreign antigens, and begins the production of antibodies that bind to those viral antigens. Antibodies are themselves large proteins, so when antibodies bind to the virion surface proteins the antibodies effectively cover the entire virion surface. This antibody cloak masks the viral proteins and physically blocks the virion proteins from interacting with their receptors. Imagine the virion surface proteins as keys and the receptors as locks into which the key must fit with great precision. In this analogy, the antibodies act like masking tape wrapping around the keys to physically prevent the keys from entering the keyholes. Antibodies with this blocking capacity are known as neutralizing antibodies because they neutralize the virion's ability to infect a cell. However, the first time we encounter a new virus, antibody production takes 5–10 days to reach effective levels, and it is during this period that we are likely to become ill. Eventually, the increasing quantities of antibodies will neutralize all the virions to help us overcome the infection and return to health. We also develop an "immune memory" that allows us to make neutralizing antibodies much more quickly if we re-encounter that same virus, and this rapid response usually protects us from getting the same disease twice. Similarly, viral vaccines generally induce neutralizing antibodies, along with immune memory, thus the vaccinated individual will now quickly be able to block viral attachment and be protected even the first time he or she encounters the target virus.

While attachment is a critical first step in the viral life cycle, the attached virion is still outside the cell. To reach the cell interior and access all the cellular raw materials and machinery, the virus needs to traverse the impermeable cell membrane that is designed to keep inappropriate materials out. Consequently, all viruses have developed a penetration mechanism that moves the virion into the cellular cytoplasm (Fig. 2.5). Viruses that lack a membrane envelope on their virion penetrate by a process known as endocytosis. Attachment of the virion to its receptor induces the adjacent cell membrane to invaginate inward bringing the virion along. As the invagination progresses the membrane region with the bound virion eventually circularizes and pinches off to form a membrane vesicle in the cytoplasm with the virion inside. Subsequent virion-triggered lysis of this internalized vesicle releases the virion into the cytoplasm. Some enveloped viruses also utilize this endocytic mechanism for their penetration step. In contrast, other enveloped viruses perform a distinctly different process known as membrane fusion. In this case, the attachment of the enveloped virion to its receptor triggers a fusion or intermingling of the virion envelope with the cell membrane. The fusion of the virion envelope with the cell membrane effectively makes the interior of the virion contiguous with the cell cytoplasm, i.e. the viral nucleocapsid is now inside the cell. By either mechanism, the virion has crossed the membrane barrier and is in the cytoplasm where it can begin the actual process of converting the cell into a viral replication factory. Note, however, that after penetration the viral genome is still packaged inside either the entire virion or the nucleocapsid structure, both of which protect and sequester the viral nucleic acid. For the genome to become accessible for transcription and replication it must be released from its packaging in a step known as uncoating. Uncoating is a complex and virus-specific series of events that depend both on properties of the virion/nucleocapsid structure and on host cell "cues", i.e. biochemical signals that help trigger the disassembly of the virion structure. This process can occur at different subcellular locations depending on the virus and results in complete or partial liberation of the viral nucleic acid. For RNA viruses, the released genomic RNA remains cytoplasmic while for DNA viruses, other than poxviruses, there is an additional transport step to direct the genomic DNA into the cell nucleus.

After uncoating, the viral genomic nucleic acid becomes accessible to the regulatory factors, enzymes, and raw materials (nucleotides and ATP) needed to produce viral mRNA. As noted above in the vaccinia virus example, some of these factors and enzymes may be brought into the cell as cargo in the virion, but many are cellular proteins the virus usurps for its purpose. Once viral mRNAs are made they are translated into proteins by the host cell ribosomes. The combined production of viral mRNAs and proteins is the gene expression phase of the life cycle. During this phase, the viral genomes initially produce the proteins they need to start genome replication (non-structural proteins). Once the replication proteins are produced in sufficient quantities the replication phase begins. Viral replication is often localized to multiple discrete sites within the cell, called replication factories or replication centers, where large amounts of the replicative proteins and new replicated genomes co-localize and accumulate. As these newly synthesized viral genomes accrue they begin to express the genes that encode the new structural proteins that will comprise the progeny virions, so the gene expression and replicative phases of the life cycle are at least partially overlapping.

The penultimate step, assembly, occurs spontaneously for some viruses as their virion capsid proteins self-aggregate and naturally polymerize into the final structure. For other viruses, there are packaging proteins that chaperone or facilitate the assembly process but may not end up in the final virion structure. In either case, the nascent virions will each bind a copy of the newly made genomes and incorporate that genome into the maturing virion. Normally a virion physically only has room to package one copy of its viral genome which prevents multiple genome copies from ending up in a single virion. New virions will continue to be produced and accumulate within the infected cell, reaching hundreds or thousands of copies before release. As we will discuss in the following subsection, filling up the cell with viral proteins, nucleic acids, and new virions has detrimental effects on cell function that contribute to the pathogenesis of viral infections.

The final step of the viral life cycle is the release of the progeny virions from the infected cell so that they may infect other cells in the host and be spread to other individuals. For some human viruses, particularly those with viral envelopes, their progeny are released by a budding process where the virions extrude through the cell membrane. Two of the prominent viruses that cause human hepatitis, Hepatitis B virus and Hepatitis C virus, use this method to leave their primary target, hepatocyte cells in the liver. In contrast, most non-enveloped viruses release their progeny by fatally lysing the host cell. As with the other steps in the viral life cycle, the molecular mechanisms that different viruses use for budding or lysis vary greatly, but in all cases, the final result is the release of the newly made virions that spread to surrounding cells to amplify the infection. All existing viruses have successfully evolved to accomplish these seven life cycle steps in some type of cell. We don't yet know the complete molecular details for most viruses, therefore, exploring the staggering variation in how different viruses perform their life cycle stages is one of the exciting pursuits of virologists, biochemists, and infectious disease researchers. However, for our purposes just understanding the basic concept of the seven universal steps that all viruses undergo to reproduce is sufficient.

Infection and Disease-How Viruses Spread and Do Nasty Things

The human body has eleven organ systems: circulatory, digestive, endocrine, integument, lymphatic, muscular, nervous, renal, reproductive, respiratory, and skeletal. These systems work continuously and so effectively that we rarely notice their efforts unless something goes awry. Individually, each of these systems consists of millions of cells that interact to provide a specific function for our body. For example, the circulatory system pumps both our blood and many cells of our immune system, the digestive system breaks down and absorbs the food we ingest, and the skeletal system provides the structural framework for our bodies. The functional integrity of these systems depends on having a sufficient number of the component cells communicating with each other in a coordinated fashion. The cells in an organ are somewhat like ants in a colony with all the different types of ants each contributing to the overall health and maintenance of the colony. If you remove or kill too many of the ants, the colony becomes unhealthy and may not survive. Similarly, as a virus reproduces and infects more and more of the cells in one or more organ systems, there are negative consequences that manifest as disease symptoms. A single infected cell that supports the viral life cycle can produce hundreds to thousands of progeny that can each infect a new cell. Each of these additional infected cells can likewise produce and release many scores of new virions, leading to millions of infected cells within only a few viral generations. As viruses hijack the cellular machinery and redirect it towards the synthesis of new virions, the millions of infected cells begin to lose some or all of their ability to repair and maintain their own proteins and nucleic acids (DNA and RNA). Additionally, the accumulation of viral components and nascent virions in replication factories within the infected cells begins to "clog up" the cell and can further reduce the cell's ability to carry out its normal function. Eventually, the infected cell's normal biochemical functions are disrupted and such cells have reduced ability to contribute to the overall activity of the organ. As more and more cells are impacted and have diminished functional capacity, the overall ability of that organ system to perform its biological tasks is decreased. Ultimately, as hundreds of thousands of cells lyse to release the newly made virions, the organ may lose so many cells that its overall functions are severely impaired. For example, viral destruction of millions of cells in the intestines causes a failure of water reabsorption which results in diarrhea. Even for viruses that don't lyse the host cells, they will usually induce an immune response that will attempt to destroy the infected cells (Chap. 5). In this case, it is our own bodies trying to stop the infection that kills our cells, but the resulting organ dysfunction is the same whether the virus kills the cells or the immune system kills the cells. To summarize, while there are other important nuances to specific viral diseases (Chap. 4), the general mechanism of viral pathogenesis is cell dysfunction and death, with the clinical symptoms determined by which organ system(s) is targeted by the invading virus.

Completing its life cycle and spreading within an infected host to generate millions of progeny is only one aspect of a successful virus. To survive in nature, a virus must not only reproduce within the infected individual but must also spread to new hosts. Paradoxically, viruses that were too virulent and killed their host very rapidly likely would have been evolutionarily unfit and unable to survive. If the host perishes before the virus jumps to a new host then that virus and its progeny are trapped in the dead host and lost for further infection. Additionally, if the virus was too lethal it might quickly wipe out its entire host population and again would be lost when there were no hosts left to infect. Ideally, a virus needs the host to survive or at least persist long enough to ensure a good chance of the virus moving to a new host before the original host succumbs. The endemic human viruses that exist today are those that have solved this evolutionary quandary. These successful endemic viruses have established an effective balance between viral reproduction with its associated morbidity and mortality, and the need for efficient transmission to the next host.

The movement of infectious agents from one host to another, termed horizontal transmission, is part of the science of epidemiology which seeks to understand how diseases spread so that we might control or eliminate them. Since most viruses do not remain infectious out in the environment for more than a few hours to days, they must generally move from host to host to continue their existence. For many common human viruses, the only suitable host is humans, so these viruses are considered endemic as they are always actively present in human populations. Such viruses move continuously from human to human which means that there are always cases of the disease somewhere in the world. The disease may migrate geographically with seasons, and it may have peaks and lulls throughout the year, but it is never entirely gone. Diseases such as influenza or measles are examples of these endemic viruses that constantly plague the human race. Interestingly, it is theoretically possible to completely eliminate viruses that only exist within the human population (see Chap. 6). This was accomplished for smallpox in the 1960s and 1970s through an intensive global vaccination campaign. Slowly the spread of the virus decreased because vaccination resulted in fewer and fewer people who were still susceptible to smallpox infection. As the number of non-immune people diminished the likelihood of transmission in communities decreased steadily. Once several years passed without a single case of smallpox worldwide the virus was declared eradicated in 1980 and it no longer exists in nature. A similar campaign has been underway for many years for poliovirus with the World Health Organization originally proposing to abolish this disease by the year 2000. Unfortunately, small pockets of cases continue to occur each year in areas that are often remote, war-torn, or impoverished, and where vaccine coverage is not complete. In 2019 there were 130 reported cases, and while this is a small number, the world still remains at risk from this virus. Optimistically, the end is near when polio should no longer exist in the world, but until then vaccination remains imperative to prevent a possible resurgence of this dreaded virus. As with smallpox, if several years go by without any polio cases, then the virus will be deemed eliminated and vaccination for this disease will be discontinued. Likewise, other common childhood diseases such as measles, mumps, and rubella (German measles) also only transmit from human to human and could be eradicated someday if vaccine compliance was more universal.

In contrast to viruses that only exist in humans, some human disease-causing viruses have other nonhuman hosts that we refer to as reservoirs. Rabies lurks in many animals, including bats, foxes, skunks, and raccoons; West Nile virus persists in birds; types of influenza are hosted by birds or swine, and HIV originated in non-human primates though it is now established in the human population. Like the HIV precursor, Ebola is another primate virus that has repeatedly jumped into humans in recent years. Animal viruses that transmit to humans are called zoonotic diseases (Chap. 11), and human infection is often associated with bites, scratches, or an insect vector that carries the virus from the animal to the human. Such animal-to-human infections are less frequent than human-to-human transmission due to our limited and incidental contact with virus-carrying animals and our insect control programs in the United States. Additionally, zoonotic infections are often confined to the initially infected individual as animal viruses are specifically adapted to their authentic animal host and generally are poorly transmissible within human populations. Nonetheless, animal virus genomes may acquire mutations that expand or change their host range in ways that now allow human to human transmission. This likely happened with the simian virus that became HIV and the animal virus that became SARS-CoV-2, allowing these nonhuman viruses to become significant human pathogens. Unfortunately, viruses with animal hosts would be much more difficult to ever eradicate as the virus would need elimination not only in humans but in the entire animal population which is a daunting project.

The third and final source of viruses that infect humans is the environment. There are a limited number of human viruses that are capable of surviving for prolonged periods outside the host. These viruses come from several different viral families, but share the property that they infect us via ingestion of contaminated food or water. This requires that they pass through the stomach and into the intestines where they will initially infect cells and begin their viral life cycle. As the stomach is very acidic and full of digestive enzymes, viruses that traverse this route have evolved to have very tough and resistant capsids, and none of them have membrane envelopes, as membranes would be easily destroyed in the stomach. When these viruses replicate they usually cause gastrointestinal distress and eventually new progeny virions are shed in the feces. Since the virions have highly durable capsids, they can remain inert but viable for days, weeks, and even months in water or soil. When humans drink contaminated water or eat uncooked food with virions on the surface they become infected and start the cycle all over again. Even touching tainted material and then inadvertently transferring virions from your fingers to your mouth can be sufficient for infection. Rotavirus is a classic example where an infected person sheds enormous numbers of virus particles in their stool and unknowingly contaminates their surroundings, for example, door handles and faucets. These shed virions persist for prolonged periods until another unsuspecting person touches the object, thereby picking up the virus on their hand and eventually transferring it to their mouth. Another common example in the United States is the hepatitis A virus that is frequently associated with ingesting contaminated fruits or vegetables. Unfortunately, hepatitis A can also be spread from person to person and by blood exchange which is common for IV drug abusers who share needles. Since one infected individual can trigger a spread to several other people, the result is that we continue to have mini-epidemics of hepatitis A each year in the United States even though there is a very efficacious vaccine available.

Several of the examples described above involve modes of viral spread by ingestion or via bites from animals and insects. However, the far more common mechanism for human viral transmission is the respiratory route. Virtually all the major childhood viral illnesses (measles, mumps, rubella, chickenpox, respiratory syncytial virus, influenza, and fifth disease), as well as multiple virus families and strains that can cause "colds" (rhinoviruses, adenoviruses, coronaviruses, and parainfluenza virus), are spread via inhalation. Remember, as viruses complete their life cycle they damage and destroy the infected cells, in this case, cells in the respiratory tract. As dysfunctional and dying cells accumulate the integrity of the airway is impacted and we begin to have congestion leading to coughing and sneezing. These mechanical responses are ideally suited to disperse the progeny viruses into microscopic droplets. Such droplets can be inhaled by nearby individuals or can land on surfaces where they are picked up on our hands and then self-transferred to the oral-nasal region. Cellular injury and demise are not only an inevitable consequence of viral reproduction that produces the clinical symptoms, but it may also be the device that facilitates the dispersal of the virus to new hosts! Since none of us can avoid breathing for long periods or completely avoid being around other people, respiratory infections plague us constantly with 1–2 such infections per year throughout our lifetimes. This is dramatically illustrated by the COVID-19 pandemic that spread worldwide in a matter of a few months (Chap. 11). It's no wonder that so many viruses have evolved to utilize this respiratory route as this is the most efficient way to ensure a high likelihood of moving from host to host.

In Chap. 4 we will examine three examples of established pathogenic human viruses and delve more deeply into their disease features, while in Chap. 11 we will explore how new viruses enter human populations. For both existing and emerging diseases, there is an important epidemiological principle known as the reproductive number (R_0). This number is a measure of the potential for transmission. Mathematically the R_0 reflects the duration of the infection, the likely number of contacts during the infectious period, and the probability of transmission per each contact with another person. In practice, R_0 specifies the number of new individuals likely to be infected by a single infected person during the period of their illness, for example, an $R_0=3$ means that each infected person will likely infect 3 others. This number can vary widely for different viruses, but for all established human illnesses, R_0 is 1 or greater as an R_0 less than 1 would lead to the eventual loss of the virus from the host population. Measles, a highly infectious respiratory virus, has a reported R_0 of 12–18 when introduced into a completely susceptible population, while Ebola, which spreads via direct contact with bodily fluids, only has an R_0 in the 1–3 range. Note that the R_0 can be diminished by reducing any of three factors: duration of disease, number of contacts, and/or transmission probability. Shortening the illness duration by antiviral drug treatment or decreasing the number of contacts by quarantining infected individuals are both effective strategies to reduce the R_0. Importantly for public health, the final factor, the probability of transmission, can be greatly reduced by decreasing the number of susceptible individuals through vaccination. If most of the persons who come in contact with the infected individual are already immune due to vaccination, then there is little likelihood of transmission hence the disease will die out quickly. This concept that disease spread can be effectively reduced by increasing the proportion of immune individuals is known as herd immunity. For most human viral diseases the vaccination coverage must be roughly 85–95% of the population for herd immunity to be effective. Thus, while unchecked viral disease caused high morbidity and mortality throughout human history, the routine vaccination schedules introduced in the last half of the twentieth century have provided enough herd immunity to make many once-common diseases now rare (see Chap. 10). Unfortunately, unless the causative virus is completely eliminated as for smallpox (and hopefully soon for poliovirus), even these now rare diseases can return with a vengeance if vaccine compliance dwindles below the levels necessary for herd immunity.

Additional Reading

1. Virus taxonomy in the age of metagenomics. Nature Reviews Microbiology, 15:161–168. 2017.
2. What is a virus species? Radical pluralism in viral taxonomy. Studies in History and Philosophy of Biological and Biomedical Sciences, 59:64–70. 2016.

Definitions

Adenovirus – A type of double-stranded DNA virus mostly associated with respiratory infections.

Antibodies – Proteins made by B cells that recognize and bind to foreign antigens.

Antigen – Any foreign substance introduced into the body which elicits an immune response.

Bacteriophage – a virus that infects bacteria.

Chickenpox virus – A double-stranded DNA virus of the herpesvirus family that causes chickenpox and shingles.

Class – A biological classification that comes between Phylum and Order.

Coronavirus – A type of single-stranded, positive-sense RNA virus usually spread via the respiratory route.

COVID-19 – The disease caused by the SARS-CoV-2 virus.

Ebola virus – A type of single-stranded, negative-sense RNA virus that causes a dangerous hemorrhagic disease.

Electron microscopy – A technique using electron bombardment rather than light to obtain high-resolution images of objects such as viruses that are too small to be seen with a regular light microscope.

Endemic viruses – Viruses that are constantly present in a specific geographical region.

Endocytosis – The process by which materials enter a cell through the invagination of the cell membrane.

Epstein-Barr virus – A double-stranded virus of the herpesvirus family that causes mononucleosis and is associated with several types of cancer. Also known as HHV4.

Family – A biological classification that comes between Order and Genus.

Fifth disease – A mild febrile disease with a rash caused by parvovirus B19.

Gene expression – The process of converting genetic information into protein. RNA polymerases transcribe genes into mRNA and then the mRNA is translated by the ribosomes into protein.

Genera (singular: genus) – A biological classification that comes between Family and Species.

Generation time – The length of time between infection of a cell by a virus and the first release of new viral progeny.

(continued)

(continued)

Genome replication – The process of synthesizing new copies of the genomic nucleic acid of an organism or virus.

Hepatitis – An inflammation of the liver commonly caused by viral infection.

Hepatitis A virus (HAV) – A virus in the picornavirus family with a single-stranded, positive-sense RNA genome that is one of three major agents of viral hepatitis.

Hepatitis B virus (HBV) – A virus in the hepadnavirus family with a double-stranded, DNA genome that is one of the three major agents of viral hepatitis.

Hepatitis C virus (HCV) – A virus in the flavivirus family with a single-stranded, positive-sense genome that is one of the three major agents of viral hepatitis.

Hepatocytes – A liver cell.

Herd immunity – The condition that occurs when a sufficiently large portion of a population is immune to a disease, thus making the person-to-person transmission unlikely.

Horizontal transmission – The spread of a virus from person to person.

ICAM-1 – A membrane protein involved in cell-to-cell interactions and used as a receptor for coxsackieviruses.

Immune memory – The ability of the immune system to remember previous exposures to pathogens. This memory allows the immune response to be much more rapid on a re-exposure to the pathogen.

Immune system – The network of bodily defenses (cells, tissues, proteins) that recognize and fight infections and other diseases.

Influenza virus – A member of the orthomyxovirus family with a segmented, single-stranded, negative-sense RNA genome. This virus is the agent of seasonal influenza.

Kingdom – A biological classification that comes before Phylum.

Lateral body – A protein-rich structural region in the poxvirus virion that is located between the viral membrane and the core.

Measles virus – A member of the paramyxovirus family with a single-stranded, negative-sense RNA genome. This is the causative agent of measles.

Mumps – An infection of the salivary glands caused by an RNA virus in the paramyxovirus family.

Negative-sense genome – An RNA virus genome that cannot be directly translated by ribosomes into protein. For protein production, the negative-sense RNA must first be copied to the complementary positive-sense RNA.

Neutralizing antibodies – Antibodies that bind to surface proteins on virions and prevent attachment of the virions to host cell receptors.

Nucleocapsid – The combination of a virus genome plus its capsid structure.

Oncogenic – Capable of causing the development of a tumor.

Order – A biological classification that comes between Class and Family.

Parainfluenza virus – A member of the paramyxovirus family with a single-stranded, negative-sense RNA genome. They are generally associated with respiratory infections.

Phyla (singular: phylum) – A biological classification that comes between Kingdom and Class.

Polio – Formally poliomyelitis, an infection of the brain or spine caused by the poliovirus.

(continued)

(continued)

Poliovirus – A member of the picornavirus family with a single-stranded, positive-sense RNA genome. This virus is the causative agent of poliomyelitis.

Polymerase – An enzyme that catalyzes the formation of a polymer from smaller subunits, e.g. DNA polymerase makes DNA from deoxyribonucleotides (A, C, T, and G).

Polysaccharides – A polymeric chain composed of sugar molecules linked together.

Positive-sense genome – An RNA virus genome that can be directly translated into protein by ribosomes.

Rabies virus – A member of the rhabdovirus family with a single-stranded, negative-sense RNA genome. This virus is the causative agent of rabies.

Replication factories – Intracellular compartments where viral genome replication and progeny virion assemble occurs.

Reproductive number (R_0) – A mathematical term indicating the contagiousness of an infectious disease. It represents the average number of people who will be infected by a single infected individual.

Respiratory syncytial virus (RSV) – A negative-sense, single-stranded RNA virus of the paramyxovirus family that causes respiratory infections.

Rhinovirus – A member of the picornavirus family with a single-stranded, positive-sense RNA genome. Rhinoviruses are typically associated with mild respiratory illness, ie. colds.

Rotavirus – A member of the reovirus family with a segmented, double-stranded RNA genome. Rotaviruses cause diarrheal disease.

Rubella – A positive-sense, single-stranded RNA virus of the togavirus family that causes German measles.

SARS-CoV-2 – A member of the coronavirus family with a single-stranded, positive-sense RNA genome. This virus is related to the original SARS virus and is the causative agent of COVID.

Smallpox virus – A member of the poxvirus family with a linear, double-stranded DNA genome. This virus causes smallpox.

Species – A biological classification that comes after Genus.

Taxonomy – The field of study that describes, characterizes, and classifies organisms into hierarchies of relatedness.

Uncoating – The step in the viral life where the virion partially or completely dissembles to release the viral genome for gene expression and replication.

Vaccinia – A member of the poxvirus family with a linear, double-stranded DNA genome. This virus is used as the vaccine for smallpox.

Variola major – The scientific name for the smallpox virus.

Vesicle – A small, fluid-filled structure enclosed by a lipid membrane.

Viral tropism – The predilection of a virus for infecting a specific type of cell or organ.

West Nile virus (WNV) – A member of the flavivirus family with a single-stranded, positive-sense RNA genome. Transmission is via mosquitoes and infection with this virus can cause severe encephalitis in some patients.

(continued)

(continued)

X-ray crystallography – A technique where X-rays are used to determine the atomic structure of a protein that has been prepared in a crystalline state.

Zika virus – A member of the flavivirus family with a single-stranded, positive-sense RNA genome. Transmission is via mosquitoes.

Zoonotic disease – A disease that is transmitted from animals to humans.

Abbreviations

COVID-19 – coronavirus infectious disease 2019
ICAM-1 – intracellular adhesion molecule 1
ICTV – the International Committee on the Taxonomy of Viruses
NS – nonstructural
R_0 – reproductive number
S – structural
SARS-CoV-2 – sudden acute respiratory virus-coronavirus-2

3

Ancient or New – On the Origin of Viruses

Keywords Last universal cellular ancestor (LUCA) • Viral evolution • Virus-first hypothesis • Cell-first hypothesis • Papillomaviruses

> *Endless forms most beautiful and most wonderful have been, and are being evolved.*
> Charles Darwin

We Don't Know Exactly Where We Came From, But We're Here Anyway

Viruses have been human pathogens throughout our evolutionary history. Even today they remain banes for humanity since we lack vaccines or treatments against many viruses that cause diseases ranging from mild to fatal. While incredibly small, viruses powerfully attack and usurp our cells to create misery and destruction; even a small number of these invading microscopic virions can defeat the entire body, leading to illness, permanent damage, or death. It seems unfair that these harmful entities populate our environment and infect us throughout our lifetimes. Did you ever wonder why these potent adversaries plague us or where they came from? What purpose do viruses serve and why are they so diverse and plentiful? Absolute answers to these questions may be unobtainable, but research over the last 50 years has pieced together more and more accurate models of how viruses arose and their relationship to other life forms. They likely are as primordial as life itself and have inexorably co-evolved with their more visible cellular brethren. To understand this fully we need to go back a long, long time.

The Big Bang theory postulates that the primordial universe existed in an incredibly hot, dense state. Around 13 billion years ago this matter suddenly began to expand. As the expansion progressed and the primordial material cooled, all the subatomic particles began to form, eventually leading to the creation of the atoms that comprise all the matter of our universe. Over many millennia, the matter coalesced into the 100 billion galaxies, the 10^{21} stars, the unknown number of planets, and all the other objects that exist in our universe today. About 4.6 billion years ago, part of the gigantic, expanding cloud of matter collapsed to create our sun. During the next 100 million years the small, insignificant planet we call home began to develop from the gas and dust remnants of the Sun's formation. Slowly the earth cooled, water formed, and the nascent oceans became chemical soups filled with small inorganic molecules such as methane (CH_4), ammonia (NH_3), and carbon dioxide (CO_2). At some point, these small molecules began to combine to form more complex precursors to biological compounds. How this actually initiated is likely unknowable, though several models that could account for these early synthetic reactions have been proposed. In the classic model proposed in 1953 by Miller and Urey, they demonstrated that applying energy to methane, ammonia, and hydrogen in water created organic amino acids, the building blocks of proteins. Miller and Urey used an electrical discharge as the energy source to mimic lightning, but other sources of energy existed on primordial earth, including meteor strikes, hydrothermal vents, and solar radiation, so there was no shortage of potential energy sources. Precursors to nucleotide bases, the basic components of DNA and RNA, likely arose by similar reactions of available inorganic compounds. As simple organic molecules grew abundant, polymerization into longer chains and more intricate organizations became feasible. Slowly, biomolecules became increasingly complex and gained the ability to self-replicate. Notably, these first self-replicating molecules likely were catalytic RNAs as even today catalytic RNAs exist that perform protein-like enzymatic functions. Such primordial RNAs eventually crossed over from being simply chemicals to being life forms, with DNA-based genomes being a later development. We don't know how this cellular life step happened or when it happened, but the first known microfossils date from 4.1 billion years ago, so we know it occurred before that time. At some point in this heterogeneous mixture of biochemicals and nascent cells, there existed a single cellular entity, the Last Universal Cellular Ancestor (LUCA) that won the evolutionary battle and became the progenitor of all subsequent life on earth (Fig. 3.1). LUCA undoubtedly lacked a nucleus, although it ultimately gave rise both to organisms lacking a nucleus (prokaryotes and

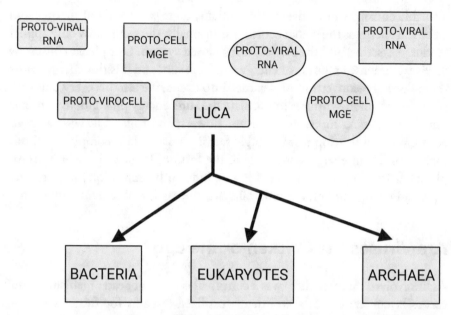

Fig. 3.1   The tree of life. On primordial Earth, there was likely a multitude of primitive replicative entities (proto-viral RNAs, proto-cells, and/or proto-virocells) that were the precursors to modern cells and viruses. Included in this mix were mobile genetic elements (MGEs) that were small segments of RNA that could jump from one location in a larger RNA to another location within the same RNA. From this complex and competing mix entities arose LUCA, the Last Universal Common Ancestor. LUCA became the dominant life form and is the seed of the Tree of Life that gave rise to all three of the modern lineages of life, bacteria, eukaryotes, and the archaea. Because all life on Earth arose from a single common ancestor, all life shares a common genetic code and uses the same basic process to convert DNA to RNA to protein. (Created with BioRender.com)

archaea) and to those with a nucleus (eukaryotes). Once unicellular eukaryotes appeared it was only a matter of time before cellular aggregates developed leading to all the diverse multicellular plants and animals existing today, including humans. Because all current life is derived from LUCA, this is why all plants, animals, and bacteria share a common genetic code and a common protein synthesis process involving mRNAs being translated into proteins via ribosomes, or so we think!

The concept that there was spontaneous formation of complex biomolecules that developed into organic life on earth is often criticized by citing the second law of thermodynamics. This law declares that all systems must move towards entropy (disorder) over time. The entropy principle seems at odds with the theory that small, inorganic molecules spontaneously became more and more complex over time to generate biomolecules and eventually life forms. However, the flaw in this criticism is that as long as sufficient energy is available entropy can be overcome. Initially, there was plenty of thermal and electrical energy as the planet cooled. Additionally, the second law only applies to closed systems, and the earth is not a closed system. Our planet constantly receives immense amounts of energy from the sun, and it is this energy influx that allows the conversion of disorder into order. We see this every day with plants using photosynthesis to translate the sun's energy plus simple nutrients into their growing roots, stems, leaves, flowers, and fruits. Likewise, animals each grow and develop from a single fertilized ovum into complex adults as long as sufficient energy is available in the form of food. So generating complexity from simple constituents is not inherently impossible, it just takes time and energy, and on a geological scale, the earth is abundant with both.

Are Viruses the Chicken or the Egg?

As life evolved on earth there was a continual capture of dead organisms in the earth's layers leading to a wealth of fossilized remains for future scientific perusal. This fossil record allows us to peer billions of years into the past to visualize not only our hominid ancestors but also the microscopic microfossils that represent the earliest cellular life forms. Unfortunately, viruses are too small and fragile to fossilize so they can't be detected this way. Without physical remains to inform us, their ancient history remains mysterious. The alternative to fossil tracing is the analysis of genome mutations which allows construction of genetic family trees for plants, animals, and bacteria. Based on known mutation rates, scientists can align these trees with the fossil record to show relatedness and critical branch points where species diverged. Unfortunately, genetic analysis is much more challenging for viruses. As noted in Chap. 2, because of their rapid life cycles and huge numbers of progeny per generation, viruses mutate much more rapidly than other organisms. Consequently, genetic analysis of their relatedness loses accuracy much sooner than for other life forms, somewhere around a million years ago. Thus, based on genomes alone it has been impossible to trace viral origins or to identify common ancestors with other life forms as these progenitors existed billions of years ago, not a mere million years. However, since scientists thrive on imagination the lack of hard data hasn't stopped them from speculating about the origin of viruses. A priori, there is a critical dichotomy in the concept of viral origination: either the ancestral precursor(s) of modern viruses arose independently from cellular life forms (the virus-first hypothesis) or cellular life forms came first and viruses derived in some fashion from cells (the cell-first hypothesis). If the first premise is correct then viruses are likely ancient and predate or are contemporaries of the progenitor life form, LUCA. In the second supposition, viruses could have arisen at any time after cellular life developed and could be much younger evolutionarily than the original cellular ancestor. In examining the two possibilities, the second scenario might intuitively seem more probable. Since viruses are absolutely dependent on cells for their reproduction, how could viruses arise before cells existed? Nonetheless, this simple conclusion is not necessarily correct. As noted above, the original precursor to cellular life might have been self-replicating, catalytic RNA molecules. It is equally plausible that these primordial RNAs might be the precursors to both cells and viruses, i.e. they were proto-virocells that diverged to form both the cellular lineage and the viral lineage. In this scenario, the progenitor viruses were not cell-dependent and could replicate autonomously with cell-dependence evolving later on. Thus, these two major hypotheses are both theoretically plausible and are not mutually exclusive as different viruses may have arisen independently by either pathway. Like the chicken and the egg conundrum, we may never have a completely conclusive answer to this fundamental question of which came first, the virus or the cell, but that doesn't stop the search for more data and a better understanding of viral origins.

Hypotheses, Hypotheses, and More Hypotheses

The virus-first hypothesis was originally proposed by the French-Canadian microbiologist, Félix d'Hérelle, nearly 100 years ago and was refined over time. In its modern-day form, this model speculates that autonomous, self-replicating RNA nucleic acids ("pre-viruses") existed prior to or during the early existence of cell-based life forms. These viral precursors were not dependent on cells for their replication which allowed them to diversify and evolve in complexity while cellular life was also developing. Exactly how far RNA-based pre-viruses evolved as independent entities is likely unknowable. Over time some may have acquired a primitive capsid structure and others may have diverged into DNA-based pre-viruses. At some point, after LUCA arose, descendent cells were invaded by various viral precursor forms. Forms that found the intracellular environment favorable for replication flourished, reproduced, and became established. Over time these forms became dependent on the host cell and lost their ability to replicate autonomously outside the confines of the cellular milieu, becoming obligate intracellular parasites that are the forefathers of modern-day viruses.

Two major pieces of evidence support the virus-first hypothesis and give considerable credence to its basic tenets. First, the vast majority of known viral genes have no cellular counterpart. If viruses had somehow originated from cells then all viral genes should bear some resemblance (known as homology) to cellular genes, and this is not the case. This lack of discernible homology between most viral genes and known cellular genes suggests that the ancestral viral genetic information arose independently of the LUCA lineage of cells. Although this genetic dissimilarity doesn't indicate whether the virus precursors appeared before or after cellular life began, it does strongly favor an independent origin before cells were invaded. The second piece of supporting evidence comes from the existence of common plant pathogens called viroids that are spread from plant to plant by insects. Viroids consist solely of a single piece of circular, single-stranded RNA. They lack a capsid, and their genomes don't encode any protein, so they aren't technically viruses. Once introduced into plant cells the viroid RNA replicates using a cellular enzyme, RNA polymerase II, which normally makes mRNA for the cell. Importantly, some viroid RNAs have retained demonstrable catalytic activity, a function that likely came from their self-replicating, ancient ancestors. Consequently, viroids look suspiciously like a remnant of an intermediate step in virus evolution where a catalytic RNA has invaded a host cell, become dependent on the cell for its replication, but hasn't yet evolved a capsid for transmission between hosts. Interestingly, there is a human pathogen, the hepatitis D virus (HDV), which resembles the plant viroids. Like viroids, the genome of HDV is also a small, circular, single-stranded, catalytic RNA molecule that replicates using the host cell's RNA polymerases rather than by a viral-encoded enzyme. Unlike viroids that produce no viral proteins, the HDV genome does encode a single viral protein called the hepatitis D antigen (HDAg) that associates with the viral RNA. A second distinction between viroids and HDV is that HDV is packaged into a virion for transmission while viroids are never encapsidated. Although HDV does not encode any virion capsid proteins, HDV is packaged when it infects a cell that is co-infected with the hepatitis B virus (HBV). HBV is a DNA virus that is one of the three major causes of human viral hepatitis, the others being hepatitis A virus and hepatitis C virus. During co-infection of cells with HDV and HBV, the HDV RNA-HDAg complex gets packaged into virions comprised of HBV proteins. These co-infected cells thus produce authentic HBV virions along with bogus virions containing the HDV genome rather than the HBV genome. HDV RNA packaged in HBV virions is infectious for new cells just as are the authentic HBV virions. Because it is defective when alone and absolutely dependent on HBV to propagate, HDV is termed a virusoid or satellite virus rather than a true virus.

Recent sequencing studies have shown that the HDV-like virusoids are widespread in a variety of animals. Given the prevalence of virusoids, perhaps these entities are another fascinating example of evolutionary intermediates that descended from self-replicating RNAs but have not yet acquired the ability to encode their own virions.

One wrinkle to the virus-first hypothesis is whether or not all modern-day viruses originated from a single ancestral cell invasion by a single pre-virus or if there were independent cellular invasions by distinct pre-viruses leading to multiple, unrelated viral lineages. In the first case, all viruses existing today would be related and would share at least some commonalities at the gene and protein levels. Since universal commonality is not seen, the prevailing belief is that distinct viral lineages arose independently in evolutionary history, a concept known as polyphylogeny. Nonetheless, these initiating invasion events must have occurred in the far distant past. For example, there are well-documented examples of seemingly very different modern-day bacterial viruses and eukaryotic viruses that share homologies in their replicative enzymes, suggesting an ancestral relatedness. Since prokaryotic bacteria and eukaryotes (plants and animals) diverged 2–3 billion years ago, the viral progenitor must have existed prior to that split, consistent with an incredibly ancient origin for modern viruses.

Even though the virus-first hypothesis has solid supporting evidence, the cell-first hypothesis also has ardent advocates. There are two major models for a viral origin deriving from cell-based life forms: the escape or progressive hypothesis and the regressive or degeneracy hypothesis. The basis of the escape hypothesis depends on so-called mobile genetic elements (MGEs) that exist in plants and animals today, both as active elements and as inactive evolutionary remnants. These MGEs come in a variety of types with distinctive properties and mechanisms for transmission, including elements known as transposons, retrotransposons, and insertion sequences. For our purposes, we can ignore the molecular and biological details of how these MGEs function. Instead, we can focus on the underlying principle that all of these elements are relatively small pieces of nucleic acid that can hop from one position to another within the genome of a cell. Similar to viruses, MGEs have a complex evolutionary history and likely preceded and/or co-evolved with cellular life. The escape hypothesis speculates that an ancestral form of MGE acquired the ability to leave its original cell and jump between cells rather than only jumping from place to place within the genome of a single cell. At some point this jumping element acquired capsid forming capacity that both protected the element and facilitated cell to cell transmission, thus becoming the progenitor of modern viruses. How MGEs might have accomplished their escape and acquired capsids is unknown. As for the virus-first hypothesis, to account for the lack of a single gene conserved among all viruses, it is necessary to postulate that there were multiple escapes by different ancestral MGEs leading to different viral lineages.

The second cell-first hypothesis, regression, proposes that viruses arose from cellular organisms that degenerated and lost their cellular features, their biochemical metabolism, and their ribosomal machinery for protein synthesis. Stated another way, viruses might be stripped down cells that lost the large majority of their properties and retained only enough of the genome to encode critical viral needs such as the capsid and essential replicative proteins. Modern plants and animals have thousands to tens of thousands of genes to provide all the functions that their cells need, while modern viruses typically have less than 100 genes. Although early life forms would have had far fewer genes than their modern descendants, the viral precursor entities could have arisen by slowly losing parental cellular genes until only a minimal set of genes remained. As these pre-virus organisms regressed in complexity, they eventually lost their independent replication ability and became parasites of their more complete brethren – like an inconsiderate relative that lives in your house, eats your food, and wears your clothes. Exactly how this genetic reduction occurred is unclear, but for either the regression and escape hypotheses to account for observed similarities between some modern bacterial and animal viruses, the critical reduction events again must have occurred before prokaryotic and eukaryotic life diverged 2–3 billion years ago.

Evidence to support the regression hypothesis comes from the identification in 2003 of the first so-called "giant" virus, *Acanthamoeba polyphaga* mimivirus (APMV). First detected in 1992, APMV was found inside amoebae isolated from a water cooling tower. Because of its huge size (easily visible by light microscopy) and since other types of bacteria are known to infect amoebae, APMV was assumed to be an unknown bacterium. It was another 10 years before investigators finally realized they were dealing with a gigantic new virus rather than another bacterial pathogen. Subsequent analysis of this new virus revealed several remarkable features. First, its virion capsid is gigantic with a diameter of 500 nanometers. This APMV capsid is even larger than some bacteria and is more than twice the size of poxvirus which had been considered one of the largest viruses. Second, its linear, double-stranded DNA genome has 1.18 million base pairs. This is approximately five times longer than the poxvirus genome, roughly 150 times longer than the genome of a more typical DNA virus such as human papillomavirus, and is even larger than the genomes of some unicellular eukaryotic organisms. Sequencing this mammoth genome revealed that it could encode over 900 predicted proteins, again 4–5 times more proteins than encoded by poxvirus. Third, while some APMV genes show relatedness to genes from other large DNA viruses, such as poxviruses and adenoviruses, many of the APMV genes have never been seen before in any known viruses. More importantly, a number of these unique APMV genes encode proteins that resemble cellular proteins involved in metabolism and protein production, two classes of activities that are not part of the standard virus repertoire!

Over the last 16 years, a variety of other giant viruses that share similar features with APMV have been discovered, confirming that this new group of viruses is widespread and not a singular anomaly. Genes related to metabolism and protein production are present in all the giant viruses, a finding compatible with the regression hypothesis. These giant viruses could represent intermediate forms that descended from a cellular forefather but had not yet lost all the cellular genes lacking in traditional viruses. At the same time, the opposite possibility of gene acquisition cannot yet be ruled out. Some scientists maintain that giant viruses may have started as much smaller and more traditional viruses that have inadvertently picked up genes from their host cells over time. MGEs in host cells could easily have jumped into invading viral genomes during infection and carried cellular genes into the virus DNA. Once inserted, some of these acquired genes may have provided increased fitness that led to a selective advantage of those viral progeny. Even acquired cellular genes that were totally inconsequential would persist as long as they didn't produce a selective disadvantage, so continual accretion of new genes into viruses may be commonplace in host cells with highly active MGEs. Future genetic analysis of yet to be discovered new giant viruses may help to resolve this dichotomy, but for now, the origin and evolution of these amazing viruses remain murky. Still, the very existence of giant viruses remains a tantalizing hint that at least some viral lineages may have developed through cellular regression.

Finally, a caveat for both the escape and regressive hypotheses is the lack of homology between a majority of viral genes and existing cellular genes. This is seen for both typical viruses and the known giant viruses. If viruses really did derive from LUCA or a cellular descendent, then we would expect detectable similarity between nearly all existing viral genes and cellular genes. However, the majority of known viral genetic information is unique to viruses and lacks a cellular counterpart which seems to argue against viruses deriving from cells. Of course, we can't rule out that virus progenitors escaped or regressed from pre-LUCA cellular life forms whose genetic lineages no longer exist on earth. In this case, the unique viral genes represent the sole remaining descendants of ancient genetic information from extinct life forms that aren't represented in any cellular organisms that inhabit the earth today. This is a difficult question to resolve, so the arguments and counterarguments continue to be debated in the scientific community.

New Data at Last

While the multiple origin hypotheses we've discussed may seem fruitless conjecture at first, hypotheses are critical elements of scientific research as they form an intellectual framework against which new results are evaluated. New findings are applied to each hypothesis to determine if the data support or refute the hypothesis. If data cannot be explained by a particular hypothesis then that hypothesis is revised or eliminated. In contrast, if a preponderance of evidence supports one hypothesis over the others then that hypothesis gains strength and the others are diminished. Clever scientists are constantly trying to develop new approaches that will provide useful data to distinguish between elements inherent in competing hypotheses. Recent pioneering studies looking at similarities in viral and cellular protein structures have provided a new window into ancient events.

As noted above, the high mutation rate of viral genomes limits estimations of viral and cellular relatedness through RNA and DNA sequence analysis. Mutations in viral genes often result in amino acid changes in the corresponding viral proteins, so simply comparing amino acid sequences between viral and cellular proteins is also limited for estimating evolutionary relationships. Fortunately, there is another structural aspect of proteins that is much more highly conserved than is their linear amino acid sequence. Proteins do not exist as simple, string-like linear chains of amino acids (the primary structure). Instead, the amino acid chains fold up into complex 2- and 3-dimensional shapes known as their secondary and tertiary structures, respectively (see Fig. 1.2). For example, a small section of the linear amino acid chain could coil into a helical form and this helix would be a region of secondary structure. Along the linear chain of a protein, there can be multiple, independent regions of secondary structure resembling beads on a string. If you take the beaded string and clump it up into a ball, the ball shape is the 3-dimensional tertiary structure. All proteins fold into tertiary structures that are critical for the functional activity of the protein – if you destroy the tertiary structure then the protein no longer functions. As increasing numbers of proteins had their tertiary structures solved, it became clear that these overall tertiary structures are themselves composed of sub-structures known as protein folding domains. Importantly, these protein folding domains are highly conserved building blocks that evolution mixes and matches in a modular fashion to construct all the different proteins. While the genomic sequences and the amino acid sequences producing a particular folding domain may vary so widely between different proteins as to seem unrelated, the overall tertiary shape of the domain will remain fairly constant. This conservation of domain structures makes it a better parameter for long-range evolutionary comparisons than are primary sequences. One way to envision this is to think about a blueprint for a house. The blueprint is the DNA sequence that specifies all the instructions to make the house (the protein). A blueprint is virtually unlimited in what kind of house it can encode, and the house itself can take on an enormous variety of shapes (its tertiary structure). However, houses all contain certain substructures (folding domains) that are conserved from house to house regardless of the house's overall shape. For example, a door is a "folding domain" in the house, i.e. a conserved structure that can be used in many different houses to perform the same entry/exit function. Residential doors are relatively standard, and while they can vary widely in color, surface texture, or stylistic details, they are traditionally rectangles with fairly conserved dimensions. Two blueprints and their respective houses may look completely different (no homology), but their doors will be fundamentally similar because they have to function as doors. All houses are comprised of common elements, for instance, doors, roofs, kitchens, bathrooms, and bedrooms, with each having specific functions. Likewise, protein folding domains often perform specific tasks, and the overall protein function develops by compiling multiple domains to achieve the desired sum of activities. Evolutionarily, complex proteins are combinatorial assemblages of modular folding domain subunits that are exchanged and reused among many different proteins, much like different shaped Lego pieces can be mixed and matched to build many varied structures.

Considering that the conservation of protein domains is much greater than the conservation of their corresponding amino acid or nucleic acid sequences, domain analysis can provide a much deeper view of evolutionary history. Revisiting our house and door analogy, a modern mansion is vastly different in size and shape than a primitive hut, but the doors on both would still be recognizable as related objects performing the same function. Similarly, the conservation of functional domain shapes allows protein tertiary structure relatedness to extend much further back in time and reveal relationships that are lost in primary sequence information, as structure is at least 3–10 times more conserved than sequence. Additionally, similar structures in different organisms typically reflect descent from a common precursor structure (divergent evolution) rather than evolution independently arriving at the same folding patterns from unrelated precursors (convergent evolution). As more and more structural information accumulated about proteins from diverse organisms in the last decade, interrogation of this data has been highly fruitful for exploring evolutionary relationships.

An elegant study in 2015 made a detailed analysis of protein domains across a wide range of organisms. The complete protein sets (known as the proteomes) from 5080 organisms were examined, with 1620 of the proteomes coming from cellular organisms and 3460 coming from viruses. Collectively, these proteomes included over 11 million proteins! From this huge pool of proteins, the study identified 1995 conserved domain structures which they termed fold superfamilies (FSFs). Of these 1995 FSFs, only 66 were unique to viruses, 1279 were unique to cellular organisms, and 650 FSFs were shared by viruses and cellular organisms. More importantly, a whopping 442 out of the shared 650 FSFs are found in viruses plus organisms from all three modern kingdoms: archaea (atypical bacteria), prokaryotic cells (bacteria), and eukaryotic cells. The observation that viruses share the majority of their protein folding domains with all three other types of cellular organisms strongly favors a very ancient origin for viruses that precedes the division of life forms into the three modern lineages. Specific analysis of FSFs in the giant viruses also revealed significant overlap with FSFs in parasitic and symbiotic organisms, consistent with reductive evolution for this group of viruses. Score one for the regression hypothesis!

Several other interesting findings also came out of the viral FSF study. Using FSF relationships to define a phylogenetic tree revealed that the earliest progenitors had segmented, single-stranded RNA genomes, supporting the general belief among biologists that RNA was the original genetic material. From these ancient progenitors, the various RNA virus families emerged first, followed thereafter by DNA viruses, with the retroviruses between them. Retroviruses (see Chaps. 8 and 9) are fascinating viruses that have RNA genomes in their virions yet convert their genomes to DNA inside the cell using a special enzyme called reverse transcriptase. It's often been speculated that retroviruses were the bridging event between RNA viruses and DNA viruses, and the FSF-based tree supports that relationship. The tree also supports a polyphyletic origin of viruses with independent emergence of the two major structural types, spherical and filamentous virions, which subsequently gave rise to all the morphological variants seen today. Consistent with this interpretation is the observation that there is no single FSF conserved among all currently known viral families, just as there is no single gene conserved. Had all modern viruses originated from a single precursor, a viral LUCA equivalent, we would expect at least some FSFs to be present in all viral families. Importantly, all of these events must have preceded LUCA and the establishment of the modern cellular lineages on the tree of life since viruses also contain genes and FSFs not found in LUCA-descended life forms. Thus, these observations support the virus first hypothesis that at least some viral lineages emerged during a pre-LUCA period. At that time there were likely many ancient life forms competing, interacting, and sharing protein-encoding genes, which explains why there is so much overlap between viral and cellular FSFs today. Still, while providing valuable new information, the FSF approach doesn't resolve the basic origin dichotomy. FSF evidence can support both the virus first and the regression hypotheses, making it more likely that both pathways may have yielded modern viral families.

Regardless of whether or not we consider viruses alive, the available data are consistent with viruses occupying a unique niche that collectively is distinct from the other three kingdoms in its evolutionary history. Because of this unique position on the tree of life prior to LUCA's appearance, the virus branch could be considered a fourth kingdom that evolved in parallel with and parasitic on the other three kingdoms once they appeared. For example, the 66 FSFs unique to viruses are predominantly found in viral proteins that are involved in pathogenicity, such as capsids and structural proteins, proteins involved in binding host receptors, and proteins that suppress host immune responses (see Chap. 5). Such functions likely evolved after the establishment of viral lineages as early viruses began to try to invade LUCA's descendants. Proteins that facilitated invasion and reproduction had a selective advantage and eventually evolved into a diverse array of pathogen-associated molecules present in today's viruses. Additionally, while 716 FSFs were identified in viral proteins, this only involved a small percentage of total viral proteins. The vast majority of viral proteins do not belong to known FSF and have no apparent homology to any cellular proteins. This highlights the distinctive evolutionary branch that viruses occupy with their unique modern proteins deriving from ancestral proteins that were not present in LUCA, and thus not shared with the other three kingdoms. As more and more novel viruses are identified and their proteins analyzed, their rich viral proteomes will offer an exciting window into the far distant past (i.e. billions of years ago) that may someday reveal even deeper insights into very early life forms.

Closer to Home

Even if the details of the origins of viruses and their cellular hosts remain obscured by time and veiled by limited available data, science is making significant progress in understanding more recent evolutionary relationships between viruses and their prey. While the virus-cell relationship appears to extend back to the dawn of life, what about our modern-day viral pathogens? When did these modern viral lineages form and when did they first infect humans? Are the common viruses that inflict us today a relatively recent invasion of the human race, or can we trace their lineage much further back to our hominid predecessors and beyond? We know from historical records that ancient civilizations were plagued by viruses such as polio, smallpox, and measles, indicating that these viruses have at least thousands of years of association with humans. Exciting new molecular studies have begun to move that clock further and further back giving us a window into the co-evolution of viral families with the plant and animal kingdoms, including a closer look at primates and hominids. Let's take an overview of some general principles and findings and then examine in more detail a ubiquitous human pathogen, papillomaviruses.

One of the key concepts in mapping viral evolution is co-divergence. This principle states that as life forms diverged into different branches, for example, the split into plants versus animals, their associated viruses would also split and continue to evolve independently along each branch. In practice, this means that if we find viruses from both branches that have relatedness at the gene and/or protein levels, then those two viruses likely had a common ancestor prior to the time of the branching. Put another way, the evolution and relatedness of viruses should follow the evolution and relatedness of their hosts. Therefore, by looking for viral similarities across plants, animals, and bacteria we can construct trees illustrating how and when viruses appeared in different species. Based on many such studies a coherent picture of overall virus evolution is emerging, although there are still many gaps and missing details. Focusing just on eukaryotic viruses, the available data again favors a polyphyletic origin with multiple, independent events leading to the currently existing viral families (Fig. 3.2). RNA viruses appear to be the oldest type, consistent with the original RNA world of the oldest life forms. RNA viruses all encode an enzyme called RNA-dependent RNA polymerase (RDRP) which is required for the synthesis of the viral RNA genomes. RDRP is an example of one of those viral proteins for which no existing cellular homolog has ever been found, implying that it derives from a pre-LUCA organism, and thus placing the origin of RNA viruses far, far into antiquity. Except for plant viroids and hepatitis D virus that both have circular genomes, modern RNA viruses have linear genomes that fall into three groups: single-stranded, positive-sense RNA (PS-RNA); single-stranded, negative-sense RNA (NS-RNA); and double-stranded RNA (DS-RNA). The earliest event in the formation of the modern RNA virus lineages was an ancestral RNA virus that

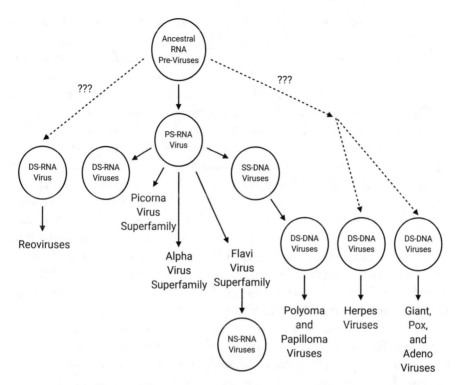

Fig. 3.2   The tree of modern virus lineages. From an ancestral pre-virus emerged the initial positive-sense RNA (PS-RNA) virus whose genomic RNA could be directly translated into proteins by cellular ribosomes. The primordial PS-RNA virus eventually gave rise to three PS-RNA superfamilies including the picornavirus superfamily, the alphavirus superfamily, and the flavivirus superfamily. Each superfamily gave rise to several of the modern PS-RNA virus families. Additionally, the negative-sense RNA (NS-RNA) viruses arose from the flavivirus superfamily. NS-RNA viruses have a genome that cannot be directly translated by ribosomes, and instead, the NS-RNA must first be converted to PS-RNA after infection. The primordial PS-RNA virus also gave rise to most of the double-stranded RNA (DS-RNA) viruses except for Reoviruses that have a different origin unrelated to the PS-RNA viruses. In addition to the DS-RNA viruses, the PS-RNA viruses also gave rise to the small single-stranded DNA (SS-DNA) viruses that became the polyomaviruses and the papillomaviruses. The larger DNA viruses have distinct lineages with herpesviruses forming one evolutionary group and the giant viruses, poxviruses, and adenoviruses forming a second distinct evolutionary group. (Created with BioRender.com)

diverged into three lines of PS-RNA viruses, the picornavirus-like superfamily, the alphavirus-like superfamily, and the flavivirus-like superfamily. Collectively these three branches contain virtually the entirety of the PS-RNA viruses that exist today. Since members of the picornavirus-like superfamily infect a wide range of both unicellular and multicellular eukaryotes, this family must have emerged before those cellular organisms diverged. In contrast, the alphavirus-like and flavivirus-like superfamilies both have a much more restricted host range indicating a later divergence from the common ancestor of all three families. The NS-RNA viruses are even more limited in host range than the alphavirus-like and flavivirus-like viruses, being found predominantly in animals with only a few members infecting plants. Recent structural analysis of an RDRP from a member of the NS-RNA group, influenza virus, showed impressive similarity to the RDRP from flavivirus, suggesting that the NS-RNA virus group arose from the PS-RNA flavivirus-like group. Because of their prevalence in animals and scarcity in plants, the flavivirus-like group transition into the NS-RNA group likely occurred after the split between plants and animals. The few modern-day NS-RNA viruses that infect plants probably arose by animal NS-RNA viruses jumping into plants and switching hosts (see Chap. 11). The final RNA group of DS-RNA containing viruses evolved by multiple, independent events from different branches of the PS-RNA virus tree. Additionally, at least one family of the DS-RNA genome type, the Reoviruses, infects a wide range of organisms from algae to vertebrates, indicative of an origin long before these organisms separated. Reoviruses also share structural relatedness with a bacterial virus family suggesting that they derived directly from this family and not through the PS-RNA virus branch that led to most other RNA virus families.

Similar to RNA viruses, modern DNA virus families have multiple origin points from diverse ancestral precursors (Fig. 3.2). DNA viruses with single-stranded genomes (ssDNA) are chimeras that arose from the replicative genes of different bacterial plasmids combining with capsid protein genes from various PS-RNA viruses. These independent combinatorial events leading to ssDNA virus progenitors likely happened several different times but must have occurred very early on in evolutionary history since ssDNA viruses widely infect organisms throughout the eukaryotic world. The picture is somewhat clearer for double-stranded DNA (dsDNA) viruses which appear to derive from three different lineages. The large dsDNA virus group, including the giant viruses, poxviruses, and adenoviruses, are all descendants of a bacterial virus family known as the Tectiviridae. Tectiviruses made the jump into eukaryotic cells then radiated out to form the different large dsDNA virus branches. In contrast, the herpesvirus family, which is also a group of large dsDNA viruses, evolved independently from a different group of bacterial viruses, the Caudovirales. Herpesvirus evolution is also a much more recent event than the other large DNA viruses, as herpesviruses are only seen in animals. Lastly, there are two small dsDNA virus families, the polyomaviruses and the papillomaviruses (see Chaps. 7 and 8, respectively, for more about these two virus families). Their evolutionary history is unrelated to either of the two large dsDNA virus lineages and instead, they are most closely related to the small ssDNA viruses. At some point, the ssDNA viruses switched their replication mode and gained a complementary DNA strand so that they began to package dsDNA into the virion rather than ssDNA. Consequently, small dsDNA viruses are closely related to PS-RNA viruses through their ssDNA progenitors while the large dsDNA viruses have a much more distal connection to the RNA world.

In summary, the polyphyletic nature of modern RNA and DNA virus families is well supported by genetic homologies and protein structural relatedness studies. The data strongly attest to current-day viral lineages arising from distinct, independent events spread out over a long period of evolutionary history. Evaluation of viruses from divergent organisms shows an evolution that generally parallels the hosts which is consistent with the co-divergence model, although it is becoming increasingly apparent that viruses can indeed hop to new host species with much greater frequency than once envisioned (see Chap. 11). Interestingly, viral structural and genetic relatedness across host species is also reflected in functional properties such as tissue tropism. For example, if a mammalian virus infects the skin, its fish virus counterpart would also likely cause skin infections. This conservation of both physical and functional viral features across widely divergent host species powerfully corroborates an ancient origin for modern viruses, and we'll explore one specific example in the next section.

Papillomaviruses and Hominid Evolution

Papillomaviruses are small, dsDNA viruses that are presumed to infect all vertebrate species, but not invertebrates, putting their origin approximately 530 million years ago when vertebrates appeared. These viruses likely had an aquatic origin in the precursors to fish, and sometime during the Mesozoic period (roughly 180 million years ago) they spread into higher vertebrates (reptiles, birds, and mammals). As mammals diversified widely around 120 million years ago the papillomaviruses were similarly distributed to all the branches, including our primate ancestors, and so have become an entrenched, endemic infection of the human lineage.

Human papillomaviruses (HPVs) have diversified into over 200 subtypes (simply named HPV1, HPV2, etc.) that have very specific skin niches that they infect. For example, some types only infect the soles of the feet (plantar types) while some only infect mucosal skin (oral and genital regions). These viruses are spread by skin-to-skin contact, including sexual activity; most of us are initially colonized by these viruses at birth and shortly thereafter via the adults who handle us as babies, but we continue to be exposed throughout our lives. Fortunately, infections are mostly asymptomatic, and the viruses simply reside in our skin with little noticeable effect. When the infections do become symptomatic they present as warts which are benign tumors caused by viral-induced hyperproliferation of skin cells. These viruses have small genomes, and modern papillomaviruses typically encode three gene sets: E1 and E2 for replication; E5, E6, and E7 which alter the host cell's internal environment to favor viral replication; and L1 and L2 which form the virion capsid. Interestingly, early progenitors of the papillomavirus group appeared to lack the E5/E6/E7 group and only had the set of genes for replication (E1 and E2) and the set for encapsidation (L1 and L2). However, the proteins encoded by E5, E6, and E7 each have potent effects on the host cell that are advantageous to the virus, so the early acquisition of these genes likely provided a positive selective pressure that led to their broad spread throughout this viral family. Unfortunately, in a small subset of the 200 plus HPVs, the E5/E6/E7 gene set has carcinogenic potential, and infection with these so-called "high risk" HPV types is the primary causative factor for cervical cancer, anal cancer, and some forms of oral cancers. In particular, HPV types 18 and 16 are the two most prevalent high-risk types associated with cervical cancer, although exactly how and why their high-risk E5/E6/E7 proteins promote cancer while most other similar HPV E5/E6/E7 proteins do not is still not entirely clear. It is also important to note that causing cancer is neither necessary nor advantageous for viral reproduction. Instead, HPV-associated cancers are a rare and inadvertent side-effect of these viral infections (see Chap. 8).

One thing we do know about the primate papillomaviruses is that they began to branch about 40 million years ago as primates radiated out into various species, including archaic hominids. Nonetheless, there appears to have been significant subsequent exchange across different types of primates over the eons. Many examples exist of modern primate and human papillomaviruses that are much more closely related than are their primate hosts to humans, so strict co-divergence is not observed. We also know that most of the modern HPV lineages were established 6–8 million years ago, around the time when hominids diverged from chimpanzees, so these pathogens have co-evolved with humans throughout our existence. Over several million years the HPVs types continued to evolve, and within some types, there is now sufficient variation to define sub-lineages. For example, highly oncogenic HPV16 has 4 subgroups, A-D, based on nucleotide sequence differences in their genomes. The A subgroup is primarily found in Asians and Caucasians, B and C in African populations, and D primarily in South/Central America. Even this divergence into related subgroups is ancient and occurred 500,000–600,000 years ago, before the initial out-of-Africa migration, and before *Homo sapiens* separated from other now-extinct hominids such as Neanderthals and Denisovans. However, these data create a conundrum. When the ancestors of modern humans eventually left Africa 60–70,000 years ago and spread into Europe and Asia they would have carried the B and C types that still exist in Africa today. How then did the A subgroup become the most prevalent in today's Europeans and Asians? Current thinking is that the progenitors of Neanderthals carried the A lineage into Europe when they were the first to populate this region roughly 500,000 years ago. When our modern ancestors reached Europe we know that they interbred with Neanderthals (2–4% of our DNA is Neanderthal), so our best guess is that Neanderthals spread the A group to modern humans through sexual activity and that the A types gradually displaced the B and C types on the Eurasian continent. Sadly, it looks like our ancestral cousins gave us an STD!

The HPV example illustrates a human virus whose origin predates mankind and whose association with humans began at the dawn of hominid evolution. Because of this ancient association, HPVs are highly specific to humans and their infections have evolved into a mostly benign relationship. This virus evokes little disease yet fosters long-term persistence and ready transmission from person to person that facilitates viral spread and maintenance in its host population. In contrast, not all human viruses have this ancient co-evolution with humans. Instead, many endemic human viruses are much more recent acquisitions by human populations that we acquired by being infected with animal viruses. Such recent viral acquisitions are often more frankly pathogenic as we haven't yet had time to co-adapt with these new agents. We'll explore other types of viral diseases and how they are spread in subsequent chapters.

Additional Reading

1. "A Phylogenomic Data-drive Exploration of Viral Origins and Evolution." Science Advances 1(8): e1500527, 2015.
2. "Phylogenetic Tracings of Proteome Size Support the Gradual Accretion of Protein Structural Domains and the Early Origin of Viruses from Primordial Cells." Frontiers in Microbiology 8:1–18, 2017.
3. "Viruses and Cells Intertwined Since the Dawn of Evolution." Virology Journal 12:169–179, 2015.

Definitions

Acanthamoeba polyphaga mimivirus (APMV) – A type of giant virus that infects amoeba.

Alphavirus – A genus of RNA viruses in the togavirus family.

Amoeba – A free-living, unicellular eukaryotic organism usually found in freshwater.

Archaea – A group of unicellular, prokaryotic organisms that are distinct from bacteria.

Caudovirales – A classification of bacteriophages at the level of Order.

Co-divergence – The process where virus lineages evolutionarily diverge as their respective hosts diverge.

Convergent evolution – The process by which two species or proteins independently develop similar properties without sharing a recent common ancestor.

Denisovans – An extinct species or subspecies of ancient humans.

Divergent evolution – The process where two species or proteins acquire differences in response to selective pressures.

Flavivirus – Positive-sense, single-stranded RNA viruses often spread via insects.

Fold superfamilies (FSF) – The highest level of grouping of proteins based on the relatedness of their structural folding domains.

Giant viruses – Extremely large viruses, typically a large as a bacterium. They have correspondingly large DNA genomes which encode many unique genes.

Hepatitis D antigen (HDAg) – The single protein encoded by the hepatitis D virus.

Hepatitis D virus (HDV) – A negative-sense, single-stranded, circular RNA virus that is dependent on the hepatitis B virus for its propagation. Because of this dependency, it is considered a satellite virus or a virusoid.

Herpesvirus – A group of viruses with large, double-stranded, linear DNA genomes and complex virions. All members of this group can establish latent infections.

Hominid – A group that includes Great Apes (chimpanzees, gorillas, and orangoutans), humans, and their immediate ancestors.

Homo sapiens – Modern humans.

Homology – Having relatedness in sequence or structure.

Insertion sequences – A short DNA sequence that can jump from position to position within a genome.

LUCA – The last universal common ancestor of all life on Earth.

Mesozoic period – The geological time from 250 to 66 million years ago.

Mobile genetic elements (MGE) – Any type of nucleic acid sequence that can move around within the genome.

Neanderthals – An extinct species of humans.

Papillomavirus – The group of small, double-stranded, circular DNA viruses that cause warts. Some members of this group have oncogenic potential.

Phylogenetic tree – A diagram showing the genetic relatedness of organisms.

Picornavirus – The group of small, positive-sense, single-stranded RNA viruses that includes poliovirus, rhinovirus, and hepatitis A virus.

Polyomavirus – Small, double-stranded, circular DNA viruses that rarely cause obvious disease although they often establish persistent infections. This family includes mouse polyomavirus, several human polyomaviruses, and SV40.

(continued)

(continued)

Polyphyletic – A group of organisms deriving from more than one common ancestor.

Polyphylogeny – Of or developing from more than one common ancestor.

Protein folding domains – Localized regions on proteins with specific structures that are evolutionarily conserved.

Proteome – The total array of all the proteins produced by an organism.

Proto-virocells – A putative ancient cellular organism that was a precursor to viruses.

Reoviruses – The group of double-stranded RNA viruses that includes rotaviruses.

Retrotransposons – A transposon that shows homology with retroviruses.

Retroviruses – The group of single-stranded RNA viruses that use reverse transcriptase to synthesize a DNA copy of their RNA genome.

Reverse transcriptase – An enzyme that uses RNA as the template and synthesizes a complementary DNA sequence.

RNA polymerase II – The cellular enzyme that transcribes DNA into mRNA.

RNA-dependent RNA polymerase (RDRP) – A viral enzyme that uses RNA as the template and synthesizes a complementary RNA sequence.

Satellite virus – A virus that is unable to produce its own capsid protein and is packaged into a virion using the capsid protein(s) of a helper virus.

Superfamily – A high-level grouping of proteins or organisms with common ancestry.

Tectiviridae – A family of bacteriophages with a linear, double-stranded DNA genome.

Transposons – Genetic elements that can jump from location to location within a genome.

Viroids – An infectious agent of plants that consists only of nucleic acid without any protein capsid. Generally transmitted via insects.

Virusoid – A viroid that can be encapsulated by a helper virus coat protein.

Abbreviations

APMV – *Acanthamoeba polyphaga* mimivirus
CH_4 – methane
CO_2 – carbon dioxide
dsDNA – double-stranded DNA
DS-RNA – double-stranded RNA
FSF – fold superfamilies
HBV – hepatitis B virus
HDAg – hepatitis D antigen
HDV – hepatitis D virus
HPV – human papillomavirus
LUCA – last universal cellular ancestor
MGE – mobile genetic element
NH_3 – ammonia
NS-RNA – negative-sense RNA
PS-RNA – positive-sense RNA
RDRP – RNA-dependent RNA polymerase
ssDNA – single-stranded DNA
STD – sexually transmitted disease

4

Of Predators and Prey

Keywords Influenza • Hepatitis C • Herpes varicella-zoster • Viral pathogenesis • Polyomavirus • Anellovirus

The single biggest threat to man's continued dominance on the planet is the virus.
Joshua Lederberg, 1958 Nobel Laureate

Viral Infections – The Good, the Bad, and the Ugly

Viruses typically sicken, injure, and even kill, or at least that's how they are usually perceived. In the previous chapters, we've considered whether or not viruses are animate entities; have examined their physical, biochemical, and biological properties; and have explored their origins and co-evolution with the kingdoms of life. However, for most of us what we really want is to understand how viruses upset the delicate homeostasis of our health to cause sickness. What are their pathogenic properties and their intimate mechanisms for infecting humans which provokes the multitude of deleterious effects we call disease? Surprisingly, while we typically equate viral infection with illness, there is more and more evidence that not all viruses are pathogenic. Instead, many cause little or no disease, with some possibly even being considered part of our "normal" flora, now being called the virome. Similarly, even many pathogenic viruses can often cause inapparent first infections that produce little or no obvious symptoms yet induce sufficient immunity to protect us from subsequent infections. Nonetheless, frank disease-causing viruses have historically been the most obvious to detect and the most highly studied in

V. G. Wilson, *Viruses: Intimate Invaders*, https://doi.org/10.1007/978-3-030-85487-4_4

our attempts to prevent and treat their resultant illnesses. Most of our common, disease-producing viruses cause acute infections where we become noticeably ill with symptoms characteristic of the particular virus, for example, diarrhea with gastrointestinal viruses. In the typical acute infection, the initial exposure seeds a small number of virions into the susceptible host where they begin to replicate. As viral numbers rapidly increase with each round of replication, more and more cells become damaged and/or killed leading to clinical symptoms. Fortunately, the presence of these viral invaders elicits a variety of immune responses (see Chap. 5) that collectively squelch the infection, eliminate all the virions, and restore us to a healthy, uninfected state. The key principle of acute infection is the complete resolution of symptoms along with the total elimination of the infecting virus from the surviving host so that there is no longer any virus present after recovery from the illness. Unfortunately, the impact of acute disease, even if the patient survives, can sometimes cause permanent residual damage, such as the pox scars from smallpox or neurological dysfunctions following viral encephalitis.

Because acute disease-causing viruses are so obvious it is no surprise that physicians and scientists focused initially on this type of virus and acquired a bias that this was how most viruses worked. However, from the viral perspective, an acute infection may not be the best strategy for survival and propagation. Acute infections are transient and require efficient mechanisms to transmit the virus from the infected host to a new host before the infection is eliminated. If the host recovers or dies before the virus can spread to the next victim then that infecting virus and its progeny, with all their genetic heterogeneity, are lost from the global pool of that virus type. Ideally, an acute disease-causing virus would produce a long-lasting, but nonfatal, illness that allows ample opportunity for the sick person to produce numerous new viruses and to interact with many other people to spread the infection. Those common acute viruses that plague us today, such as many respiratory and gastrointestinal viruses, are the evolutionary victors that successfully established an infectious process that is conducive to their spread and persistence as endemic pathogens in human populations. Yet these acute viruses are likely the few highly noticeable survivors of the evolutionary competition and only represent a small minority of human viruses.

In contrast to acute infections where viruses are operating in a "hit and run" mode, perhaps a more effective alternative strategy is to establish a persistent infection where the virus is never cleared from the infected individual. While transient acute disease may or may not occur upon exposure to a persistent virus, after the initial infection resolves persistent viruses will remain in the host and not be eliminated completely as for acute viruses. There are two types of persistent infections, chronic and latent. In the chronic type, the persistent virus replicates unceasingly and virions remain detectable in the infected people. While these individuals show no clinical symptoms and are completely healthy, they continually produce low levels of new viruses that are infectious for other people, a state that can continue for decades. This provides an ideal scenario for the virus where it can persist for many years in a single host and be passed unwittingly to many potential new hosts. Unfortunately, this chronic, low-level virus production is not always harmless and may lead to long-term clinical effects resulting in serious disease many years after the initial infection. This long-term effect can result from viral-induced cell damage and/or chronic inflammation and immune attacks on the infected cells. In either case, this late-term disease or death of the host is insignificant for the virus's survival as it will have already spread widely long before the initial host succumbs. Whether or not there are persistent viruses that truly cause no disease effects is still being debated. This has important clinical implications as some estimates suggest that most humans have 8–12 persistent viral infections.

For latent infections, the virus typically "hides" in some unique site where the virus can persist indefinitely without replicating. Since little or no new virus is produced, these latent viruses are very difficult to detect and patients are generally not infectious during the latent stage. Additionally, these latent infections are less likely to cause long-term damage than chronic viruses since there is little viral activity during latency. Instead, the dormant latent viruses can sometimes be reactivated months to years after the initial infection. When reactivated, the awakened virus begins to replicate and produce new virions which now give rise to a new occurrence of acute disease. During these acute outbursts, the host is also infectious and can spread the virus to new susceptible individuals. Eventually, the acute outbreak resides and the virus returns to latency, a pattern that may repeat throughout the life of any infected person. So, as for chronic infections, latent infections provide the virus with a long-term, stable host where it likely has many opportunities for spread over the lifetime of the host. The last decade revealed a plethora of new human persistent viruses that have no obvious disease presentation. The abundance of these new viruses indicates that persistence with little or no frank disease is likely a far more prevalent viral strategy than the acute presentation. In the following sections we will explore representative viruses that illustrate fundamental principles of acute, chronic, and latent infections with known disease associations, as well as consider some of the more recently identified persistent viruses whose disease potential remains uncertain.

Influenza – More Than Just Bad Air

More than any other acutely infectious virus, influenza has been the major epidemic threat for the twentieth and twenty-first centuries. The influenza virus is a rare triple threat with significant potential for host mortality; effective transmission, including across species; and high mutability. This lethal combination results in 3–5 million cases worldwide each year with 300,000–650,000 deaths annually, based on the Centers for Disease Control (CDC) estimates. These numbers include roughly 40,000 deaths per year in the United States, a staggering number compared to other acute viral infections, except for the recent SARS-CoV-2 pandemic. The young (less than 5 years old) and the elderly (over 65 years old) are particularly at risk for fatal infections, though some yearly strains are quite effective at causing lethal infections in healthy adults. For example, the so-called Spanish influenza pandemic of 1918 killed somewhere between 50 and 100 million people as it spread out of Europe and covered the world. Most victims were less than 65 years of age, and deaths were common in the 20–40 age group. Luckily, no such massive killer pandemics of influenza have occurred since the 1918 outbreak though several lesser pandemics in the last 60 years each caused nearly a million deaths.

The word influenza means "influence" in Italian, a reference to adverse meteorological conditions or "bad air" that was once believed to cause the disease. Nonetheless, the disease is much older as the symptoms were accurately described by Hippocrates over 2400 years ago. Unfortunately, since many respiratory diseases have symptoms that overlap those of true influenza, it is difficult to attribute historical disease events definitively to true influenza prior to around 500 years ago. In 1493, after the arrival of Columbus in the Antilles, a possible influenza epidemic nearly decimated the indigenous population and was likely the first spread of this virus into the New World. Since that time there have been many localized epidemics and pandemic outbreaks that were likely influenza, though the virus itself was not identified until 1933. Consequently, it's only in the last 90 years that we've had the formal scientific tools to study this virus and elucidate its history, properties, and clinical implications.

Influenza virus has a negative-sense RNA genome classified in the family of Orthomyxoviridae. Influenza viruses are widespread in mammals, birds, reptiles, fish, and even lower vertebrates, indicative of an ancient history for this type of virus. An important feature of influenza viruses is that they can be transmitted across host species much more readily than most other viruses. Because these viruses can jump species effectively, their genetic relatedness doesn't always parallel the evolutionary relationships of their hosts, making these a marked exception to the general co-divergence model. For example, a fish influenza virus is the closest relative to one of the major human influenza virus types; this would not be expected considering how long ago fish and humans diverged. Because of this propensity for species jumping, wild birds are an important reservoir for influenza disease that can jump to humans. Birds travel widely across vast geographic regions and often in large flocks, so there is ample opportunity for distinct wild bird species to exchange and transmit different influenza viruses. Avian influenza viruses of wild birds can also be readily transmitted to their domesticated cousins, often chickens, starting the chain of transmission into other farm animals such as swine, horses, and dogs, or even into humans. If an animal influenza virus is effective both at causing human disease and being transmitted from human to human, then it can become an endemic human strain.

Another consequence of species jumping is co-infection where two distinct influenza viruses infect a single host. This is often a human-avian mixed infection, but could also be any combination of different influenza viruses infecting a single bird or other animals. Unfortunately, mixed infections can produce genetic scrambling that yields hybrid progeny viruses (Fig. 4.1). The influenza virus has a segmented genome meaning that its genetic information is not a single, unbroken RNA molecule, but instead consists of 7–8 individual RNA molecules depending on the specific virus. Conceptually, each of these RNA pieces in the virion is a viral chromosome, just as we humans have 23 separate DNA chromosomes that form our genome. Each viral chromosome encodes different proteins, and when reproducing, every new virion must get 1 copy of each of the viral chromosomes to produce a complete, functional virus. Imagine a mixed infection where a human and avian influenza virus are both infecting a human cell. Within the infected cell, both the human and avian parental viruses will replicate many copies of their own chromosomes. When packaging these RNA pieces into the new progeny virions, there is a random selection from this mixed pool of chromosomes such that any virion can get some chromosomes of the avian virus type and others of the human virus type. For example, in a particular progeny virion, its chromosomes 1 and 2 may derive from the human virus while the remaining chromosomes come from the avian virus. This mixing process is termed reassortment and the resulting hybrid virion is a reassortant.

Reassortment is only possible with segmented viruses and is essentially like sexual reproduction in that the "offspring" virions have hybrid genetic information contributed by both parent viruses. Like sexual reproduction, the mixing of genetic information can result in offspring that have new properties

Parental Virus A   Parental Virus B

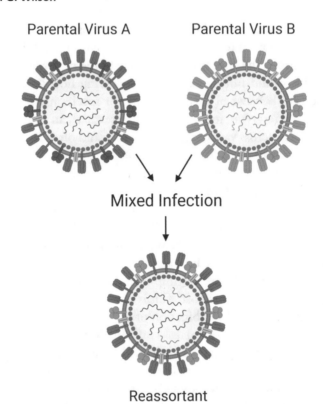

Mixed Infection

Reassortant

Fig. 4.1 **Viral reassortment.** Shown are two related influenza viruses, Parental Virus A and Parental Virus B. Both parental virions contain the 8 different RNA segments that constitute the complete influenza virus genome (blue RNA segments for virion A and red RNA segments for virion B). The surface proteins of these two parental viruses are antigenically different as represented by the different colors for these proteins. During a mixed infection, the 8 genome segments from each parental virus will be replicated, and the cells will fill with a mixture of many red and blue copies of each segment. As new virions assemble, each progeny virion must package one copy of RNA segments 1–8. For each of these 8 RNA segments packaged there is a random choice between a blue copy and a red copy. Some progeny virions may contain all blue copies or all red copies identical to the parental viruses. However, most virions likely possess a mixture of some red and some blue copies and these are the reassortants. The reassortant shown has 6 blue RNA segments derived from Parental Virus A and 2 red segments derived from Parental Virus B, but every combination of red and blue mixed reassortants is possible. In the example shown, the mixed RNA segments in the reassortant encode surface proteins from each parent virus resulting in a reassortant that is antigenically different from either parent virus. Such changes in reassortants can lead to novel viruses with properties very different from the parental viruses. (Created with BioRender.com)

different from either parent, a so-called genetic shift. In many cases, these reassortants are similar to or less fit than the parental viruses and pose no new health problems. However, occasionally they attain heightened virulence, enhanced transmission properties, and/or increased replication capacity for human hosts that potentially could lead to a devastating global pandemic such as the 1918 influenza outbreak. The reassortants may also have novel surface antigens against which human populations have little or no existing immunity, and for which entirely new vaccines would have to be developed. Consequently, abrupt genetic shifts can produce deadly new versions of influenza viruses that could be extremely dangerous to human populations. Because influenza transmits via the respiratory route it is difficult to avoid infection, and the disease spreads rapidly and widely each year. These properties make the influenza virus a great danger to humans with the potential for a massive, lethal, global pandemic that would be difficult to control before hundreds of thousands to millions of people perished. To protect against the spread of hazardous reassortants into human populations, the CDC and the World Health Organization (WHO) routinely monitor animal influenza viruses around the world to search for new hybrids. The discovery of such hybrids necessitates massive kill-offs of potentially infected animals, such as the purges of fowl that occur frequently in the Orient. By culling the susceptible animal populations the hope is to curtail the spread of the reassortant so that the viral lineage quickly dies out and has no chance to spread into humans.

In the absence of dangerous reassortants, the normal flu season has two main types of human influenza in circulation, type A and type B. The two other known types of influenza virus are found infrequently (type C) or not at all in humans (type D). Type A is the most frequent cause of human infections and is also the most diverse type with numerous strains designated by the HN nomenclature. H and N are both viral proteins found on the surface of the virion. H stands for hemagglutinin, and this protein binds to the host cell receptor to mediate viral attachment and entry. N is the viral neuraminidase protein that is critical for the release of new virions from infected cells. Currently, across animal and human influenza viruses, there are 18 known H subtypes and 11 known N subtypes. Typically, we designate influenza virus strains by the H and N subtype pair they carry, such as the H3N2 strain that caused the Hong Kong flu in the late 1960s. Only a few of the many possible H and N combinations are found in influenza viruses commonly associated with human disease, but reassortment could produce any possible combination, so new human pathogenic subtypes are likely to arise.

For acutely infecting viruses like influenza, there are only two possible outcomes of infection: survival or death. Infection begins with the entry of the virus into the eyes, nose, or mouth by one of three routes: direct contact with an infected individual, airborne inhalation, or hand-to-face from fomites (objects contaminated with viruses). When an infected person coughs or sneezes they exhale microscopic aqueous droplets that carry thousands of viral particles that can drift directly into the eyes, nose, or mouth of people nearby to infect cells and start the disease. Alternatively, the virion-infected droplets will eventually fall out of circulation onto whatever objects are in the vicinity. Aerosol deposited virions can persist from minutes to days depending on the surface and the environmental conditions. Additionally, if the infected person gets viruses on their hands then they can directly plant viruses onto whatever they touch. When an uninfected individual unknowingly touches a contaminated surface with their hand and then touches their eyes, nose, or mouth they can self-inoculate. The infecting virus replicates rapidly and spreads into the respiratory system. Within 1–2 days after infection there is an abrupt onset of symptoms including fever, chills, headache, muscle aches, and loss of appetite, followed by the respiratory symptoms of dry cough, sore throat, and nasal congestion. This constellation of symptoms is due both to viral-induced tissue damage and to a vigorous innate immune response (see Chap. 5) releasing interferon and other defense molecules known as cytokines and chemokines. Slowly over the first week antibodies and anti-viral T cells begin to develop. In most healthy people these combined bodily defenses are extremely effective in stopping viral reproduction, resolving the infection within 7–14 days, and completely eliminating the influenza virus from the host body. Complications and fatalities are of greater concern for persons with underlying health issues, and for the very young and the elderly who have less effective immune responses and more limited respiratory capacity.

Our immune system is highly effective in curtailing influenza virus infections and preventing persistent infections by this virus. Unfortunately, recovery doesn't protect against repeat infections as it does for most other acute viruses. What is it about influenza that leaves us vulnerable to new infections year after year and which requires new vaccinations yearly to help prevent these new infections? The answer lies in another genetic property of influenza known as genetic drift. Unlike genetic shift which is a major, dramatic change in genetic content due to new chromosomes acquired via reassortment, genetic drift is a more gradual change in the viral genetic information caused when replication errors generate mutations that accumulate in the genome. All viruses can undergo genetic drift, usually a slow process, but the influenza virus mutates at a remarkable rate. During viral reproduction, the genomic RNA replicates into new copies using the viral RNA-dependent RNA polymerase enzyme. This particular influenza virus enzyme has very low fidelity, and it makes approximately one mistake every 10,000 nucleotides during the synthesis of the new RNA pieces. Since the entire influenza genome is roughly 14,000 nucleotides long, this means that virtually every progeny genome has at least 1 mutation due to replication error. This enormous rate of mutation allows the virus to change its proteins so rapidly that antibodies generated against the current year's strain may have no protective effect on strains prevalent in subsequent years. Consequently, recovery doesn't provide life-long immunity, or even very short-term immunity, as next year's prevalent strain may be quite different antigenically from last year's strain. This rapid antigenic change necessitates that manufacturers change the vaccine yearly to target the strains predicted for circulation in the upcoming flu season. However, predicting the correct strain for the vaccine remains a difficult and somewhat dicey process since influenza is a quickly moving target. As a result of these issues, flu vaccine efficacy rarely ever exceeds 40–60%, unlike most other common vaccines whose efficacies exceed 90%. The holy grail of influenza vaccinology is to create a vaccine against some relatively immutable part of the virion. Such a target would generate long-lasting and broad protection against influenza so that yearly vaccination would no longer be necessary. Until then, this acute virus will continue to afflict us year after year.

Hepatitis C – The Great Deceiver

John is a 55-year old male in seemingly good health. During a routine annual physical his blood test results showed elevated liver enzymes typical of liver inflammation or damage. Although John never experienced acute hepatitis, follow-up testing revealed that he had a chronic hepatitis C virus (HCV) infection. John's case is a fairly typical presentation for this silent, but potentially deadly infection. Like most others with this disease, John was unaware of his infection while spreading the virus unknowingly to others. This is an ideal scenario for a virus since it persists life-long in its chronically infected victims and stealthily moves on to new prey over many decades. Because of this insidious pattern, we didn't even suspect this virus existed until the 1970s as it silently spread globally. By 2018 there were an estimated 3.5 million Americans chronically infected with HCV and a staggering 150 million people worldwide (2% of the world population)!

By the mid-1970s, both hepatitis B virus (HBV) and hepatitis A virus (HAV) were identified as the major causes of viral hepatitis and diagnostic tests were developed for each. Once HAV and HBV could be ruled out in patients it became clear that there were some cases of viral hepatitis caused by neither of these viruses. This led to the nomenclature of Non-A, Non-B hepatitis (NANB) as a placeholder designation for an undiscovered viral agent. These NANB infections were particularly prevalent in the burgeoning blood transfusion patient population arising in the 1970s. Since the agent was unknown there was no way to screen blood donors resulting in the blood banks becoming contaminated. Unknowingly contaminated blood helped spread the disease to transfusion recipients, often leading to silent chronic infections. Sadly, it would take close to 15 years before the hepatitis C virus was discovered in 1989. HCV turns out to be a small (9600 nucleotide genome), positive-sense RNA virus in the family Flaviviridae (which also includes West Nile virus and Zika virus). Much of the delay in identifying this new virus was due to HCV being refractory to standard viral cultivation methods; it cannot be grown in cell cultures like most common viruses, so many standard approaches for isolating new viruses failed. Ultimately HCV was found by cloning the viral genome out of chimpanzees experimentally infected with blood samples from patients with NANB hepatitis. Consequently, its initial discovery and subsequent characterization have been mostly via molecular and genetic analysis rather than standard virological approaches. Nonetheless, over the last 30 years, there has been tremendous progress in the diagnosis and treatment of these infections, and the once intractable disease is now curable in over 90% of infected individuals treated with anti-HCV drugs. In recognition of this epic discovery and its impact on clinical care, Drs. Harvey Alter, Michael Houghton, and Charles Rice received the 2020 Nobel Prize in Physiology or Medicine.

While HCV can and does cause acute hepatitis, most patients (~80%) have an initial silent infection with little or no noticeable symptoms. Importantly, for over 70% of infected individuals the virus establishes a silent, but chronic infection of the liver that persists for the lifetime of the unfortunate victim. The HCV target cell in the liver is the hepatocyte, a cell that comprises 70–85% of the liver's mass. Hepatocytes manufacture several blood proteins, including serum albumin and some clotting factors, as well as synthesizing cholesterol and bile salts. Furthermore, hepatocytes are essential for the metabolism and detoxification of external compounds such as drugs and environmental chemicals, as well as certain endogenous compounds such as steroids, making their adequate functioning critical to our overall health. This predilection for hepatocytes by HCV is likely manifested because of complex and not yet fully understood interactions between virion surface proteins and cell surface proteins that serve as receptors, co-receptors, and auxiliary factors. Only hepatocytes (and possibly a few other limited cell types) appear to have the array of surface proteins necessary to facilitate HCV binding and entry, so viral infection is limited to these cells.

HCV infection of hepatocytes converts them to permanent viral production factories. Unlike most acute viruses which ultimately lyse their host cells for the release of progeny virus, HCV does not kill the hepatocyte. Instead, new virions are released from the infected cells by budding from the cell surface in a process that does not lyse or kill the cells. Thus, HCV-infected hepatocytes can live a normal life except that they continually excrete new virions that either infect other hepatocytes or enter the bloodstream. Because of this chronic release, HCV-infected persons persistently have virions in their blood (called viremia) and bodily fluids (e.g. urine, saliva, semen, breast milk, and vaginal secretions), making these biofluids sources of infectious virus for transmission to other people. The highest viral levels are typically in blood, making direct exposure to contaminated blood the most common route of transmission. Historically, viral transmission was due to transfusions or the use of blood products prior to our ability to screen for HCV. Currently, infection is commonly associated with IV drug abusers who share needles or with people who partake in risky, unprotected sex, though medical personnel and first-responders exposed to infected blood are also at risk. Additionally, around 10% of HCV-positive people don't have any of these high-risk factors in their backgrounds, so other less intimate means of transmission must exist.

At the start of an HCV infection, the innate immune system (Chap. 5) would normally be activated to limit viral replication, but HCV has multiple mechanisms to inhibit our innate antiviral responses. This allows the virus to establish a persistent infection with little initial symptomology and only low-level inflammation. However, the cellular immune system does recognize these infected cells over time and attempts to kill them to stop the infection. This immune response leads to hepatocyte cell death, but the liver is highly regenerative and makes new cells to replace the destroyed cells. As these new cells arise they quickly become infected due to the high levels of virus in the liver, leading to a chronic cycle of infection, immune-mediated killing, and excessive regeneration. Even antibodies are of limited help in this infection as HCV is another virus that is highly mutable due to replication errors, similar to the influenza virus. This genetic diversity for HCV is so great that within a single infected individual there are distinct viral subpopulations that are different enough at the genome sequence level to be designated as quasispecies. This genetic heterogeneity translates into enormous antigenic variation in the virion proteins within every infected person. We do make neutralizing antibodies against the viral surface proteins, but these antibodies are rarely sufficient to neutralize all the diverse and changing virion types that are present. Consequently, there are always unneutralized and infectious circulating virions so the re-infection of new hepatocytes continues unabated by our antibody response. This combination of extreme genetic heterogeneity and non-lytic reproduction by the virus, coupled with the liver's regenerative powers, effectively thwarts our immune system and allows the virus to establish life-long liver infections. Over several decades this unrelenting virus-immune system battle leads to liver damage that ultimately manifests as scarring, cirrhosis, or liver cancer, with victims succumbing to these liver diseases and not directly from the viral attack. Evolutionarily, persistence with minimal clinical effects is a wonderful life cycle strategy for the virus as it allows the host to stay healthy for decades while continuously being able to transmit the virus to other people!

Another corollary of HCV's persistent viral life cycle is that vaccine development has been ineffective. Most vaccines work in whole or part by inducing antibodies that bind virions and block their ability to interact with cell receptors. Such neutralizing antibodies don't destroy the virions but do render them noninfectious and help the virions to be cleared by other mechanisms. If our own immune system is stymied by the virus and cannot protect us from initial infection or effectively clear an active infection, then how would a vaccine evoke a protective response? The answer is that to date we haven't been able to devise a protective vaccine strategy, although many creative approaches to enhance and/or augment our immune responses are still being tested. Fortunately, antiviral drug therapy has been much more successful, and chronic HCV infection is now a treatable disease with a high cure rate. Three classes of anti-HCV drugs now exist, each inhibiting a different viral protein. The first class, available since 2011, are protease inhibitors that target the viral NS3 protein. The NS3 protein is a protease enzyme that cleaves other viral proteins. Several HCV proteins are initially synthesized as large inactive precursors that must be cut into smaller functional proteins by NS3. This process is critical for viral reproduction, and drugs that inhibit the protease activity of NS3 severely cripple the production of progeny virions in infected cells. The other two classes of HCV drugs inhibit viral proteins NS5A and NS5B, respectively. NS5A is a multifunctional protein that is critical for viral replication and assembly of new virions, while NS5B is a viral RNA-dependent RNA polymerase that synthesizes new copies of the viral genome; inhibition of either of these proteins is highly detrimental to viral reproduction. By using combinations of these different classes of drugs viral reproduction can be completely eliminated, resulting in effective cures in around 90% of chronically infected patients. This is a remarkable achievement for a disease that was virtually untreatable prior to 2011. So while this virus successfully deceives our immune system and remains refractory to vaccine development, scientific understanding at the molecular level has led to effective treatment and hopes for the eventual elimination of this disease.

Herpes Varicella-Zoster Virus – Now You See It, Now You Don't

Herpes varicella-zoster virus (HVZ), the agent of chickenpox, exemplifies the second type of persistence, a latent infection rather than a chronic, active infection like the hepatitis C virus. After initial infection, HVZ can remain hidden in a quiescent state with little or no virion production. This inactive state helps the virus avoid immune recognition and provides a mechanism for these viruses to establish permanent residence in human populations. The latent stage can continue for life and the virus never reawakens in many individuals. Unfortunately, for some people, the virus does reactivate after decades of hiding to cause a new disease, shingles. Shingles lesions are full of viruses, so the victim of reactivation has the potential to spread the chickenpox virus to new generations of uninfected people. Imagine, before the HVZ vaccine existed, huge numbers of humans throughout history unknowingly carried this insidious viral passenger for most of their lives.

HVZ is one of the eight known human herpesviruses, all of which cause persistent infections. Like other members of the herpesvirus family, HVZ has a large, double-stranded DNA genome of around 125,000 base pairs encoding greater than 70 proteins. Contrast this to influenza and HCV that have genomes roughly 10 times smaller and which only encode 10–14 proteins and 10 proteins, respectively. What every HVZ protein does is still uncertain, but many of these additional proteins are known to alter the host cell environment to favor viral reproduction, to function in counteracting the host immune response, and to comprise the complex virion particle. The virion particle for this virus has three components: the capsid surrounding the viral DNA, a so-called tegument layer composed of 10–15 viral proteins, and an outer lipid envelope acquired from the cell yet punctuated with 11 types of embedded viral proteins. These surface viral proteins are critical for attachment of the virion to host cells and for subsequent penetration of the host cell membrane to release the virion into the cytoplasm. During this penetration process, the virion loses its envelope which liberates the tegument and nucleocapsid. Released tegument proteins can immediately function to subvert the cell and block immune responses without any need for the synthesis of new viral proteins. Delivery of preformed viral proteins is a very effective strategy to facilitate viral survival during the critical period between entry and new viral gene expression, a time when the virus would be very susceptible to interdiction by host defenses. Eventually, the viral DNA is transcribed into mRNAs that are translated into new proteins that perform viral DNA replication and assembly into new progeny virions.

Clinically, the connection between chickenpox and shingles (also known as zoster) took decades to recognize and to confirm as there are significant differences in the presentations of these illnesses. In its classic acute presentation, HVZ is an extremely infectious virus that spreads via aerosols and direct contact with lesions. Before the chickenpox vaccine was introduced in 1995, this disease was typically acquired in childhood and spread easily through households and schools. After initial viral replication in the oral region, progeny virions infect lymphocytes and disperse throughout the body as these cells circulate in the bloodstream. The incubation period ranges from 10 to 21 days as the virus replicates and spreads from the initial site of infection to its ultimate target in the skin. Reproduction of the virus in infected skin cells damages these cells and gives rise to the characteristic reddish rash that usually first appears on the face, back, or chest and then travels to the limbs. Fever, headache, fatigue, and loss of appetite also accompany the rash, often preceding the rash by a day or two. Over about a week the rash progresses to itchy blisters that eventually scab over and heal up as our immune system defeats this invader. Although in most children the disease is fairly mild, complications such as secondary infection or varicella pneumonia are possible and rare fatalities can occur.

Recovery from chickenpox eliminates the circulating virus and generally provides long-lived immunity. Unlike the influenza virus and HCV, HVZ does not mutate rapidly as its DNA replication process has high fidelity and a low error rate. Consequently, all circulating HVZ is nearly identical, so one exposure is usually sufficient to generate neutralizing antibodies that prevent any subsequent reinfection. Unfortunately, by the time our host defenses have marshaled sufficiently to curtail the initial infection, this crafty pathogen has already sequestered itself from our immune system by hiding in our neurons (Fig. 4.2). Our skin is highly innervated with millions of sensory neurons that provide sensations such as touch, temperature, and pain. These skin sensory neurons are unusual cells that have an extremely long projection known as the axon. The main bodies of these cells cluster in groups known as ganglia that are located adjacent to the spinal column. From each cell body, its single axon extends to the skin, making these axons as long as 2–3 feet! Some studies show that HVZ replicating in skin cells can enter the adjacent sensory neurons whereby the infecting nucleocapsid particles move back up the axon to the nucleus of the cell body in the ganglia. Alternatively, there is evidence of ganglia infection directly via infected circulating lymphocytes, thus either or both of these mechanisms may be at play. Regardless of the route, the HVZ genomic DNA eventually locates within the nucleus of the infected neuron where it persists in a circular form, usually at around 5 copies per cell. For reasons that

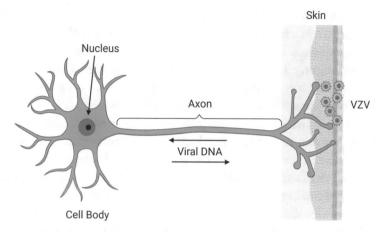

Fig. 4.2 Varicella-Zoster virus latency. The figure depicts a single neuron innervating a region of the skin. The cell nucleus is located within the cell body, and a single long projection called an axon stretches from the cell body to the tissue being innervated. Cell bodies from many individual neurons group together in structures near the spinal cord called ganglia (not shown). When varicella-zoster virus (HVZ) infects the skin it replicates productively in the epithelial cell where it produces chickenpox lesions. Some virions may enter the axon where their viral genomic DNA is transported within the axon to the cell body nucleus. The HVZ DNA remains in the cell nucleus in an inactive state without viral reproduction. This is a permanent condition that persists for the lifetime of the infected person. As immunity wanes with aging, viral DNA in one or a few cells may reactivate and travel back down the axon to the original region of skin that the axon innervated. Upon reaching the skin the virus will start the productive replication cycle that produces new virions and causes the skin eruption of shingles. The location where shingles occurs on the body is random and depends on which neuronal cell suffered the reactivation event. (Created with BioRender.com)

have not yet been well elucidated, the viral DNA remains inactive in the nucleus of these cells, expressing few if any proteins. Without viral protein expression, viral DNA replication does not occur, new virions are not produced, and these neuronal cells are not harmed by the viral DNA. Also, without viral protein expression, there is nothing to mark these cells as infected. Consequently, the immune system doesn't recognize these cells as harboring viruses and doesn't mount an attack to destroy these latently infected cells. Without an immune response, the viral DNA simply resides in the cell nucleus for the life of the neuron, and this type of cell generally lives for our entire lifetime. The result is that anyone who has had an HVZ infection now has many sensory neurons with persistent viral DNA that is never cleared and which remains with that person throughout his or her life.

We are not aware of this latent HVZ DNA hidden in our nerve ganglia, and it isn't infectious in this state since virions are not produced; the only way to detect the viral DNA would be to do an invasive biopsy of the ganglia. If quiescent persistence in nerve cells was the only outcome there would be no clinical consequence of the latent infection, and we'd simply be carrying harmless and nearly undetectable extra DNA in some of our nerve cells. Unfortunately, by mechanisms that are poorly defined, occasionally the viral DNA in one or a few cells starts expressing its genes which leads to the normal replicative cycle. The produced progeny virions pass back down the axon to infect whatever skin region that particular neuron innervates, and viral replication in these skin cells results in the characteristic lesion of shingles. Since the infectious pathway for reactivation is directly from the axon to the skin and not the original systemic dissemination via the blood-borne route, there are usually only localized patches of lesions in shingles rather than the generalized lesions seen throughout the skin during the initial chickenpox infection. The painful, blistery shingles patch can be located almost anywhere on the body depending on what neurons had the viral reactivation event. We believe that reactivation of latent HVZ may occur periodically throughout a lifetime, but in our younger years the prior immunity from chickenpox squelches the new infection before it progresses to lesions, and the individual never realizes that reactivation even happened. As we age and our immunity to HVZ dwindles, particularly T-cell immunity, we may drop below a threshold of protection. At that point, our immunity is now insufficient to curtail the reactivation of the latent virus, which is why shingles generally occurs in older persons many decades after the initial infection. An estimated nearly one-third of Americans will develop shingles in their lifetime. This large number could be reduced through available shingles vaccines that stimulate anti-HVZ immunity and help block viral reactivations. Likewise, preventing chickenpox through the childhood vaccine also eliminates shingles since latent HVZ is never established in the vaccinated individual. When shingles does occur the lesions are full of infectious virus and generally take a few weeks to resolve. The virus in these lesions can readily spread via direct contact to uninfected people to give them chickenpox, so keep those unvaccinated youngsters away from grandma or grandpa when they have a shingles outbreak!

This pattern of initial acute infection followed by decades of latency before localized reactivation may seem a strange lifestyle for a virus, but there is a significant advantage to this mode of attack. As mentioned previously, a virus that fails to spread to another host becomes lost from the evolutionary pool. Failure to spread could be due to killing the host too quickly or simply running out of other susceptible hosts to infect before the original host's immune system destroys the virus. Imagine when animal or human populations were sparse and an acute infection type virus entered into a new group of hosts. If the virus spreads quickly it soon infects everyone in the group. At this point, the virus would be lost as all recovered individuals have anti-HVZ immunity and can't be reinfected, and there are no susceptibles left to infect. In contrast, a latent virus such as HVZ can run its acute course in the new population then evade the immune system by lurking for years hidden in the nerve ganglia before reappearing. By the time the virus reactivates there will be new births and perhaps whole generations of new susceptibles ripe for infection and another round of spread. Such a cycle can persist indefinitely and allow passage of this type of virus for eons and eons with no effective way to interdict the generational spread until we developed a vaccine.

The attack, hide, attack approach used by HVZ is a very ancient and successful viral strategy. The herpesviruses family infects not only vertebrates like humans but also invertebrates such as oysters and abalones. As discussed in Chap. 3, this dissemination across such diverse groups denotes a very ancient origin for this type of virus. Interestingly, the abalone herpesvirus is neurotropic like HVZ, indicating that this property of the virus likely existed more than 500 million years ago before vertebrates and invertebrates diverged. The primordial HVZ is more recent and arose in African primates around 100 million years ago. Consistent with this dating, African monkeys today carry the simian varicella virus (SVV) while New World monkeys lack this virus. Thus, SVV must have arisen after South America drifted away from the African continent around 120 million years ago. Remarkably, SVV is 70% identical to human HVZ and therefore both SVV and HVZ likely shared a direct progenitor. As early primates carrying the ancient SVV progenitor diverged and the hominin lineage developed, early hominins were already carrying latent infections. Over time reactivations kept the virus present through generation after generation. This allowed the progenitor virus to co-evolve with the hominin lineage to become the modern HVZ that spread out of Africa with human migration. The net result is that humans all over the world harbor modern HVZ, and a human generation free from this pathogen has never existed. Unless we can somehow prevent new transmission until all the carriers of the latent virus die off we will never totally free the human population from one of its oldest adversaries.

Polyomaviruses and Anelloviruses – Predators or Passengers?

Some studies suggest that humans typically carry 8–10 persistent viruses, and this is likely a significant underestimate. HCV and HVZ are examples of two types of persistent infections, chronic and latent, respectively, which can cause clinically obvious disease both on initial infection and after many years of persistence. But what about more subtle persistent infections that might or might not lead to any obvious disease? Does our virome contain hidden passengers, so-called "orphan" viruses, whose effects on our health are minimal or so stealthy that we haven't yet connected their presence with a subsequent disease process? Exciting findings with polyomaviruses and anelloviruses certainly suggest that this may be the case.

Polyomaviruses were first discovered in 1953 as an infection in mouse colonies, giving rise to the prototype virus known as mouse polyomavirus (MPyV). The name comes from the Greek words for many (poly) and tumors (oma) because the virus could cause tumors in many anatomical locations in mice. In the early 1960s, scientists discovered a primate polyomavirus contaminating the early poliovirus vaccines. This virus became known as simian virus number 40 (SV40) and raised considerable consternation about a human virus with potential cancer-causing ability being injected into vaccine recipients. Both MPyV and SV40 became important models for understanding virus biochemistry and molecular biology, which are covered in-depth in Chap. 7. The polyomaviruses are small viruses with double-stranded, circular, DNA genomes of around 5000 base pairs that encode less than 10 proteins. Their genome is packaged into a simple capsid made up of 2–3 proteins and lacking an envelope. They are absolutely dependent upon the cellular DNA replication machinery to replicate their viral genomes, but they do make 2–3 fascinating viral proteins that can reprogram the host cell's environment to the virus's advantage as we'll see in Chaps. 7 and 8.

The first two human polyomaviruses were discovered in 1971 and were named BK polyomavirus (BKPyV) and JC polyomavirus (JCPyV) after the initials of the patients who harbored them. Over the ensuing decades, it became clear that both these human polyomaviruses were highly widespread as persistent infections in human populations with between 50% and 90% of adults being infected. However, neither seemed to cause apparent disease in immunocompetent individuals, so initial infection and persistence are largely unnoticed. Only under immunosuppressed conditions did either of these viruses reactivate to cause significant disease. Reactivated BKPyV causes kidney disease while reactivated JPyV causes a fatal demyelinating disease of the brain known as progressive multifocal leukoencephalopathy, PML for short.

The human set of polyomaviruses remained at two for 36 years until 2006 when two additional human viruses, KIPyV and WUPyV, were discovered in patients with respiratory illnesses. Again both viruses were named after the initials of the patients from which they were isolated. Two years later a fifth human polyomavirus was isolated from a skin cancer called Merkel cell carcinoma, and since then nine additional human polyomaviruses have been discovered, six of them from human skin. To bring more consistency to the naming of these viruses they are now designated human polyomaviruses (HPyV) 1–14, though the original names are still widely used. This sudden explosion of human polyomavirus discovery is due to new methods for identifying unknown viruses by their genomes and molecular signatures rather than culturing them from clinical samples. These methods allow researchers to screen healthy individuals and their samples more broadly and not just look at sick individuals. Whether or not there are still more than these 14 human polyomaviruses lurking in our bodies remains unknown. It is clear though that we all harbor many more of these inconspicuous viruses than was ever imagined even a few years ago.

The rapid expansion of the human polyomavirus family means that most of the newly discovered types are not well studied and their role in human disease is largely undefined. There are, however, several features of these viruses that are becoming well established. First, all of these viruses are highly ubiquitous in human populations with estimates ranging from 40% to 90% of adults carrying one or more polyomaviruses. Second, most of the 14 known types have no or only a tenuous connection to a disease state, at least in immunocompetent individuals. Interestingly, persistent polyomaviruses have been found in diverse tissues such as skin, kidney, bladder, prostate, gastrointestinal tract, and brain. While models to explain this multi-organ persistence have been postulated, there is still a dearth of solid evidence about how these viruses maintain themselves in our bodies and avoid immune eradication. Third, the infections begin in childhood and continue throughout life, as the percentage of people showing viral positivity increases with age. For example, one study with HPyV8 showed that around 20% of children age 2–3 years were positive for this virus while the percent positive jumped to nearly 90% in the 60–69 year age group. Other studies in infants and young children support the conclusion that infection can occur very early in life, contracted primarily from the parents and close caregivers. The virus is shed from the skin and is also found in blood, saliva, and most bodily excretions such as urine, feces, and respiratory secretions. Given this diverse array of viral shedding, it is no surprise that direct person-to-person contact is a major route of transmission. Unfortunately, it appears that these viruses can also be spread through infected food and water, as well as contaminated surfaces, so no one is really safe from polyomaviruses, and our odds of becoming infected increase the longer we live. All of these features suggest a long co-evolution with humans that resulted in a biological truce: as long as polyomaviruses don't bother us much we let them hang around and pass them on to our offspring and other contacts. It's even possible that polyomaviruses could be beneficial early in life by helping to prime and develop our immune systems without exposing us to serious disease consequences.

Another poorly understood group of orphan viruses is the family *Anelloviridae* (from the Latin word anellus which means ring). Like polyomaviruses, anelloviruses have a small, circular DNA genome, though in this case, it is single-stranded DNA not double-stranded. The first members of this family were discovered in the late 1990s in the blood of patients with post-transfusion hepatitis, yet subsequent studies found that anelloviruses are not likely to cause hepatitis. Once the first members of this family were identified it became possible to develop molecular detection methods to screen samples from other individuals, both healthy and sick. Numerous studies around the world over the last 20 years have shown two remarkable outcomes. The first is that this family is highly diverse with over 60 species infecting humans and with unusually high genetic heterogeneity within each species. Secondly, members of this virus family are extraordinarily prevalent as silent, persistent infections with most healthy adults having virus present in their blood, often at millions of virions per milliliter! Other studies report finding these viruses in breast milk and amniotic fluid, suggesting that early infection from one's mother, either in utero or soon after birth, is a major mechanism for disseminating these viruses from generation to generation. Similar to polyomaviruses, anelloviruses are found not only in blood and breast milk, but also in saliva, tears, sweat, skin, urine, feces, the respiratory tract, and genital secretions, so acquiring infection from other close contacts is likely if you were not initially infected by your mother. The virus also survives well in sewage, rivers, and lakes, so environmental exposure is another potential source of infection. Given its ubiquitous distribution in both the environment and the human body, it's not surprising that almost 100% of adults worldwide show evidence of infection.

You would think that it would be easy to find a disease connection with viruses that infect nearly everyone. Yet hundreds of studies have attempted to link anellovirus infection to human disease with disappointingly conflicting results. Beyond hepatitis, scientists and physicians have attempted to associate anelloviruses with respiratory infections, multiple sclerosis, autoimmune diseases, diabetes, periodontal disease, and various cancers with no clear-cut results. Without a consistent and uniformly observed correlation with a disease, there is no widely accepted conclusion about the pathogenic nature of these viruses. While it is clear that anelloviruses make up a large family that is genetically highly diverse and pervasive in human populations, their direct role in human disease and/or normal biology remains enigmatic. There are, however, tantalizing hints that members of this virus family might subtly influence diseases or other infections by modulating our immune system. This may be very difficult to establish because it could depend on many factors such as the specific anellovirus or combination of anellovirus species causing infection, the age of acquisition of infection, the overall level of virus production in an individual, personal genetic factors, or co-infection with another pathogen. Teasing out the role, if any, of anelloviruses in human disease will be an exciting adventure that we should all keep an eye on in the next decade. Conversely, could anelloviruses also be useful for priming and developing our immature immune systems in childhood as suggested for polyomaviruses? There have been fewer attempts to address this question, so for now, this is only speculation.

In looking at the range of viral predators we've seen that they span from acute to persistent infections. Some viruses cause nothing but disease and misery while others have little or no apparent effect on our health. Studies were initially skewed towards viruses that elicited obvious and significant disease, but now we have the tools and the insight to realize that these acute viruses are in the minority. Instead, dozens to hundreds of human viruses exist that can infect us and persist with little or no apparent disease. These orphan passengers are often acquired very early in life and may form part of our commensal virome. Such infections may contribute to certain diseases or may benefit us if they promote a healthy immune system. While much remains to be learned about virus-human interactions and the effect on our overall health, at the very least we know that hundreds of human viruses surround us and constantly infect us from birth to death.

Additional Reading

1. Biology, evolution, and medical importance of polyomaviruses: An update. Ugo Moens, Andi Krumbholzb, Bernhard Ehlers, Roland Zell, Reimar Johne, Sébastien Calvignac-Spencer, Chris Lauberg. Infection, Genetics and Evolution 54:18–38. 2017.
2. Human anelloviruses: an update of molecular, epidemiological and clinical aspects Sonia Spandole, Danut Cimponeriu, Lavinia Mariana Berca, Grigore Mihaescu. Archives of Virology 160:893–908. 2015.(continued)

Definitions

Acute infection – An infection that generally develops suddenly and persists for only a short period, typically days to weeks.

Amniotic fluid – A protective liquid that surrounds the growing fetus.

Anellovirus – A type of virus with a single-stranded, circular DNA genome. They are highly prevalent as persistent infections producing little or no disease symptoms.

Autoimmune disease – A disease process where the immune system attacks the body's own tissues rather than a foreign substance.

Axon – The long projection from a nerve cell that carries signals from the cell to some other part of the body such as the skin or muscles.

Bile salts – Organic biomolecules produced in the liver that aid in the digestion of fats and certain vitamins.

Biofluids – Any biological fluid such as sweat, urine, breast milk, blood, or the fluid in blisters and cysts.

Cell culture – The process of growing cells in the laboratory using artificial media.

Chemokines – A large group of proteins that are part of the cytokine family. Chemokines attract white blood cells to sites of infection.

Cholesterol – An organic compound that is an important component of cell membranes.

Chronic infection – An infection that persists for an extended period, typically months to a year. There may be little or no symptoms associated with much of the chronic period.

Cirrhosis – Scarring of the liver that is usually associated with chronic infections or alcohol abuse.

Cloning – The process of generating identical copies of a DNA or organism. In molecular biology, this refers to inserting a target DNA fragment into a vector (typically a plasmid or virus) that can be propagated to amplify the linked target DNA.

Co-infection – The infection of a single cell with two or more viruses.

Commensal virome – The entire collection of viruses that are normally in or on the human body.

Cytokines – A large family of immune system proteins that include interferons, chemokines, and growth factors. Cytokines control the growth and activity of immune system cells and blood cells.

Diabetes – A disease characterized by an inability to make or respond to insulin.

Encephalitis – An inflammation of the brain.

Epidemic – A disease with a widespread occurrence that affects a particular population, community, or region.

Epithelial cell – A type of cell that comprises the tissue the covers our body surface as well as lining many internal structures.

Flaviviridae – The taxonomic family of single-stranded, positive-sense RNA viruses that are transmitted via insects and which includes the West Nile virus.

Flora – The microbial organisms that inhabit the human body.

Fomite – An inanimate object that carries microbial organisms, e.g. a door handle.

Ganglia – A collection of nerve cell bodies located adjacent to the spinal cord.

Genetic drift – The slow accumulation of mutations that leads to changes in the properties of a virus or other organism.

(continued)

(continued)

Genetic shift – An abrupt change in the genetic composition of an influenza virus (or other segmented viruses) that results in a dramatic change in the properties of the virus.

Hemagglutinin (H) – A protein on the surface of the influenza virion that functions in the binding of the virus to the cell and the subsequent penetration into the cell.

Herpes varicella-zoster virus (HVZ) – A member of the herpesvirus family with a double-stranded, linear DNA genome. This virus causes chickenpox (varicella) and shingles (zoster).

Homeostasis – A stable, healthy state of an organism where the metabolic and physiologic processes are functioning optimally.

Hominin – The taxonomic group comprised of humans, chimpanzees, and our close ancestral relatives.

Immunocompetent – Possessing a normally functioning immune system.

Immunosuppressed – Possessing an immune system with reduced functionality.

Inflammation – An invasion of white blood cells, especially in response to an infection, that causes swelling, redness, heat, and pain.

Innate immune response – The branch of the immune system that uses pathogen recognition receptors to detect foreign molecules unique to pathogens. Detection of these foreign molecules triggers a defense response that produces interferon and other cytokines.

Interferon – A type of cytokine that is produced and excreted by cells in response to infections, especially viral infections, and induces an antiviral state in recipient cells.

Latent infection – The presence of a virus (or other pathogens) in the body in a state where no new virions are produced. Typically there are no disease symptoms produced during latent infection.

Lymphocytes – White blood cells, either T cells or B cells, that are part of the immune system.

Merkel cell carcinoma – A type of skin cancer associated with infection by Merkel cell polyomavirus.

Multiple sclerosis – A chronic disease of the brain that causes progressive damage to nerve cells.

Neuraminidase (N) – A surface protein enzyme on the influenza virus virion that functions to help new virions be released from the infected cell.

Neurons – A nerve cell that transports signals from the brain to other parts of the body.

Neurotropic – Preferentially affecting or attacking the nervous system.

Orphan virus – A virus that is not specifically associated with any disease.

Orthomyxovirus – The family of single-stranded, negative-sense RNA viruses that includes the influenza viruses.

Pandemic – A disease outbreak that spreads across countries or the world.

Persistent infection – A virus infection where the virus is not eliminated after the initial exposure and instead the virus remains associated with the infected individual for months to years. Persistent infections may (chronic) or may not (latent) produce new viruses during the persistent phase.

(continued)

(continued)

Progressive multifocal leukoencephalopathy (PML) – A progressive, fatal demyelinating disease of the brain associated with the JC polyomavirus.

Protease – An enzyme that can cut proteins.

Protease inhibitors – A compound that can block the action of proteases.

Quasispecies – A term used to describe viral populations that have extremely high genetic variation even within a single infected individual.

Reassortant – A virus whose genome is derived from two different parent viruses during a co-infection. The process of generating a reassortant is called reassortment and it only occurs for viruses with segmented genomes.

Segmented genome – A viral genome that is composed of separate pieces of nucleic acid rather than a single nucleic acid molecule.

Sensory neuron – A nerve cell that transmits sensations, e.g. heat, to the central nervous system.

Serum albumin – A protein prevalent in the blood which helps to maintain the integrity of the circulatory system.

Shingles (aka zoster) – The skin disease resulting from reactivation of latent herpes varicella-zoster virus.

Simian varicella virus (SVV) – The close relative of herpes varicella-zoster virus found in Old World monkeys.

Simian virus 40 (SV40) – A double-stranded, circular DNA virus of the polyomavirus family.

Steroids – A class of organic molecules with a related chemical structure that has a wide range of biological functions.

T cells – Lymphocytes that function in adaptive immunity, often by recognizing and killing infected cells.

Tegument – a protein-containing region of a herpesvirus virion between the nucleocapsid and the viral envelope.

Viremia – The presence of viruses in the bloodstream.

Virome – the total collection of viruses found in an organism.

Zoster – Another name for shingles.

Abbreviations

BKPyV – BK polyomavirus
CDC – Centers for Disease Control
H – hemagglutinin
HAV – hepatitis A virus
HCV – hepatitis C virus
HN – hemagglutinin and neurominidase
HPyV – human polyomavirus
HVZ – herpes varicella-zoster virus
IV – intravenous
JCPyV – JC polyomavirus
KIPyV – KI polyomavirus
MPyV – mouse polyomavirus
N – neuraminidase
NANB – non-A, non-B hepatitis
PML – progressive multifocal leukoencephalopathy
SV40 – simian virus number 40
SVV – simian varicella virus
WHO – World Health Organization
WUPyV – WU polyomavirus

5

Immunity, Immunity, Immunity

Keywords Intrinsic immunity • Innate immunity • Adaptive immunity • T cells • B cells • Antibodies • Interferon • Cytokines

> *If we think of the immune system as a machine, then we are far from even knowing all of its parts.*
> Bruce Beutler, 2011 Nobel Laureate

The Big Three

There's an old adage about the most important thing in real estate being location, location, location; but when it comes to protecting ourselves from a viral infection, it's immunity, immunity, immunity. That phrasing is appropriate because we are now aware of three distinct layers of immune defense (the big three): intrinsic, innate, and adaptive (Table 5.1). Without this elaborate, dynamic, and extraordinarily flexible defense system we would rapidly succumb to almost any viral infection, so understanding and appreciating these three elements of immunity are fundamental to understanding our normal interactions with the viral world. Intrinsic immunity consists of pre-existing proteins within our cells that have direct anti-viral activity. Since these proteins are already present before infection, they are immediately available to combat incoming viruses that enter our cells. In contrast, innate immunity is slightly slower than intrinsic immunity as many of the anti-viral proteins for the innate system are not premade and are synthesized only after viruses are detected. Still, this is a very rapid system that activates within minutes to hours after viral

© The Author(s), under exclusive license to Springer Nature Switzerland AG 2022
V. G. Wilson, *Viruses: Intimate Invaders*, https://doi.org/10.1007/978-3-030-85487-4_5

Table 5.1 Summary of the Three Major Immune Defense Systems

System	Components	Response time
Intrinsic	Macrophages	Immediate
	Viral Restriction Factors	
Innate	Pathogen Recognition Receptors	Minutes to Hours
	Cytokines/Interferons	
Adaptive	T cells/B cells	Days to Weeks
	Immunoglobulins	

entry into cells and provides remarkable protection early in the infectious process. Lastly, adaptive immunity is much slower and typically takes 5–10 days to reach an effective level. Though slow to develop, adaptive immunity is incredibly specific, is critical for resolving infections, and produces long-lasting immune memory that protects us from future reinfection with the same virus. Each of these three layers has unique features and properties, yet there is abundant communication between the layers to regulate and coordinate their various mechanisms for eliminating the viral infection. This multilayered defense system has sophisticated crosstalk and is the product of eons of evolution in our constant battle with pathogens. Because this combined system exists we often defeat the common viral pathogens we encounter and develop permanent immunity. Alternatively, if we are unable to completely defeat a virus then we may establish a truce and restrain the virus as a persistent infection that has little, or at least long-delayed, clinical effects. This truce allows both the individual and the virus to survive, propagate, and pass on their favorable genetic traits to the next generation. But the cold war arms race never ends as viruses continually evolve new strategies to circumvent our immune defenses.

A critically important aspect at every level of our immune defenses is the ability to discriminate between ourselves and a foreign pathogen. Our bodies are composed of millions of cells expressing tens of thousands of proteins, carbohydrates, lipids, and other complex biomolecules. When a virus particle enters the body, how do we recognize that this interloper is not one of the myriads of normal molecules that constitute our anatomy? We can't just continuously or indiscriminately turn on our immune defenses as such inappropriate activation would damage our own cells as happens in autoimmune diseases. This problem is analogous to the situation that the military has when monitoring American airspace. Our airspace contains numerous planes and helicopters; these include commercial, private, governmental, and military aircraft that are constantly coming and going over American soil. There are even objects such as weather and passenger balloons floating in our skies. How are we able to track all these objects and distinguish them from a potential foreign invader that is seeking to do us harm? The answer is that we have a sophisticated system that uses multiple tools to locate, identify, and follow everything in the sky. This includes radar to monitor position and speed, radio communication to help identify manned aircraft, flight plans, identification markings and transponders on planes and helicopters, and visual intercepts by fighter planes if necessary. Collectively these tools allow us to guard our airspace and to distinguish any outside entity that has no legitimate presence in our territory. If an aerial object is identified as foreign or is present in a prohibited location then we label it a threat and take appropriate action. Similarly, our immune system must constantly surveil our entire body from brain to bloodstream in search of pathogens, and evolution has endowed us with intricate mechanisms that detect molecular signatures unique to these invaders. Like an enemy aircraft, if an incoming organism displays non-self molecules or is found in an inappropriate cellular location then it is deemed dangerous and our immune system will mount an attack.

Another general principle of our immune system is that many of our defenses divide into two broad types, humoral versus cellular. Humoral simply means of or relating to bodily fluids. When pertaining to immune defenses, humoral signifies some type of soluble molecule that is released outside the cell, typically a protein. Classic examples of humoral defense molecules are interferon and antibodies, the primary mediators of innate and adaptive immunity, respectively. Both of these molecules are proteins released from cells into the intracellular spaces and/or the bloodstream where they can act distantly from the site of production. In contrast to these molecular effectors, cellular immunity refers to cells themselves that perform some type of defense function. Examples include white blood cells such as macrophages and T cells. Macrophages are circulating defense cells that can directly ingest viral particles and destroy them, while T cells attack virally infected host cells. Attacking T cells destroy the host cells to stop them from acting as factories that continue to produce new viruses. The details of the humoral and cellular branches of the three immune defense layers are covered in the subsequent sections.

It is also important to understand that our immune system is highly self-regulated such that activation in response to pathogens is quickly dampened once we eliminate the threats. As presented below, immune activation leads to inflammation and cell death as a means to stop viral production and spread, and some of our disease symptoms are the result of this self-inflicted damage. While this may seem an odd strategy, damaging and killing some of our cells is the cost of winning the war against a viral invader. Just as in warfare some soldiers will die to protect the nation, so too some cells will die to protect our bodies. It is far better to suffer some limited self-inflicted damage than to suffer greater damage or death from an unchecked virus. As long as the immune response is only transient then we recover quickly once the virus is controlled.

This immune shut down requires feedback mechanisms and control loops that deactivate the humoral and cellular pathways and reset the system to armed readiness rather than open combat. Numerous details about our immunity and its control systems remain unknown, but at least in broad strokes, this chapter will explain how we detect viral pathogens and combat them through the combined efforts of the intrinsic, innate, and adaptive branches of our immune system.

Intrinsic Immunity – Always There When We Need It

Viral infections can move very quickly. Imagine a virus such as influenza that can reproduce in as little as 6 hours, yielding thousands of progeny viruses for each infected cell. Even a small exposure to 100 virions would lead to billions of progeny viruses within a day or so. Each of these progeny viruses is infecting and killing our cells leading to massive damage if left unchecked. Intrinsic immunity attempts to stop and delay this initial onslaught by utilizing pre-existing proteins and cells. The cellular branch of intrinsic immunity consists primarily of macrophages. The word macrophage derives from the Greek language and literally means "big eater". Macrophages are a type of white blood cell that is prevalent throughout the body. They constantly surveil the extracellular environment where they can engulf particulate matter, such as viruses, in a process known as phagocytosis. Ingested viruses are then degraded and destroyed inside the macrophage by cellular digestive enzymes. Phagocytosis is a nonspecific process such that any type of virus can potentially be ingested and destroyed. If incoming viruses are destroyed by macrophages before they reach their target host cells, then an infection can be completely avoided. This likely helps control some exposures to viruses where we encounter only a small number of viral particles. Unfortunately, when exposed to larger numbers of virions, some inevitably reach their target cells before they are found by the macrophages so infection proceeds.

The humoral branch of intrinsic immunity consists of antiviral proteins located on or inside our cells. These proteins are collectively known as viral restriction factors (VRFs). Conceptually, you can think of each cell as a home and VRFs as makeshift weapons used to defend the home such as a golf club, a baseball bat, or anything else you have on hand. Since these proteins ("weapons") are normally present in or on our cells, they are constitutively poised for immediate response to viral invasion. Each VRF interdicts some step in the viral reproduction process from entry to egress. By restricting critical steps in the viral life cycle, the VRFs attempt to reduce rampant viral reproduction and limit the early rapid spread of infection. However, successful human viruses generally have countermeasures that diminish the effectiveness of VRFs. Consequently, viral reproduction is not completely stopped, only reduced, and infection often still proceeds towards disease. Nonetheless, any reduction in viral reproduction early on decreases the number of cells infected and gives the innate and adaptive immune defense time to activate and function.

The total number of VRFs that humans possess is not known although currently around 50 proteins have been identified with viral restriction capability. This number is far more than we can individually describe here, so we'll focus more on general principles of VRF function using a few well-characterized examples of specific VRFs. A key feature of humoral intrinsic immunity is that the individual VRFs do not have broad antiviral activity against all virus families. Unlike the macrophages that can essentially "eat" any virus, each VRF has only a specific set of virus types that it is active against. This limited range is likely because VRFs are cellular proteins that have some normal biological function as their primary mission. Just as a baseball bat's real purpose is for sport, it can certainly be an effective weapon in some circumstances, for example, against an unarmed intruder. Through evolutionary selective pressure to survive viral infection, some cellular proteins acquired a secondary antiviral activity and became VRFs. However, the VRF's original biological function must be maintained or the cells would be defective. Consequently, the antiviral activity of each VRF relates to its normal function, with each VRF only targeting those viruses whose life cycle can be affected by this function. To summarize, all VRFs are cellular proteins that also acquired the ability to counteract viruses whose life cycle brings them into contact with the VRF. Because viruses are highly diverse in their intracellular locations and their interaction with host proteins, there is no single cellular protein that can have restriction activity against all viruses. Many VRFs only attack a few viruses, and even the most broadly active VRFs still only interdict a limited subset of the human viruses. The fact that VRFs each have a limited viral target population may partially explain why there are so many VRFs; we needed to evolve many of these factors to ensure that every virus family has at least one VRF that can potentially stymie its infection.

An excellent example to illustrate VRF specificity is the VRF called tetherin. Tetherin has antiviral activity against ten families of enveloped viruses, including both DNA and RNA genome viruses. Tetherin is a host protein embedded in distinct patches in our cell membranes called lipid rafts. Available evidence suggests that tetherin has diverse and important functions for our cells that include organizing portions of our cell membranes, modulating the release of cellular vesicles, and participating in cell signaling pathways. For many enveloped viruses, the new virions assemble adjacent to these rafts to capture our membranes for incorporation into their viral envelope. Through its position in the cell membrane rafts, tetherin evolved to bind to these enveloped viruses while still performing its normal functions. As its name implies, tetherin interacts with these nascent virions at the membrane and tethers the virions to the infected cell surface. Keeping the new virions anchored to the cell that produced them is a mechanism to limit spread by reducing subsequent infection of new cells and diminishing new virion production. Viruses without envelopes, or enveloped viruses whose new virions don't co-localize with tetherin in the raft regions, are unaffected by this particular VRF. Thus, the biological location of tetherin as a membrane raft protein dictates which virus types it can potentially inhibit.

The second feature of VRFs is that there is considerable redundancy with multiple different VRFs that can restrict the same viral family. As systems engineers will attest, any effective and stable system will have built-in redundancy to ensure that failure of one part doesn't cause a system-wide catastrophe. If a particular virus's life cycle step was only blocked by a single VRF, then viral mutation might easily generate a virus that could bypass or escape the VRF action. Having two or more unrelated VRFs that block the same viral life cycle step makes it evolutionarily more difficult for a viral mutant to arise that can simultaneously bypass these redundant VRFs. Additionally, for any defense system, blocking an invader at sequential steps or stages of the attack increases the likelihood of defeating the attacker. When there are VRFs that block a virus at two stages in its life cycle, the dual attack will make viral reproduction increasingly difficult. For example, one VRF might block viral entry and another VRF might inhibit genome replication. During an actual infection where a cell might be attacked by hundreds of viral particles, a few virions might slip through the entry block and reach the cell interior. Without further defenses, then these few infiltrating virions will initiate gene expression and replication leading to new virions and subsequent spread. A second VRF that targets viral replication now thwarts these remaining invaders and prevents viral reproduction. Importantly, having VRFs that operate at more than two steps just increases the odds that the cell will effectively block any invading virions from producing progeny. An excellent example of both sequential and redundant VRF defense concerns the herpesvirus family. We know of distinct VRFs that function at four different stages of the herpesvirus life cycle: entry, gene expression (mRNA production), translation (protein production), and virion release. Additionally, there is redundancy at the gene expression step with at least four VRFs (KAP1, SAMHD1, RNaseL, and APOBEC3) that can each attack herpesvirus mRNA production. Similar situations with multiple VRFs have evolved for many other viral families. This need for redundancy and activity against multiple viral life cycle steps likely contributes to why evolution has created so many VRFs.

At this point, maybe you are wondering why we ever get sick from viruses when we have all these wonderful VRFs in place. Again, it all comes down to evolution, and viruses evolve much faster than humans. Remember that viral generation times are hours to days compared to decades for humans. So no matter what defenses we've evolved over the eons, modern human viruses are the ones that have successfully evolved countermeasures to our defenses. These successful viruses can each circumvent their specific VRFs, at least to an extent that allows enough viral reproduction for the spread and survival of the virus. Interestingly, geneticists have determined that the rate of evolution for our VRF genes is higher than for our genes in general. This higher rate is consistent with these genes being under greater selective pressure in the race to respond to viruses that change more rapidly than we do. Alas, with VRFs alone this is a race we can never win and our only hope is to stay close to our viral rivals. The inherent limitations of this VRF system likely provided evolutionary pressure to develop the more broadly acting innate and adaptive immune systems.

There are two other interesting effects of VRFs. First, some VRFs contribute to viral tropism, the ability of viruses to only reproduce in certain cell types (see Chap. 2). If two different cell types both have the appropriate viral receptor on their surface, why might viral reproduction only occur in cell A and not in cell B? One explanation is that cell B contains a high level of a VRF that inhibits this virus while cell A lacks this VRF. Without the inhibitor VRF, cell A cannot stop the infection from progressing through the complete viral life cycle. An excellent example of this phenomenon concerns a protein called EWI-2wint that restricts the hepatitis C virus (HCV). Most cells express EWI-2wint on their surface which blocks entry of HCV even though those cells have the HCV receptor. Consequently, EWI-2wint expressing cells are resistant to HCV and do not easily become infected. In contrast, hepatocytes in the liver lack EWI-2wint and cannot prevent HCV entry and subsequent reproduction. Thus, the absence of the EWI-2wint restriction activity makes hepatocytes very susceptible to HCV infection and at least partially explains the tropism of this virus for this cell type. Similarly, a second important role for VRFs is in host range restriction. Most viruses tend to be highly adapted to their native host species and don't easily transmit across species. At least some of this cross-species restriction likely involves VRFs. As noted above, our human VRFs have only a limited effect on human-adapted viruses because human viruses have evolved strategies to overcome human VRF activities. On the other hand, our human VRFs should be highly restrictive against animal viruses since animal viruses haven't adapted to human cells and human VRFs. Consequently, our human VRFs help prevent us from being infected by incidental exposures to non-human viruses. An excellent example of this cross-species protection is the human MxA protein, a VRF that originally had activity against human influenza. Unfortunately, most circulating human influenza strains are now largely unaffected by this VRF due to the acquisition of viral mutations. However, avian influenza virus strains are extremely sensitive to human MxA protein restriction which greatly protects us from disease if we are accidentally exposed to avian influenza. While humans can still get avian influenza infections, the infections are generally mild and don't readily spread to other humans. So while less important to our antiviral defense than innate and adaptive immunity, intrinsic immunity still makes significant contributions to our health and safety by partially restricting human viruses and more strongly restricting non-human viruses.

Innate Immunity – Locked and Loaded

Although the scientific community discovered and appreciated innate immunity much later than adaptive immunity, we now recognize that it plays an early and essential role in protecting us from viral and bacterial infections. Studies in mice powerfully demonstrate the importance of this defense system. Experimentally knocking out their innate immune mechanisms makes mice quickly succumb to viral infections that are normally harmless to these animals. Such a complete absence of innate immunity would likely be fatal in humans as well, since contracting almost any pathogen could lead to rapid death. Humans have been identified with gene mutations that result in partially defective innate immunity, and as expected they do suffer from increased susceptibility to some types of infections. Fortunately, there is built-in redundancy in this system, just as there is for intrinsic immunity, so completely losing innate immunity due to genetic mutations is unlikely. Given such an important role, how does innate immunity function, and how does it work to protect us?

Before delving into mechanistic details, let's consider the overall strategy of the innate defense system and how it differs from the intrinsic system. The intrinsic VRF system is primarily an individual cell defense strategy where each infected cell tries to save itself by blocking an infecting virus from reproducing. Again, it's like a homeowner using a baseball bat to threaten an intruder; it might successfully ward off the intruder and protect their home, but using this weapon doesn't raise a more general alarm that alerts the neighbors or directly calls in the police. Consequently, if the VRF fails and the infected cell produces progeny virions, then all the neighboring cells are at risk of infection. A superior defense system would not only detect and attack the intruder but would also warn the neighbors and call the police; this is exactly what the innate defense system evolved to do. By alerting its neighbors, the innate system can warn other cells that viruses are in the vicinity. This allows neighboring cells to activate their internal defenses for rapid response in case they too get infected. By summoning the police (the adaptive immune response), innate immunity activates the highly specific system that ultimately arrests the infection and gives protection from any subsequent attack by the same virus. It is this coordinated and integrated interplay of the innate and adaptive defenses that allows us to survive most viral infections.

Mechanistically, innate immunity has three basic steps: detecting pathogens, signaling danger to other cells, and activating antiviral defenses in cells

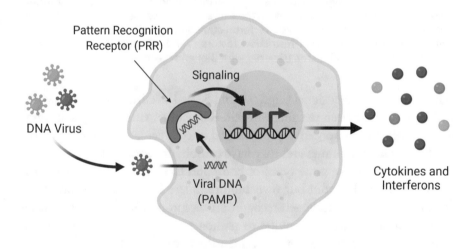

Fig. 5.1 Innate immunity. Depicted are the recognition and signaling steps of the innate response by a cell infected with a DNA virus. After attachment and entry of the viral particle, virion uncoating releases viral DNA into the cytoplasm. The cytoplasm is normally free of DNA, so the viral DNA is a pathogen-associated molecular pattern (PAMP) that warns the cell of a viral infection. The viral DNA is bound by cytoplasmic receptors (pattern recognition receptors – PRRs) that specifically recognize DNA. The binding of viral DNA to the PRR initiates a signaling cascade that transmits to the cell nucleus to turn on the production of various proteins. Among the proteins produced in response to viral infection are cytokines and interferon that are released from the infected cell as danger signals to warn surrounding uninfected cells that a virus is present. (Created with BioRender.com)

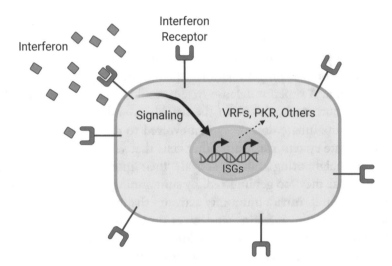

Fig. 5.2 Interferon signaling. Interferon release by an infected cell can diffuse to nearby cells or enter the bloodstream and become more systemic. Cells with an interferon receptor can bind the circulating interferon which activates a signal cascade in the receiving cell. The signal generated by interferon binding to its receptor transmits a message to the cell nucleus to induce the expression of dozens to hundreds of genes that are collectively referred to as interferon-stimulated genes (ISGs). Among the ISG protein products are the viral restriction factors (VRFs) and protein kinase R (PKR) which acts as an antiviral factor. (Created with BioRender.com)

that receive the danger signal (Figs. 5.1 and 5.2). The critical first step involves discriminating between self and non-self. Remember, virions are just biomolecules minimally composed of proteins and nucleic acid, though some virions also have lipid membranes and may have carbohydrates as well. Still, our cells are replete with these same types of compounds, so how do we distinguish virion biomolecules from our own biomolecules? Evolution has solved this dilemma by creating pathogen recognition receptors (PRRs) that specifically bind molecular signatures unique to pathogens. The PRRs fall into several families named after the originally identified member of each family. For example, the Toll-like receptor (TLR) family is named for the Toll protein and the Rig-like receptors (RLRs) are named after the retinoic acid-inducible gene 1 protein (RIG1). Each PRR family has multiple members so the total number of known pathogen receptors is currently over three dozen; note that only some of the PRRs recognize viruses with the others recognizing bacteria and other types of pathogens. The TLRs are positioned in our cell membranes so that they can bind and detect pathogens at the cell surface or in internal membrane structures called endosomes; endosomes are vesicles through which many viruses transit during entry into the cells. The remaining PRRs are in our cytoplasm to detect pathogens inside our cells.

With all these multiple PRRs spread throughout our cells, what exactly are the distinctive pathogen signatures that each PRR recognizes? It turns out that the signatures vary widely depending on the type of pathogen, so collectively we term them pathogen-associated molecular patterns or PAMPs for short. For viruses, their PAMPs are primarily nucleic acids, either RNA or DNA. This may seem surprising given that our cells also contain RNA and DNA, but the secret lies in where the viral nucleic acid locates in the cell or in biochemical differences between viral and cellular nucleic acids. Let's consider DNA viruses as an example. When DNA viruses infect a cell they penetrate the outer cell membrane and the virion uncoats. Uncoating releases their genomic DNA either into the cytoplasm or endosome vesicles before the DNA traffics to the nucleus for viral gene expression and genome replication. While this viral DNA is biochemically the same as our own DNA, in this case, the key difference is in intracellular location. All of our cellular DNA is either inside the nucleus or inside mitochondria and should never be in the cytoplasm or inside endosomes. Hence the mere presence of DNA in either endosomes or the cytoplasm is a danger signal for a likely invasion by a virus. Consequently, we evolved PRRs in both locations that simply bind any DNA to trigger the innate immune response. Likewise, there should also be no RNA in endosomes, so when RNA from RNA viruses traffics through endosomes it will be bound by specific PRRs to activate the innate system. The situation is trickier for RNA viruses that don't traffic through endosomes and instead locate directly into the cytoplasm. Since the cytoplasm also contains various host RNAs such as messenger RNAs (mRNAs), transfer RNAs (tRNAs), and ribosomal RNAs (rRNAs), the mere presence of viral RNA in this location cannot be the activation signal. Instead, PRRs have evolved that recognize features of RNA that are unique to viral RNA and are not found in cellular RNAs. The simplest example is PRRs that recognize long double-stranded RNA. Our cellular RNAs are single-stranded molecules while all RNA viruses must go through a double-stranded RNA intermediate as they replicate their genomes. Certain PRRs have evolved that only bind the double-stranded RNAs that are uniquely characteristic of the presence of a virus and will not bind single-stranded RNAs. This discrimination allows activation of these PRRs only by RNA viruses and not inadvertently by our cellular RNAs that are predominantly single-stranded or covered with proteins. Other PRRs use different tricks for distinguishing between viral and cellular RNAs, but the result is that each PRR can recognize some distinct and unique molecular feature of viral nucleic acids. Collectively, this PRR network provides extensive surveillance of the cell surface and intracellular milieu and can detect any type of virus.

Once viruses are detected by the appropriate PRR, the next step is sending out a signal to the surrounding cells to warn them that viruses are in the vicinity and that they should prepare their defenses. PRRs are normally inactive and simply reside quietly in their assigned locations until an appropriate PAMP is encountered. The binding of a PAMP to its respective PRR activates the PRR and starts a multi-step protein signaling cascade within the cell. The overall process is like a chain reaction where each step activates the subsequent step with the final step activating key cytoplasmic proteins, notably transcription factors called NF-κB and interferon regulatory factors 3 and 7 (IRF3 and IRF7). All three of these proteins reside in the cytoplasm until activated by the PRR signaling cascade. Once activated, all three proteins migrate into the cell nucleus where they bind to our DNA and activate the expression of the cytokine genes (by NF-κB) and the interferon genes (by IRF3 and IRF7). The cell now begins to produce several cytokines and interferons; interferons are a type of cytokine, but for our discussion, we'll keep them separate. These newly made cytokines and interferon proteins are the humoral signals for the innate immune system. There are many different cytokines and multiple types of interferons with a very complex array of biological functions. As with the VRFs, there is extensive redundancy and overlap in function, so sorting out all the effects of each cytokine has been a slow and difficult process. Rather than try to explore this vast subfield of immunology in minute detail, let's instead focus on some general concepts. Once exported by the producing cell, cytokines can diffuse to nearby surrounding cells as well as enter the bloodstream to have more distant effects throughout the body. Functionally, the production and release of cytokines and interferons transmit a warning to other cells that a viral pathogen is present, analogous to a blaring home alarm siren that signals danger to the entire neighborhood.

If cytokines and interferons are the danger messengers, how do cells receive these messages and what do they do with them? The receiving part is fairly clear as we know that our cells have a specific surface receptor for each of these messenger molecules. The receptors are transmembrane proteins embedded in our cell membranes. One end of the receptor can bind external cytokines or interferons and the other end of the receptor is in the cytoplasm. As for the PAMP-PRR system, binding of these cytokine messenger molecules to their specific surface receptors triggers a signal cascade through the cytoplasm that ultimately causes many host genes to be expressed. For viral-induced cytokines, there are numerous and complex effects on host cells, but for our purposes, we'll focus on just three: inflammation, cell killing, and priming of adaptive immunity. Cytokine-induced inflammation is an important defense that results from an influx of white blood cells and macrophages being attracted to the infected cell by the released cytokines. These recruited cells attempt to destroy infected cells, resulting in localized tissue damage. Cytokines also cause blood vessels in proximity to the infected cells to dilate which improves the flow of cells and nutrients, but the resultant tissue swelling can be painful and may cause further local tissue damage. On a whole-body level, some cytokines induce fever which is actually helpful in that many viruses replicate less well at elevated temperatures, so fever can reduce the viral load. Consequently, while inflammation is a helpful defense response to limit viral reproduction and spread, it does contribute to some of the symptoms that we associate with disease.

Instead of inflammation, other cytokines induce cell death through processes known as apoptosis and autophagy. Although these two processes are biochemically and molecularly distinct, you can think of them simply as cell suicide mechanisms. Both apoptosis and autophagy are normal physiologic processes used to remove unneeded cells, but each can also be used to destroy virus-infected cells and prevent those cells from producing thousands of progeny virions. Inducing suicide in infected cells may seem harsh, but most of these cells would eventually die anyway from the viral infection. Consequently, it is much better if the initially infected cells are quickly killed before viral reproduction to save many other cells from infection and death. In addition to these direct attempts to limit infection through inflammation and cell death, another highly important function of cytokines is to activate the adaptive immune system. As presented in the next section, adaptive immunity is critical for resolution and recovery from viral infection. In our home defense analogy, adaptive immunity is the police officers that come in to remove the intruder permanently and to monitor for that intruder if he ever tries to return. Cytokines help prime this system and provide early information about the type of infection so that adaptive immunity can begin to produce and send in the appropriate troopers.

Lastly, interferons are a class of cytokines that have evolved to play a crucial and central role in the early stages of anti-viral defense (Fig. 5.2). As intrinsic immunity became less effective due to viral adaptation, innate immunity likely developed as an important second line of defense. While all cytokines are important contributors, interferons are a major weapon against viruses before adaptive immunity takes over. Most cells in the body have interferon receptors. Like the other cytokines, interferon binding to its receptors sends a signal cascade from the cell membrane to the cell nucleus. In the nucleus, the end effect of the interferon signal is that hundreds of genes turn on to make their respective proteins and create an "anti-viral" state in the receiving cells. What every one of these newly expressed proteins does for cell defense is not yet defined, but a number of them do have well-understood roles in antiviral activity. One subset of interferon-induced genes is those encoding many of the VRFs. While these VRF proteins are typically expressed constitutively in our cells, interferon causes increased VRF gene expression resulting in higher levels of these protective proteins. So even if VRFs are only partially effective under normal conditions, raising the amounts of these proteins to a higher level should provide more defense. Thus, elevating VRF levels in uninfected cells prepares them in advance for possible infection and contributes to a state of anti-viral readiness in the cells. Since interferon produced by a single infected cell can spread to many surrounding uninfected cells, this early warning system alerts the entire neighborhood of cells to marshal their defenses long before they ever encounter the virus.

In addition to the VRFs, there are many other genes whose expression is increased by interferon. Several of these interferon-regulated genes make proteins with well-defined anti-viral activity, and we'll examine the protein kinase R (PKR) protein to illustrate the concept (Fig. 5.3). PKR is a protein with an enzymatic ability to modify other proteins by attaching a phosphate group to the target protein. PKR is normally expressed at low levels and exists in an inactive form. Upon interferon stimulation, cells express PKR more abundantly, though PKR remains inactive. To become activated, PKR must bind double-stranded RNA, a molecule that is scarce in our cells. However, long double-stranded RNA is a signature of RNA viruses as they create this molecule during their genome replication. Even DNA viruses can generate double-stranded RNA during their reproduction through the hybridization of

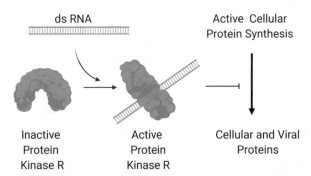

Fig. 5.3 **Protein kinase R.** Protein kinase R (PKR) is one of several anti-viral proteins induced by interferon. In response to interferon, cells begin to make larger amounts of PKR, but the enzyme exists in an inactive form in the cytoplasm of the producing cells. For activation, PKR requires long, double-stranded RNA (dsRNA), a molecule not usually available in the cytoplasm. However, if the cell becomes infected, particularly with RNA viruses, then long, dsRNA becomes abundant as the viral RNA genomes copy themselves as part of the replication process. When dsRNA is present it binds to PKR and converts PKR from an inactive protein to an active enzyme. Active PKR now modifies a second protein that is critical for cellular protein synthesis, resulting in the inhibition of protein synthesis. Without protein synthesis, the cell cannot produce new virions, thus helping to limit the spread of the virus. (Created with BioRender.com)

complementary mRNAs. Consequently, double-stranded viral RNA is generated in interferon-stimulated cells that are subsequently invaded by a virus. Inactive PKR then binds the double-stranded RNA which converts PKR into an active enzyme. Active PRK modifies a host protein that is critical for the initiation of protein synthesis, and this modification inactivates the protein-synthesizing machinery in the cell. Without protein synthesis new viral proteins cannot be made, thus stopping virion production. Stopping protein synthesis is also detrimental to the cell itself, but again it is better to sacrifice the infected cell quickly rather than let it turn into a virion factory whose output can infect thousands of other cells. The beauty of this system is that interferon merely increases PKR levels, but doesn't activate the enzyme. Therefore, cells that encounter interferon are merely poised to respond to viruses but are not harmed unless they become infected, generate double-stranded RNA, and activate PKR. Since interferon can spread widely, this two-step process prepares uninfected cells for virus attack yet spares them from PKR damage unless they become infected. Many other interferon-induced anti-viral proteins work similarly, so this two-step process seems to have evolved as an effective strategy to prevent collateral damage in uninfected cells. Collectively, all the various cytokine- and interferon-induced effects dampen viral reproduction and spread, preventing viruses from completely overwhelming our bodies in the first few days after infection. Yet as for the intrinsic system, most human viruses have evolved countermeasures against innate immunity. Because of these viral countermeasures, viral infections are not completely halted by innate immunity and disease still occurs. Resolution and recovery from most viral infections ultimately require the third level of immunity, the adaptive immune system.

Adaptive Immunity – The Gift That Keeps on Giving

The biological theme that runs through innate immunity is the use of receptors to bind target molecules. In biological lingo, any target molecule bound by a receptor is called a ligand; PAMPs are ligands for PRRs and interferon is the ligand for interferon receptors. The result of the specific interaction between a ligand and its cognate receptor is the activation of a biochemical signaling cascade. Ultimately, these cascades transmit a signal to DNA in the cell nucleus where the signal changes gene expression patterns to increase or turn on the production of specific proteins. In the case of innate immunity, many of these newly expressed proteins have an anti-viral activity to help protect cells from infection. Similarly, adaptive immunity also makes extensive use of receptor-ligand binding to initiate its protective effects. A key difference though is that innate immunity responds very rapidly, on the order of minutes to hours from virus detection to establishment of the antiviral state by cytokines and interferon. In contrast, adaptive immunity takes days to weeks to reach its full activity. If innate immunity was not active and preventing uncontrolled viral reproduction, we would all succumb to almost any viral infection long before adaptive immunity became available.

Adaptive immunity primarily involves two types of white blood cells: B cells and T cells, also known as B and T lymphocytes. T cells are further subdivided into cytotoxic T cells, helper T cells, and suppressor T cells. The cytotoxic T cells constitute the cell-mediated arm of the adaptive immune response due to their ability to target and destroy virally infected cells. As their names imply, the helper and suppressor T cells provide regulatory functions that help control the activity of both cytotoxic T cells and B cells. B cells supply the humoral arm of adaptive immunity through the production and secretion of antibodies; antibodies, also called immunoglobulins, are proteins that are highly specific in their ability to recognize and bind to their targets known as antigens. Both cytotoxic T cells and antibodies are critical for completely eliminating viral infections and allowing recovery from illness. Also, both T and B lymphocytes have subpopulations of so-called "memory" cells that provide rapid response and protection from any subsequent re-exposure to a virus. Humans have approximately 2 trillion lymphocytes, about the same number of cells as in the brain or liver, reflective of the adaptive immune system's biological importance. First, we'll consider the roles of T cells and then revisit B cells.

As mentioned in the previous section, virus detection by the pathogen recognition receptors (PRRs) of the innate immune system ultimately results in cytokine release. Among other things, some cytokines bind to receptors on lymphocytes to prime the adaptive immune response and prepare the naïve B and T cells for activation. Activation of naïve T cells requires assistance from another group of cells collectively known as antigen-presenting cells (APCs) (Fig. 5.4). When APCs encounter foreign antigens, including virions, the antigens are taken up by an internalization process called phagocytosis. While antigens can be different types of molecules, for simplicity we'll focus on protein antigens such as the proteins that make up the virions. Inside the APCs, enzymes digest the viral proteins into small fragments called peptides. These peptides travel to the surface of the APCs for display in conjunction with a host protein complex called the major histocompatibility complex (MHC). It is these MHC-peptide complexes that are the critical activators for the T cells. As the APCs circulate, each of their MHC-peptide complexes is scanned by

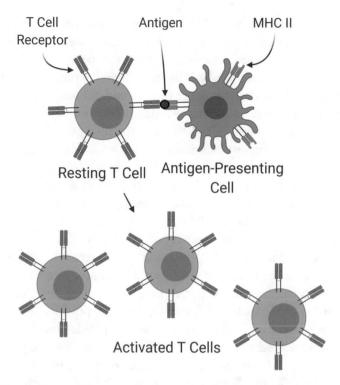

Fig. 5.4 T cell activation. Activation of a resting T cell requires interaction with an antigen-presenting cell (APC) that displays a foreign antigen (for example a viral protein fragment) bound to its major histocompatibility complex II (MHC II). For the interaction to occur, the T cell receptor must recognize the foreign antigen associated with the APC MHC II. The formation of this cell-to-cell complex via antigen recognition triggers the T cell to activate and stimulates its proliferation. (Created with BioRender.com)

the T cell population. Every individual T cell has a receptor called the T cell receptor (TCR) that recognizes both the MHC component and one unique ligand. Since we have trillions of T cells, this cell population can recognize trillions of different ligands, hence the incredible ability of the adaptive immune system to respond to virtually any foreign antigen. Out of all these trillions of T cells, each cell with a TCR that can bind a presented peptide will form a complex between that single T cell and the APC. Formation of this stable complex triggers signals from the APC that activate the naïve T cell to mature, reproduce, and begin to function in anti-viral defense.

The key player in antiviral defense is the cytotoxic T cell. When activated as above, a cytotoxic T cell will reproduce itself to make hundreds of identical T cells. Each of these new T cells can be activated by the APCs causing them also to reproduce. The result is an army of tens of thousands of identical, activated T cells whose TCRs each recognize the specific viral peptide that initiated the process. Each activated T cell is like a laser-targeted missile that will now seek out and destroy any cell expressing the viral peptide. Since the target is a viral peptide, normal uninfected cells won't express this peptide or be harmed by this T cell army. However, virally infected cells will express viral peptides on their surface, and T cells will recognize these peptides through the TCR. When the activated T cell binds to the virally infected cells this triggers the release of enzymes from the cytotoxic T cell that kill the infected cell. As we saw for innate immunity, killing our own infected cells is a necessary sacrifice to stop the production of new virions. If you think of infected cells as factories that are manufacturing enemy munitions, then blowing up these factories is the only way to stop production and end the continuous resupplying of the enemy. This specific ability of cytotoxic T cells to recognize and destroy infected cells is critical for resolving infections since it will eventually eliminate all virus-producing cells from the body. Keep in mind though that this process is slow compared to intrinsic or innate immunity because of the many steps involved. First, APCs presenting viral peptides have to develop, then they have to encounter naïve T cells with a TCR that will bind the peptide. This APC-T cell interaction activates the T cell and stimulates it to proliferate. Next, this initial batch of activated T cells needs to reproduce and these new progeny cells also become activated. This process repeats over and over to generate a large population of activated T cells. Since there may be a substantial number of infected cells, it can take days to weeks to produce enough activated cytotoxic T cells to eliminate all the infected cells. Fortunately, intrinsic and innate immunity are working hard during this period to restrict viral reproduction and keep us alive while this T cell army is generated!

The activation process of helper T cells and suppressor T cells is similar to cytotoxic T cells, though their functions are much different from cytotoxic T cells. Suppressor T cells do what their name implies, suppress the immune response once the infection resolves. This is important so that all our armed defenses don't run amok and cause unnecessary damage, but suppressor T cells don't contribute to the resolution of viral infections so this cell type won't be further discussed. More important to our story, helper T cells do play a critical role in infection defense through their ability to help coordinate the overall adaptive immune response. Once activated, the helper T cells release various other cytokines, some of which promote the activated cytotoxic T cells and activated B cells to proliferate. Additionally, helper T cells are directly required for the initial activation of B cells.

B cells provide the humoral arm of adaptive immunity through the production and secretion of antibodies; each B cell produces one unique antibody. Like T cells, there are trillions of B cells giving us a huge repertoire of possible antibodies. Also like T cells, each B cell carries a receptor (the B cell receptor or BCR) that will recognize a unique antigen. In the case of the B cell, its receptor is actually a form of the antibody that the cell can produce. Circulating B cells are in an inactive or naïve state and do not produce antibodies until they encounter an antigen that is recognized by their BCR (Fig. 5.5). For example, the binding of a virion surface protein to the BCR triggers a process whereby the cell internalizes the protein, digests it, and returns a viral protein fragment to the B cell surface in a complex with MHC. The B cells are now primed but require a second event to become activated. The second event is the binding of the B cell peptide-MHC complex by a helper T cell. For this to happen the TCR must recognize the viral peptide that the B cell is presenting which ensures that both the T cells and B cells are responding to the same viral invader. This two-step activation process confers additional specificity to the response and helps prevent false B cell activation.

Once activated, B cells mature into plasma cells. Plasma cells proliferate rapidly, in part through stimulation by the helper T cells cytokines, and secrete large quantities of their unique antibody; estimates suggest that a single plasma cell can produce millions of antibody molecules per hour. These released antibodies will circulate and eventually find and bind to the viral protein that contains the peptide that started the process. For the resolution of viral infection, the important antibodies are the ones that bind virion surface proteins and "neutralize" the virus. Remember, virion surface proteins are critical for interacting with host cell receptors to allow viral attachment to the cell and subsequent entry into the cells. If antibody molecules bind to the virion surface they can coat the entire virion and physically block the virion proteins from interacting with the host receptors. Think of the receptor as a glove and the virion as the hand that has to fit snuggly into the glove. Imagine wrapping the hand with bulky padding, like a boxer preparing for a fight, and then not being able to get the hand into the glove. Neutralizing antibodies are the padding that blocks the virion-receptor interaction. During an infection, neutralizing antibodies will eventually coat all the free virions and prevent them from being able to infect new cells. The neutralized virions aren't directly destroyed by the antibody binding, just rendered harmless since they cannot enter cells to produce new virions. Coating the virions with antibodies does make them more susceptible to phagocytosis, and the neutralized virions are eventually "eaten" by phagocytic cells and eliminated. In addition to neutralizing virions, some of the anti-viral antibodies can also bind to virus-infected cells and induce cell death, supplementing the cytotoxic T cells in destroying the viral factories.

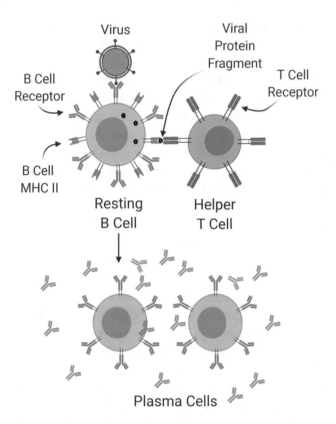

Fig. 5.5 B cell activation. Resting B cells express both B cell receptors (BCRs) and major histocompatibility locus II (MHC II) proteins on their surface. The BCR is a membrane-bound form of the specific antibody that a B cell can produce. The binding of a virion surface protein to the B cell receptor triggers step 1 of the B cell activation process. Bound virions are internalized by the B cell, degraded into protein fragments, and a fragment recognized by the BCR is returned to the B cell surface for display with the MHC II. Step 2 of the activation requires a helper T cell with a T cell receptor (TCR) that can recognize the protein fragment displayed by the B cell. The binding of the T cell to the B cell via the MHC II-fragment-TCR linkage activates the resting B cell and converts it into an antibody-secreting plasma cell. The antibody produced is directed against whatever viral protein was responsible for the initial interaction between the virus and the BCR. (Created with BioRender.com)

Similar to the response time for T cells, B cell activation and antibody production take days to weeks to reach a level sufficient to neutralize all the free virions. Nonetheless, eventually there are sufficient numbers of activated B and T cells to combat most viral infections. The cytotoxic T cells eradicate virally infected cells to stop new virion production, and neutralizing antibodies nullify existing virions to prevent any new cells from becoming infected. This combinatorial adaptive attack, in conjunction with ongoing intrinsic and innate responses, ultimately thwarts viral reproduction and spread, eliminating the infection and allowing us to recover.

An additional feature of adaptive immunity that is extremely important for fighting disease is the existence of immunological memory. When both T and B cells are activated, a small subset of each cell type becomes so-called memory cells. Memory cells don't participate in the fight against the triggering virus or other pathogens. Instead, memory cells remain in reserve for future infections by the same organism and can persist for decades. If a virus is re-encountered sometime after the original infection then the memory T and B cells will recognize the virus through their respective TCRs and BCRs. The binding of the virus to the memory cell receptors now triggers an extremely rapid activation and proliferation by the memory cells, much faster than the original process. Instead of 1–2 weeks to generate large numbers of cytotoxic T cells and antibody-producing plasma cells, the memory response occurs in hours to days. The rapid accumulation of cytotoxic T cells and neutralizing antibodies is usually sufficient to block and eliminate the virus before it even establishes enough infection to cause disease symptoms. This memory system constantly surveils our bodies for known pathogens and prevents us from suffering through the same infection multiple times. The memory response is also the principle involved in vaccination. Vaccination introduces pathogen antigens to our body to produce an adaptive immune response and generate memory cells. These vaccine-induced memory cells are available to protect us if we ever become exposed to the actual pathogen we were vaccinated against. Vaccines will be discussed in more detail in Chap. 10.

To conclude, immune defenses are ancient systems that began to evolve with early life forms as necessary protection from other hostile life forms and viruses. Intrinsic and innate immunity provide a critical rapid response to viral infections that helps limit the viral spread in the first few days after infection. Without this immediate response, we would be overwhelmed by viruses and most infections would likely be lethal. Adaptive immunity, which arose around 500 million years ago, is a beautiful, elegant, and powerful evolutionary advance that has enabled animals to thrive and survive the constant viral onslaught. While slower than intrinsic and innate immunity, adaptive immunity is highly specific and provides the mechanisms to effectively eliminate infecting viruses. Adaptive immunity also provides the invaluable immune memory that allows us to become resistant to reinfection with the same virus, thus saving us from having the same diseases repeatedly during our lifetimes. Collectively, the three layers of our immune system are incredibly complex and absolutely essential for our health and survival in our daily conflict with viruses. When these systems meet a virus that can effectively subvert all three layers then we face life-threatening diseases like Ebola or HIV. As we continue to encounter novel and changing viruses (see Chap. 11), it is only our immune systems that stand between us and disastrous pandemics that could devastate human populations.

Additional Reading

1. An Introduction to Immunology and Immunopathology. Clinical Immunology, Vol. 14(Suppl 2):49. 2018. Jean S. Marshall, Richard Warrington, Wade Watson, and Harold L. Kim.
2. Immunology's Coming of Age. Frontiers in Immunology, Vol. 10:684. 2019. Stefan H.E. Kaufmann.

Definitions

Adaptive immunity – The branch of the immune system that includes B cells and T cells. This branch produces antibodies and cytotoxic T cells that are critical for resolving viral infections and also provide long-lasting immunological memory that prevents reinfection by previously encountered pathogens.

Antigen-presenting cells (APCs) – A specialized immune system cell that displays antigens on its cell surface for interaction with other immune system cells.

APOBEC3 – A cellular protein involved in editing mRNA that also serves as a viral restriction factor.

Apoptosis – The process of programmed cell death used to remove unneeded cells during growth and development. Also used to destroy virally infected cells.

Autophagy – Self-digestion of cells in the body to remove damaged or infected cells.

Avian influenza virus – A member of the orthomyxovirus family with a segmented, single-stranded, negative-sense RNA genome that infects birds.

B cell receptor (BCR) – A transmembrane protein on B cells that binds antigens to stimulate B cell activation and antibody production.

B cells – The cells of the adaptive immune system that produce antibodies. Also called B lymphocytes.

B lymphocyte – A type of white blood cell that produces antibodies. Also called a B cell.

Carbohydrate – The class of biomolecules that includes sugar, starch, and cellulose.

Cellular immunity – The portion of the immune system that uses specific cells (phagocytes and T cells) to destroy pathogens and pathogen-infected cells.

Cognate receptor – The specific receptor that matches with a particular ligand.

Cytotoxic T cells – A type of lymphocyte that can kill pathogen-infected cells, cancer cells, and foreign cells.

(continued)

(continued)

Endosomes – A network of intracellular structures involved in trafficking and sorting biomolecules. Some viruses use the endosomal network for part of their life cycle.

EWI-2wint – A cell-surface protein with viral restriction activity against hepatitis C virus (HCV).

Helper T cells – The subpopulation of T cells that secretes regulatory factors that influence the activity of other immune cells, especially B cells and other T cells.

Host range restriction – The adaptation of a virus to a particular host species that limits the ability of viruses to infect other species.

Humoral immunity – The immune protection provided by antibodies and other extracellular immune system proteins.

Hybridization – The process of complementary nucleic acid single strands coming together to form base-paired, double-stranded regions.

Immunoglobulins – The group of proteins that function as antibodies. Humans have five types of immunoglobulins: A, D, E, G, and M.

Immunological memory – The ability of the adaptive immune system to remember and more quickly react to antigens and pathogens that were previously encountered.

Interferon regulatory factor (IRF) – A protein transcription factor that regulates the production of interferon mRNA.

Interferon-stimulated genes (ISGs) – Any gene whose expression is turned on by interferon.

KAP1 – A cellular protein involved in transcriptional regulation that also has viral restriction activity.

Ligand – Any molecule that binds to a receptor.

Lipid rafts – Regions in cell membranes that are enriched for cholesterol and which affect membrane function.

Macrophages – Specialized cells that ingest and destroy pathogens.

Major histocompatibility complex (MHC) – A multi-protein complex on the cell surface that displays foreign antigens for detection by T cells.

Memory cells – Long-lived B and T cells of the adaptive immune system that develop following initial exposure to foreign antigens and pathogens. These cells provide a very rapid adaptive immune response following reexposure to the initiating substance.

MxA – An interferon-induced protein with antiviral activity, especially against the influenza virus.

NF-κB – A cellular protein that is activated by the innate immune system and functions to turn on the expression of cytokine genes.

Pathogen recognition receptors (PRRs) – A diverse group of cellular proteins that are part of the innate immune systems. These proteins function to bind PAMPs which triggers the production of interferons and cytokines.

Pathogen-associated molecular patterns (PAMPs) – Any of a diverse group of molecular features that are unique to pathogens and are either not found in cellular biomolecules or are not normally found in specific subcellular locations.

(continued)

(continued)

Peptides – Short chains of amino acids, often derived from the digestion of proteins.

Phagocytosis – The process of a cell engulfing and ingesting bacteria, viruses, other pathogens, or debris.

Plasma cell – A mature and fully differentiated B cell that is producing antibodies.

Protein kinase R (PKR) – An interferon-induced protein with antiviral activity that acts by inhibiting protein synthesis to prevent the production of viral proteins.

Ribosomal RNA (rRNA) – A group of RNAs that form part of the structure of ribosomes.

RIG1 protein – A pathogen recognition receptor (PRR) that recognizes double-stranded RNAs.

Rig-like receptors (RLRs) – The family of pathogen recognition receptor proteins related to RIG1.

RNase L – An interferon-induced protein with antiviral activity that acts by digesting RNAs.

SAMHD1 – A cellular enzyme involved in deoxynucleoside metabolism that also has viral restriction activity, especially against retroviruses.

Suppressor T cells – The subpopulation of T lymphocytes that dampen the immune response to return the immune system to its resting state once a pathogen is eliminated.

T cell receptor (TCR) – A protein complex on the surface of T cells that recognizes and binds antigens presented by the major histocompatibility complex (MHC).

Tetherin – A cellular protein located in lipid rafts that has viral restriction activity against several enveloped viruses.

Toll protein – A pathogen recognition receptor (PRR) that recognizes bacterial lipopeptides.

Toll-like receptors (TLRs) – The diverse family of pathogen recognition receptors (PRRs) related to the Toll protein.

Transmembrane protein – A protein that spans a membrane and has portions of the protein exposed on both sides of the membrane.

Vaccination – Use of an innocuous material to induce immunity to an infectious agent.

Viral restriction factors (VRFs) – Any constitutively present cellular protein with intrinsic antiviral activity.

White blood cells – The group of blood cells that includes the B and T lymphocytes and several other cell types such as monocyte, basophils, neutrophils, and eosinophils.

Abbreviations

APCs – antigen-presenting cells
APOBEC3 – apolipoprotein B editing complex 3
BCR – B cell receptor
EWI-2wint – EWI (glutamic acid [E]/tryptophan [W]/isoleucine [I]-motif containing protein)-2 without its N-terminus
IRF3 – interferon regulatory factor 3
IRF7 – interferon regulatory factor 7
ISGs – interferon-stimulated genes
KAP1 – KRAB-associated protein 1
MHC – major histocompatibility complex
MxA – myxovirus resistance protein A
NF-κB – nuclear factor kappa-light-chain-enhancer of activated B cells
PAMPs – Pathogen-associated molecular patterns
PKR – protein kinase R
PRR – pathogen recognition receptors
RIG1 – retinoic acid-inducible gene 1
RLR – Rig-like receptors
rRNA – ribosomal RNA
SAMHD1 – SAM and HD Domain Containing Deoxynucleoside Triphosphate Triphosphohydrolase 1
TCR – T cell receptor
TLR – Toll-like receptors
VRFs – viral restriction factors

6

Viruses That Shaped Our World

Keywords Smallpox • Polio • HIV • AIDS

A virus can change the fate of the world; power has
nothing to do with being tiny or giant!
Mehmet Murat ildan

The previous chapters focused primarily on the virus-host interaction at the individual level, however, viral infections can have even more powerful effects at the population level. The death of any person is tragic, but the death of thousands is a powerful sociological event that can have wide-ranging effects. Many examples exist of viruses that changed history, and this chapter will explore several well-documented cases from both distant and recent times. In each instance, specific conditions developed that introduced and/or facilitated rapid viral spread into naïve populations, usually with devastating consequences. In times past, the decimation of large numbers of people led to the eventual collapse of empires and the demise of certain cultures. In more recent times, new epidemics galvanized sociological, political, and scientific action to confront and contain the viral spread. Yet even with all our modern scientific advances and medical technologies, there is nothing to prevent novel viruses such as SARS-CoV-2 from emerging to cause worldwide devastation (see Chap. 11). We can only hope that learning from each encounter will help prepare us for new threats and keep us ever vigilant for future pandemic outbreaks.

Smallpox – Mankind's First Viral Conquest

The scientific name for the smallpox virus is variola major, a DNA virus with a very large and complex genome. Most readers of this chapter likely have never seen someone with smallpox because the last case in the United States was in 1949, and the disease was eliminated worldwide by the late 1970s. Many readers probably are not even vaccinated against this disease as smallpox vaccination stopped in the United States in 1972 and was discontinued in most other countries within the next decade. Yet it is doubtful that there is any viral disease with a greater impact on world history than smallpox. No precise numbers are possible, but estimates suggest that smallpox killed 300–500 million people in the twentieth century alone, several times more than were killed in World Wars I and II combined. Older historical records describe smallpox devastation across every continent, often with massive decimation of local populations. Sometimes called the "red death", this virus is one of the most deadly and feared disease agents due to its extremely contagious respiratory spread, its high mortality rate, and the scarring pox lesions it left on the survivors. Initial symptoms of smallpox were a sudden high fever coupled with head and body aches, followed by the telltale rash that started first with red spots on the tongue and in the mouth. These spots quickly progressed to pustulant sores filled with millions of virions. As the oral lesions broke open, a red rash began on the face and then spread downward to the arms and legs covering the entire body in 24 hours. Within 3–4 days, each lesion became a large, raised, hard pustule teeming with virus. Over the next 10 days, the pustules burst or scabbed over, often causing scarring at each pustule that collectively could be quite disfiguring. Scabs sometimes took up to 4 weeks to completely heal over, resulting in the patient being infectious throughout this period due to viruses in the lesions. Internally, numerous organs were affected, with pneumonia often cited as a primary cause of the high mortality rate. Other issues, such as disruption of normal blood coagulation and "toxic" effects possibly related to immune system dysregulation by the virus, also likely contributed to the high incidence of fatality. Unfortunately for scientific research, since the last smallpox case in the world occurred in 1978, modern technologies and knowledge haven't been extensively applied to smallpox pathogenesis. With only limited historical information available about the disease process, a complete understanding of its killing mechanisms is lacking.

Even without a clear picture of how smallpox kills there is ample speculation about why. Most researchers believe the high fatality rate for smallpox reflects its more recent introduction into human populations. High lethality is often a trait of viruses when they are first introduced into new populations since these naïve populations may have no natural resistance to a new virus. More direct evidence for the relatively recent introduction of smallpox into humans comes from an analysis of the DNA sequences of human and animal poxviruses. A comparison of sequences allows the construction of family trees that show relative age and relationships among these viruses. From these studies comes the conclusion that the closest relatives to human smallpox are camelpox and taterapox, a poxvirus of the naked sole gerbil found in Africa. All three of these poxviruses appear to have evolved from a common ancestor around 3000–4000 years ago. We know that the naked sole gerbil is found only in central Africa, including the Horn of Africa. Camels were first introduced into the Horn region from 3500 to 4500 years ago, coincident with human populations growing in number and density. One speculation is that the mixing of these three species, humans, gerbils, and camels, allowed the spread of the progenitor virus into all three of these new hosts with subsequent divergence into three distinct, host-specific poxviruses. We don't know the first human who became infected with the smallpox progenitor, or what happened to him or her, but at some point, this new virus established itself as a contagious human pathogen. Once established in humans, the virus that we now call smallpox became endemic and likely spread along human migration routes throughout the world.

Ancient historical records are abundant with possible smallpox cases and epidemics, though there is no actual scientific proof about which pathogen caused these outbreaks. For example, several Egyptian mummies from over 3000 years ago, including Pharaoh Ramses V, have pox-like scarring that is highly suggestive of smallpox. These early Egyptian cases are consistent with the proposed timing and geography of smallpox entry into human populations. However, there is no definitive proof that the Egyptian lesions resulted from smallpox as opposed to some other disease since scientific studies have been unable to find virions or viral DNA in tested samples. Similarly, the Plague of Athens in 430 BCE and the Roman Antonine Plague from 165 to 180 CE, are both speculated to be massive smallpox epidemics based on the writing of Thucydides and Galen, respectively. This later plague is estimated to have killed up to one-third of the Roman population, including Emperor Marcus Aurelius, and may have helped start the decline of the Roman Empire. However, since no evidence remains of the actual pathogen, we only have the descriptions by these ancient writers on which to base the diagnosis. Over the next five centuries, 200–700 CE, descriptions characteristic of smallpox presentation can be found in writings from Africa, the Middle East, India,

Europe, and Asia, suggesting dissemination of this pathogen along human exploration and trade routes. These writings inevitably relate horrible epidemics that slew vast numbers of people as the disease entered new territories. Still, portrayals of these epidemics and depictions of the symptoms are not conclusive proof that smallpox was the agent of disease. Most experts agree that the first definitive account of smallpox symptoms was by the Persian physician, Al-Razi, in 910 CE. His careful observations of the lesions and the physical symptoms clearly delineated smallpox from measles and other diseases that all caused skin rashes and fevers.

In the second millennium CE, the historical record is much better and allows a more convincing illumination of variola's journeys across the Old World and eventually the New World. While smallpox had already reached Europe, the Crusades helped spread it widely and established it as an endemic disease throughout this continent. Over the next several centuries numerous smallpox outbreaks in Europe killed hundreds of thousands of individuals. It was during the fifteenth century in Great Britain that the term smallpox was first coined to distinguish the characteristic smallpox skin lesions from the so-called "great pox" lesions of syphilis. During the next 300 years, smallpox outbreaks were responsible for the deaths of six reigning monarchs: Mary II of England (1694), Joseph I of the Holy Roman Empire (1711), Luis I of Spain (1724), Peter II of Russia (1730), Louis XV of France (1774), and Maximilian III of Bavaria (1777). The impact of these leaders' deaths, as wells as the deaths of thousands of their followers, on the social, cultural, and political fabric of Europe is hard to quantify. Who knows what potential great artists, scholars, scientists, politicians, and leaders were lost among the multitudes that perished from variola infections? Every family's smallpox tragedy was societies' loss, yet two notable things did arise from the deadly rampages of this virus. First, as epidemics swept through populations the most susceptible people succumbed while those with more natural resistance recovered. Over centuries this natural selection led to a more resistant general population and a decline in smallpox death rates among Europeans and other Old World population groups. Second, the deaths and disfigurements from smallpox were so egregious that they spurred the first attempts to develop prophylactic treatments that ultimately formed the concept of vaccination.

Historical records suggest that attempts at smallpox prevention were widespread in many cultures. The origins of the oldest practices are obscure, but by the 1500s one method was common in Asia and the Middle East, particularly in China. The Chinese approach was to take dried smallpox scabs, grind them into powders, and blow the powder into the nostrils of children: the right nostril for boys and the left nostril for girls. Being full of virions, these dried powders could produce a protective immune response, though frequently they instead caused full-blown smallpox infection with the usual high risk for death and disfigurement. Eventually, an alternative approach known as variolation was developed. In variolation, pus material was collected from a smallpox lesion and then scratched into the skin of the recipient. This inoculated material produced a localized smallpox infection on the skin that again was sufficient to evoke protective immunity. Smallpox is a respiratory virus with natural infection starting through inhalation. Introducing a small amount of the smallpox virus into the skin instead of the respiratory system greatly reduced the risk of a severe systemic infection, making this approach much safer than the original Chinese method. Still, historical accounts suggest that one in fifty to one in a hundred recipients died from this procedure, making skin variolation far from an ideal prophylactic. Variolated individuals who developed generalized smallpox were also infectious and could spread the virus to others, adding to the danger of this approach.

The next step in the evolution of smallpox prevention occurred in Great Britain in the late 1700s. Variolation was introduced to Great Britain in the early 1700s by Lady Mary Worley Montagu who discovered the practice while in Istanbul with her husband, the British Ambassador to that country. She became a strong proponent of variolation and helped establish the practice among the British royalty and upper class. However, since the procedure was both dangerous and expensive it never became widely adopted among the masses. It was Edward Jenner, a British royal physician, who in the late 1700s popularized a much safer alternative involving a cattle disease known as cowpox. Jenner noted that milkmaids rarely got smallpox, and he correctly guessed that milkmaids who were infected by cowpox were somehow protected from smallpox. Jenner knew nothing of viruses or how immunity worked, but we now know that the cowpox virus is closely related to the human smallpox virus. Cowpox infection produces only very mild disease in humans yet generates an immunity that provides cross-protection against smallpox. Jenner modified traditional variolation by extracting material from cowpox lesions on cows and using this material to inoculate the skin of recipients, rather than using pus from human smallpox lesions. This new approach provided similar immunity, but without the danger of contracting a smallpox infection from the inoculation. Since his starting material came from cows, "vacca" in Latin, he called his procedure vaccination. The success of vaccination at protecting from smallpox established the concept of vaccination as a potentially general method for disease prevention. From this fortuitous origin as an attempt to control one horrid disease, vaccination was eventually adapted and applied to numerous other diseases. Perhaps more than any other medical advance, modern vaccination has helped conquer or control dozens of diseases, not only viral but bacterial as well. We now live in a world that is relatively free from illnesses that were once common and frequent killers, and we owe a debt to Jenner and the other pioneers in vaccine development.

While smallpox remained a brutal disease in Europe until widespread vaccination became common in the nineteenth and twentieth centuries, the introduction of smallpox into the New World by Europeans was catastrophic. In contrast to the Old World populations that slowly developed the ability to survive smallpox over centuries of infections, the New World populations were completely naïve having never encountered variola before the arrival of Spanish explorers. The Spanish subjugated the Caribbean islands of Haiti, Cuba, and the Dominican Republic in the early 1500s and likely introduced smallpox carried by their imported West African slaves. In 1518 smallpox swept through these islands and is estimated to have killed up to half the native populations. Similarly, Hernando Cortez and Spanish conquistadors unknowingly brought smallpox to the great Aztec empire of Mexico during their attempted conquest of this territory. Militarily the Spanish were unable to defeat the superior Aztec forces, and Cortez was eventually forced to retreat in 1520. However, by the time he returned in 1521 smallpox had decimated the Aztec population, killing the Aztec leaders and many warrior chiefs. With the Aztecs dying and leaderless, Cortez's small force was able to seize the capital of Tenochtitlan (now Mexico City). More significantly, the Aztecs took this sweeping pestilence as a sign that the Spanish god was superior to their own gods, and a culture of over 12 million people meekly surrendered to a few hundred Spanish. In less than a generation, the entire rich Aztec culture was largely destroyed and replaced by Spanish traditions.

Once introduced by the Spanish, smallpox swept through Central America into South America laying waste to native populations and destroying centuries of culture. By the time the Spanish explorer Pizarro reached what is now Peru in 1532, the mighty Inca Empire had already been devastated by smallpox, including the death of the Inca Emperor, and their civilization was easily defeated. Later that century the Portuguese explorers carried smallpox into Brazil, again leading to the complete extermination of many indigenous peoples. Similar to the outcomes in Central and South America, smallpox introduction into North America was a major factor in the destruction and defeat of many indigenous tribes from the Atlantic to the Pacific, and greatly contributed to successful European colonization. Even before the Pilgrims landed in 1620, a smallpox outbreak in the preceding 3 years had killed almost 90% of the native population of Massachusetts, an event the Pilgrim leader called an act of God to facilitate their colonization. By the late 1600s, smallpox had spread from the northeast to the Great Lakes region, and both the Mohawk and the Iroquois tribes suffered massive epidemics leading to the loss of over 50% of their population. Over the next 100 years, smallpox continued to spread westward along with settlers and soldiers, wreaking havoc among each indigenous peoples it encountered.

In addition to its natural spread, there is even good historical evidence that smallpox was used as the first bioweapon during Pontiac's Rebellion of 1763. Writings by both Colonel Henry Bouquet and militia commander William Trent describe intentionally giving smallpox-infected blankets to the natives in hopes of starting outbreaks that would greatly weaken their forces. Similarly, the British forces may have used this tactic against the Continental Army in the Revolutionary War both during the siege of Boston by George Washington and the siege of Quebec City by Benedict Arnold. The British forces purportedly sent newly variolated civilians to mingle with the Continental troops in hopes of starting smallpox outbreaks among the colonists who were not themselves routinely variolated. Smallpox did break out among Arnold's forces leading to their ultimate retreat from Quebec province. Had it not been for smallpox driving the Continental forces south, much of what is now eastern Canada might have ended up as part of the United States.

After the Revolutionary War, the continued westward spread of settlers brought smallpox destruction to the Plains Indians and eventually to the West Coast Indians. Everywhere smallpox ranged it left behind devastated tribes with death rates reported from 30% to 95%. In roughly 300 hundred years, from Columbus' arrival in 1492 to the year 1800, it is estimated that the native New World population (North and South America) declined from 72 million to less than one million. This decline was not solely due to smallpox as influenza, measles, and warfare certainly contributed to numerous deaths. However, the effects of smallpox were dramatic and horrific as the native populations and their cultures were literally extinguished by this invisible marauder. Even though mostly unintentional, the introduction of smallpox into the naïve New World population greatly facilitated European colonization and helped transform the New World into primarily a European culture.

After rampaging across the Americas, smallpox remained a serious and ruinous disease worldwide throughout the next century and a half. During this period, Jenner's vaccination strategy gradually become more widely accepted as a relatively safe and very effective method for preventing smallpox. Over time, the field of virology emerged and laboratory methods for growing viruses were developed. Eventually, poxvirus stocks for vaccines were cultivated in labs rather than taking material directly from cowpox lesions.

Different viral stocks were prepared in various labs and the lab strains were passaged in eggs or cultured cells. Strains were often shared between labs over many decades with limited record-keeping and without the ability to carefully characterize the viral stocks. Eventually, in 1939 it was discovered that the standard strain commonly used for smallpox vaccination was not in fact cowpox. To distinguish this new strain it was renamed vaccinia, and vaccinia continued as the vaccine source material throughout the rest of the twentieth century. One of the intriguing mysteries of virology is that to this day we still don't fully know the origin of vaccinia. Modern DNA sequencing reveals that even relatively modern stocks of vaccinia from different sources are not identical and contain complex differences in their genomes. What we call vaccinia appears to be a collection of related strains that arose through mutation and recombination during decades of manmade passaging. Thus, vaccinia is not a natural virus that ever existed in the wild and is instead a laboratory-developed strain. Surprisingly, modern vaccinia is more highly related to horsepox than cowpox, so somewhere in the past horsepox must have been introduced into early vaccine stocks. Without adequate records, exactly how and from where the vaccinia strains arose will likely never be deciphered, though perhaps we should just be thankful that it became such an effective weapon against smallpox. Even without any understanding of the scientific basis for immunity, demonstrating that an injection of innocuous material could protect from subsequent disease was a revolutionary concept that completely changed medicine in the twentieth century.

Mandatory vaccination with vaccinia helped eliminate smallpox from the United States and Europe by the early 1950s, showing just how effective a sustained national approach could be. The 1950s also saw a heat-stable, freeze-dried version of the vaccine developed which greatly facilitated vaccine dissemination into third world areas lacking refrigeration. Given the demonstrated success of vaccination in the United States and Europe, along with the technological advancements in the vaccine, in 1966 the World Health Organization (WHO) launched an aggressive campaign to eradicate smallpox worldwide. Smallpox has no nonhuman reservoirs, consequently, if all human disease cases could be eliminated then the virus would cease to exist in nature. Worldwide vaccination efforts in Brazil, Africa, Asia, and Indonesia eventually succeeded, and the last recorded natural smallpox case occurred in 1977. After 3 years with no other cases reported anywhere in the world, WHO declared in 1980 that the planet was finally smallpox-free. A virus that terrorized humans for nearly 4000 years, killed more people than any other virus, and reshaped world history was finally eliminated through science and hard work by thousands of healthcare workers and volunteers. This remains the only human virus that has been eliminated globally although we are nearing that goal for polio. Today, the only known smallpox virus stocks remain at two sites sanctioned by WHO: the Centers for Disease Control (CDC) in Atlanta and the State Research Center of Virology and Biotechnology (VECTOR) in Koltsovo, Russia. The stocks are maintained for research purposes, though some argue that the potential danger from release, accidental or intentional, is too great and that even these stocks should be destroyed so that smallpox can never arise again.

Polio – A Force for Change

While the smallpox virus brutally ravaged the last 3000–4000 years of human history, poliovirus remained a minor threat until the last 200 years. Although its origins are unknown, an ancient Egyptian carving from around 1580–1350 BCE depicts an individual with a withered lower leg characteristic of paralytic polio, suggesting that polio has been an endemic human disease for at least that long. However, polio never caused large-scale epidemics and there are very few historical accounts until the 1700s. The writer and poet, Sir Walter Scott, was lamed by childhood polio in 1773, but other well-documented cases are few. Perhaps the first modern description of the disease appeared in 1789 in a volume entitled "Diseases of Children" by Michael Underwood, a British physician and one of the first members of the Royal College of Physicians. Initially called infantile paralysis, what we now call polio was at best an infrequent and minor disease of that era. During the early 1800s, there was a modest rise in polio outbreaks with small pockets of the disease reported in various European nations and the United States. These outbreaks were still insignificant by smallpox standards as they typically involved less than 20 cases with little or no mortality. As the 1800s progressed into the early 1900s, there was a marked increase in cases, surprisingly in the more highly developed nations. Major epidemics occurred in Scandinavia in 1905, followed by the massive 1916 New York City outbreak with roughly 26,000 cases and 7000 deaths. Suddenly polio had gone from an obscure disease to a major public health threat. The disease was particularly egregious because of its high incidence in children and its ability to cause different degrees of permanent paralysis in many survivors. Death was usually the result of paralysis of the muscles used for breathing, a particularly gruesome and brutal demise. With no understanding of the causative agent or means of spread, numerous theories abounded. Approximately 80,000 cats and dogs were killed in New York City during the 1916 outbreak because of an erroneous belief that pets might be spreading the disease. Likewise, other vermin such as insects and rodents were also falsely blamed. Additionally, places where children congregated, such as movie theaters and pools, were closed in futile attempts to stop the epidemic.

The 1916 outbreak gradually subsided, but polio had now established itself as an endemic disease in the United States. Polio plagued the country with cases and deaths every year, and with significant spikes in cases of polio every 3–5 years. From 1916 through 1942 there were 180,000 cases of polio in the United States with a little over 33,000 deaths. Based on the estimated US population of 135,000,000 in 1942, this amounted to about 1 in every 1000 people who were infected with polio during the 1916–1942 period. This high prevalence meant that every community and every family likely knew someone affected by this dreaded disease, and the whole country feared the summer season when polio was most rampant.

To understand the evolution of polio as a disease in the United States we need to understand the nature of this virus and its transmission. Poliovirus is a member of a large family of small, non-enveloped viruses known as Picornaviruses. The family name itself (Pico-rna- virus) is descriptive of the viruses in this family as Pico means very small and rna indicates that this group of viruses all have RNA as their genomes. This family has 47 genera, many of which contain viruses that infect animals but not humans. Poliovirus is in the genus *Enterovirus* and is highly related to other *Enteroviruses* known as Coxsackieviruses and Echoviruses. In fact, the likely direct precursor of polio was a type of Coxsackievirus of the C subgroup. Acquisition of mutations in the Coxsackievirus genome led to a change in host receptor usage by the virus with eventual divergence into a distinct new virus that we call poliovirus. Unlike Coxsackieviruses which use a host protein called ICAM-1 as their receptor, poliovirus diverged to use a human protein called CD155.

All three of these *Enteroviruses*, Coxsackievirus, Echovirus, and poliovirus, replicate primarily in the gut and spread by fecal-oral transmission. Viral entry is primarily via the mouth where the virus infects cells in the tonsils, eventually releasing new infectious virus into the saliva. Swallowed saliva introduces the virus into the digestive system where it thrives. Poliovirus, like *Enteroviruses* in general, has highly resistant capsids that can survive the harsh digestive conditions of the stomach. Virions can successfully pass through the stomach into the small intestines where the major viral infection and reproduction occurs in cells lining the intestines. Copious quantities of virions are shed in the feces where they can contaminate water supplies, the hands of the infected individuals, or the hands of caretakers in the case of infants and small children. Viruses on the hands are easily spread to other objects where they can then be picked up by the hands of other people and inadvertently introduced into the recipient's mouth. Similarly, any foods that become contaminated by handling are infectious if ingested. This hand-to-mouth mechanism is quite effective in spreading poliovirus among households and other close contacts of an infected individual.

One of the surprising features of poliovirus is that it is actually a quite benign virus that causes little or no disease in the vast majority of infected individuals. Around 95% of infections are either asymptomatic or present with mild gastrointestinal symptoms caused by the viral-induced damage to the small intestines. These mild gastrointestinal symptoms are no different from those caused by dozens of other viruses and would arouse no great concern. Consequently, most infected people shed the virus and pass it on to others without ever knowing that they are infected with poliovirus and without ever being diagnosed. It is only in the minority 5% that more consequential symptoms develop that identify the victim as having poliovirus. In this small subgroup, the virus escapes the intestines into the bloodstream where it causes more systemic effects, including infection of the central nervous system. These individuals can develop high fevers, more severe gastrointestinal effects, headaches, and signs of neurological involvement such as neck and back pain and/or behavioral changes. Of the 5% that show these more severe symptoms, roughly 1 in 5 develop paralytic polio to some degree. Paralytic polio involves infection of the central nervous system and direct attack by poliovirus on motor neurons that control our muscles. Motor neurons express the CD155 protein that serves at the poliovirus receptor which likely accounts for the ability of poliovirus to readily infect this cell type. The lytic destruction of these motor neuron cells by the virus can lead to anything from localized limb weakness to total paralysis depending on the location and number of neurons affected. Some patients did eventually recover most function, though many were left with permanent disabilities in their limbs. Before medical innovations, many of the most severely affected patients perished from suffocation due to paralysis of the breathing muscles. If you think of this 5% symptomatic group in terms of the 1916 New York outbreak, then the 26,000 reported cases potentially represented over half a million total infections in a city with around five million inhabitants, or approximately 1 in 10 people infected!

How then did a virus with this type of transmission go from being historically insignificant to a major public health crisis in the twentieth century? While there are several proposed explanations, including a possible mutational change to greater virulence, the most generally accepted model is more sociological than virological. For much of human history, populations were smaller and less concentrated, so the spread of non-respiratory viral infections was more limited. As urban areas developed and population density increased, oral-fecal virus spread became easier since early cities often lacked adequate sanitation, safe foods, and clean water. Under such conditions, the fecally shed virus transmission was rampant, and most individuals were likely to be exposed to poliovirus during infancy. Infants typically possess maternal antibodies that pass to the newborn during pregnancy and via breastfeeding. Mothers who themselves had a prior polio exposure would have passed antipoliovirus antibodies to their offspring, and these antibodies would have provided at least partial protection. Thus, when these partially protected infants encountered poliovirus during early life they had antibodies in their bloodstream that neutralized any circulating virions. Consequently, the virus was prevented from reaching the brain so they rarely suffered paralytic complications. Instead, these infants would have had mild to inapparent gastrointestinal cases, though they still would have shed virions, contributing to the environmental burden of poliovirus in their communities. The 1900s saw a dramatic sociological change within the United States and Europe with the growing realization that sanitation was important in preventing disease and improving living conditions for the masses. As public health measures were implemented to ensure food and water safety and to promote increased personal hygiene, the risk of poliovirus exposure during infancy declined. Unfortunately, poliovirus didn't go away, it just became an infection that was more likely encountered in later childhood when kids left the family confines and interacted more at school and in other group situations. By this later age, the maternal antibodies had long since vanished, along with any partial protection, so now infection could run its normal course unimpeded, resulting in the risk of paralytic disease that became the hallmark of this virus. Paradoxically, it was the very conscious movement to improve public sanitation and hygiene in our country that transformed poliovirus disease from a mostly silent and benign infant infection to one of the most dreaded childhood viruses in modern history.

As often happens in the United States, dread inspires innovation, and poliovirus was no exception. One of the primary causes of poliovirus death in the early 1900s was suffocation from paralysis of the breathing muscles. In 1928 a trio of Harvard faculty, Philip Drinker, Louis Shaw, and James Wilson, created a device to assist the breathing of these paralyzed children. This mechanical respirator device became known as the iron lung, and it was basically a sealed box with a vacuum pump. Victims of paralytic polio were placed in the device so that their torsos were enclosed and sealed. When the air was pumped out of the box the patient's chest expanded and air was drawn into their lungs. When the vacuum was released air rushed into the box compressing the chest and forcing exhalation. By repeated cycling of this process, artificial breathing was established that could be maintained indefinitely if needed. Fortunately, many children eventually recovered from the paralytic stage and could leave the iron lung, so this simple device kept alive thousands of children who otherwise would have perished during their paralytic interval. Subsequent modifications led to a much less expensive design that enabled more widespread production and accessibility to these machines. The dramatic lifesaving effect of the iron lung is best exemplified by examining the polio death rates before and after the widespread use of this device. In the period from 1916 to 1930, about 1 in 4 polio victims died, while from 1931 to 1940 that ratio dropped to just over 1 in 10. Even more dramatic was the effect during the infamous American polio outbreak that started in 1944 and lasted into the mid-1950s. This epidemic caused almost 400,000 cases, twice as many cases in just 10 years as in all the preceding years of the century. Yet even with this enormous case number, only 22,000 people perished, or about 1 in 20, thanks largely to the iron lung and better palliative care.

The second innovation attributed to poliovirus was the creation of the modern charitable foundation with a specific disease focus. While charities certainly existed before the 1900s, they tended to be more general like the Red Cross, and more likely funded by a few wealthy philanthropists. All that changed in 1938 when President Franklin Delano Roosevelt established the National Foundation for Infantile Paralysis in response to the ongoing polio problem in the United States. President Roosevelt also had a strong personal interest as he became wheelchair-bound from polio in 1921 when he was 39 years old. (Note that there is now speculation that President Roosevelt suffered from Guillain-Barré syndrome rather than polio, but he and his doctors certainly believed that his permanent paralysis from the waist down was due to polio.) Appointing his close friend, Basil O'Connor, as the foundation's president, they set out to create a national network for fundraising to support the delivery of healthcare assistance and to further poliovirus research. Almost immediately after the foundation's creation, a famous entertainer of the era, Eddie Cantor, called for a national fundraising campaign that he called the "March of Dimes". This inspirational innovation urged everyone in the country to send a dime to the White House and launched the idea of targeting hundreds of thousands of small donors rather than a few wealthy individuals. The March of Dimes campaign became an annual event during the Christmas season and was wildly successful over the years at raising funds for the anti-polio cause. Once the American people embraced this new funding approach, many other groups adopted this same strategy leading to the creation of a variety of successful disease-specific charities that still exist today. In 1976 the foundation formally changed its name to the March of Dimes Birth Defects Foundation and later in 2007 to simply the March of Dimes Foundation.

The creation of the National Foundation for Infantile Paralysis in 1938 was particularly fortuitous as the massive 1940s polio outbreak was just a few years in the future. Funds raised by the March of Dimes campaigns were instrumental in reducing deaths by providing iron lung machines to hospitals and clinics around the country. The increased nationwide availability of these machines was likely the most significant factor in the lower death rate compared to previous outbreaks. But even more importantly, these funds supported the intensive research efforts by Jonas Salk, Albert Sabin, and others to find cures or vaccines against polio. By the mid-1950s, the foundation had invested over $25 million in research and was rewarded by Salk's inactivated polio vaccine. However, neither the Salk vaccine nor the subsequent Sabin vaccine would have been possible without the groundbreaking work of Harvard virologist, John Enders. Among his many achievements, in 1948 Enders and colleagues were the first to grow poliovirus in human and monkey cells in culture. Before Ender's work, poliovirus could only be produced in live monkeys, a process that was not suitable for generating stocks of the virus for vaccine development. Enders and two other scientists received the 1954 Nobel Prize in Physiology or Medicine for this important breakthrough, and Salk soon applied their methods to produce copious quantities of poliovirus for his vaccine studies. By 1953 Salk had developed a procedure to inactivate purified poliovirus by treating it with formalin. This treatment destroys the infectivity of the virus but allows it to retain its virion shape and to induce neutralizing antibodies when injected into humans or animals. Like circulating maternal antibodies, vaccine-induced anti-poliovirus antibodies stopped viral spread to the central nervous system and prevented motor neuron infection. A huge field trial of Salk's killed-virus vaccine on two million children was launched in 1954 with spectacular success, leading to the spring 1955 announcement that paralytic polio could be completely prevented. Salk refused to patent his vaccine, and this life-saving advance became freely available throughout the world. In the United States, large-scale vaccine production and a nationwide vaccination campaign were immediately implemented with astounding results. From 1950 to 1955 there was an average of 37,100 polio cases and 1750 deaths per year. By 1960 the total number of cases dropped to 3190, a ten-fold reduction, and with only 230 deaths that year.

At roughly the same time that Jonas Salk's vaccine was revolutionizing the fight against poliovirus, Albert Sabin was pursuing an alternative vaccine strategy. While Salk's vaccine was an injectable, killed virus preparation, Sabin was developing a live poliovirus vaccine that could be administered orally rather than by a shot (What child wouldn't prefer this!). The secret to live-virus vaccines is to develop an attenuated, or weakened, form of the virus that causes only mild disease while generating immunity against the virulent form of the virus. Live, attenuated vaccines are typically superior to the killed vaccine because their mild infection promotes long-lasting immunity of both the cellular (T cells) and humoral (B cells and antibodies) types, while killed vaccines mostly induce an antibody response. An additional issue with Salk's killed vaccine is that it doesn't prevent poliovirus infection. Individuals vaccinated with the Salk vaccine can still catch poliovirus, can still get a gastrointestinal infection, and can still shed viruses in their feces. Importantly, what the Salk vaccine does do is prevent viral escape from the gut and spread to the central nervous system, so it is very effective at protecting recipients from the paralytic sequela. While the individuals are themselves protected from the serious consequences of polio infection, the vaccine does nothing to halt viral spread in communities. In contrast, the orally administered Sabin vaccine produced a strong gastrointestinal immunity. When individuals treated with the Sabin vaccine later encountered virulent poliovirus, this vaccine prevented viral replication in the gut and eliminated spread in the feces, thus protecting both the individual and the community. A second unexpected benefit of the Sabin vaccine was the property of secondary vaccination. When given the oral poliovirus strain, recipients transiently develop a mild gastrointestinal infection where the vaccine virus replicates and sheds in the feces just like the virulent virus. This shed vaccine virus could be picked up by other close contacts, identical to the way that wild-type polio is acquired. If they weren't already immune, infection of household members or other close contacts with the attenuated virus resulted in them becoming vaccinated also, thus spreading protection beyond the original recipient of the vaccine.

Since Salk's vaccine was developed first and used immediately in the United States, Sabin's vaccine trials couldn't be conducted there. Instead, in a surprising act of cooperation during the 1950s Cold War, Sabin collaborated with Russian scientists to conduct a massive trial in the Soviet Union on millions of children. The dramatic success of this trial led to Soviet production and distribution of the attenuated vaccine throughout Eastern Europe and Japan, quickly leading to the near elimination of polio cases in these regions. The United States soon followed with their own production, and the superiority and convenience of the Sabin vaccine resulted in it largely supplanting the Salk vaccine by the end of 1962. Still, both vaccines continued in use for the next four decades and together succeeded in eliminating the polio threat in the United States. By the 1970s the number of US polio cases had dropped from tens of thousands a year to an average of fewer than 20 cases per year with only a small number of deaths. The last known community-acquired case in the US occurred in 1979, and by 1991 polio was eliminated from the Western Hemisphere, a remarkable achievement against a frightful virus. Sadly, during this period a new issue arose known as vaccine-associated paralytic polio (VAPP). The attenuated poliovirus used as the vaccine is essentially a mutant form of the wild-type virus. The mutations reduce the vaccine strain's virulence while still allowing it to infect humans and replicate in the gastrointestinal system. However, during replication new mutations can occur, and sometimes these new mutations restored the attenuated strain to full virulence. The vaccine recipient then has wild-type polio replicating in their gut and is at risk for paralytic polio, as well as now shedding wild-type polio out into the community. The risk of this happening is quite small, about 1 in every 2.7 million vaccine recipients, yet when tens of millions of people received the vaccine each year these VAPP cases did begin to occur. Because of VAPP, the United States stopped using the oral polio vaccine in 2000 and now only uses an updated version of the original Salk inactivated vaccine, although the oral vaccine is still used in some countries. Even though there is no endemic polio left in the Western hemisphere, vaccination must continue unabated since polio could be reintroduced at any time by travelers from areas of the world where the virus still exists.

The phenomenal success of the vaccines in eliminating polio from Europe and the Americas led the World Health Organization (WHO) to declare a global war on polio in 1988. WHO's objective was to eliminate this virus entirely from our planet by the year 2000. As of 2021, this mission has not succeeded, though it is creeping closer and closer each year. Like smallpox, poliovirus has no natural host except humans and can only survive in the environment for a few weeks at best. If all human cases are eliminated then within a short time poliovirus will cease to exist except for laboratory stocks. In 2018 there were only 33 wild-type poliovirus cases and 104 VAPP cases in the entire world, so we are painfully close to the final eradication of a second major human pathogen. The remaining barriers to final victory are more sociological than scientific, as the natural cases only come from Afghanistan and Pakistan. Pockets in both these countries suffer from poverty, warfare, superstition, lack of sanitation, and suspicion of Western medicine, all factors that make vaccination of the affected populations challenging. Nonetheless, there is unrelenting hope that a soon-to-be-born generation will never experience polio. Then the virus that terrorized the first half of the twentieth century, that ushered in the modern vaccine era, and that gave us the concept of public philanthropy to support the fight against diseases will be simply a small entry in history books.

HIV/AIDS – A New Pandemic for the Twentieth Century and Beyond

By the end of the 1970s, we were entering a golden age of medicine. Smallpox and polio were completely defeated in the United States; vaccines were available for the major childhood illnesses, including the viral diseases of measles, mumps, rubella; and a proliferation of new antibiotics was marketed to thwart bacterial infections. Infectious diseases were fading from life-threatening horrors to mere nuisances that could be readily prevented or controlled. This freedom from fear of infectious diseases, coupled with the sexual revolution of the 1960s and the ready availability of birth control, created a sociological climate that was ideal for the spread of sexually transmitted diseases (STDs). In America and Europe, the sexual revolution was paralleled by the gay rights movement that saw increasing demands for gay rights and public acceptance for LGBT (lesbian/gay/bisexual/transgender) peoples. In the United States, especially on the east and west coasts, thriving and vibrant gay communities developed in many of the major cities. After decades of being closeted, facing severe cultural discrimination, and even criminal prosecution, much of the gay community reveled in their emerging freedom. For some, this freedom meant satisfying their sexual appetite through casual sex with multiple partners. Little did the gay community imagine that lurking in the background was an obscure African primate disease that would soon dramatically impact them and change the world forever. While smallpox and polio were fading historical curiosities for most young adults of that era, a devastating new epidemic was poised to explode, sowing death and dismay among an unsuspecting population.

The first intimation that something new was amiss came in the late 1970s and early 1980s when unusual opportunistic disease patterns were observed by physicians. So-called opportunistic diseases are caused by organisms that are usually benign in healthy individuals and only cause disease when a person has some underlying medical condition, such as immunodeficiency. Physicians on the East and West coasts began to notice a surge in opportunistic infections in otherwise seemingly healthy young men. Prominent among these emerging opportunistic diseases was pneumonia caused by the fungus *Pneumocystis carinii* (now called *Pneumocystis jirovecii*), a disease seldom seen in healthy individuals. By 1981 the Centers for Disease Control (CDC) had formally noted this strange occurrence. In the June 5, 1981 issue of the CDC *Morbidity and Mortality Weekly Report*, 5 cases of *P. carinii* in homosexual men were described. Similarly, the New York Times ran a report on July 3, 1981, describing 41 cases of Kaposi's sarcoma (KS) in homosexual men, mostly in New York and San Francisco. Kaposi's sarcoma is a rare skin cancer that previously was found almost exclusively in elderly men of Eastern European and Middle Eastern descent. Finding this type of cancer in 41 fairly young men without the typical heritage was surprising and disturbing. Cases of both KS and *P. carinii* continued to increase throughout the year, and by year's end 159 confirmed cases were identified, almost all in gay men. Epidemiological investigations of these initial cases revealed sexual relationships between many of the victims, suggesting a possible sexually transmitted agent. Confusingly, additional cases of opportunistic infections were showing up in hemophiliacs and immigrants from Haiti, both groups without a sexual history suggestive of an STD. However, studies on these various patient groups revealed that they all suffered from an underlying immunodeficiency, and by mid-1982 the CDC had coined the term Acquired Immunodeficiency Syndrome (AIDS).

While providing a unifying acronym, the term AIDS implied nothing about the causative etiology of this disease. There was rampant speculation in the press, the general public, and the scientific community about possible causes ranging from homosexual activity, illicit drug use, environmental toxins, or dietary factors, not to mention the fringe allegations that AIDS was all a government plot to decimate the gay community. As researchers and physicians pondered the cases and the growing pool of data on the affected individuals, a consensus grew that AIDS likely arose from an infectious agent that could be spread through sexual activity or blood transfer, possibly a virus.

Preceding and concurrent with the AIDS epidemic there was intense research focused on discovering new viruses that might play roles in more poorly understood human diseases such as some cancers (see Chaps. 7 and 8). In 1980, Dr. Robert Gallo's group discovered a new virus associated with an uncommon human cancer known as adult T-cell leukemia, and they named the agent the Human T-cell Leukemia Virus (HTLV). Soon thereafter this same group isolated a related virus, HTLV-2, and proposed it as the causative agent of AIDS, a claim quickly disproved though they were on the right track. The breakthrough occurred in 1983 with the isolation of a third related new virus from an early-stage AIDS patient; this new virus was originally named Lymphadenopathy-Associated Virus (LAV) by Luc Montagnier's group in France and HTLV-3 by the Gallo group. With the candidate virus in hand, diagnostic reagents were rapidly developed to screen individuals for this virus.

Multiple studies soon showed that all AIDS patients were infected with LAV/HTLV-3, leading to the general acceptance that this virus was the cause of or at least an important factor in the development of AIDS. However, further characterization of HTLV-3/LAV showed that it had significant genetic differences from HTLV-1and 2, leading to the 1986 proposal to rename the AIDS virus as Human Immunodeficiency Virus (HIV). Ultimately Dr. Montagnier was established as the actual first discoverer of HIV, and he was awarded the 2008 Nobel Prize in Physiology or Medicine for this momentous achievement.

After HIV was established as the causative agent of AIDS, the question of where it came from and how it spread became paramount. Once accurate and sensitive detection reagents became available, it was possible to go back and examine stored clinical samples for evidence of HIV infection. These retrospective studies provide both geographic and temporal information about the viral spread, and such studies were and continue to be quite illuminating about the origin and dissemination of HIV. The early and prolific presentation of AIDS among the gay population in the United States led to a mistaken belief that this infection was somehow uniquely associated with this group and their sexual practices. However, the scientific and medical communities soon began to piece together evidence of a broader, silent epidemic. For example, during the early to mid-1970s in New York City there appeared a malady known as "junkie pneumonia" or "the dwindles" that was almost certainly an AIDS-related issue. Unfortunately, junkies and other marginalized street people often received little or no medical care and little public health attention, so this malady was unappreciated until much later. There is speculation that HIV reached New York between 1971 and 1975 where it spread effectively among the intravenous (IV) drug abuser population through the sharing of used needles for drug injection. This speculation is supported by the clear evidence of HIV antibodies in stored serum samples from New York IV drug users as early as 1978, consistent with the virus arriving in New York before that year. Similarly, there were three children born in 1977 from IV drug-abusing mothers who were retrospectively confirmed to have HIV. Clearly, the virus was not confined to homosexuals, did not originate in that group, and was circulating long before it became known as the "gay plague".

Although not specifically a gay disease, the early identification and widespread penetration of HIV within the gay population triggered massive social activism, and ultimately some significant societal changes. The gay rights movement was already quite strong, and the burgeoning awareness that this silent killer was widely disseminated among gay men gave even greater urgency to their demands for recognition and social acceptance. Faced with no cures and no effective treatments, the gay community was at the forefront of the battle to demand a solution to the HIV/AIDS epidemic. Groups like ACT UP (AIDS Coalition to Unleash Power) became prominent voices to expand public awareness, pressure the government to provide more funding to AIDS research, and educate young people about how to avoid infection. Throughout the 1980s and 90s, ACT UP was a leader in the call for local and national actions to combat the spread of HIV. One of the demands that came out of this activism was the "right to try" movement. Normally new drugs go through a long, slow vetting process of three stages of clinical trials that can take years to complete. In phase I, drugs are tested in a small group of healthy volunteers to determine if any serious adverse effects occur at different dosages. In phase II the new drug is tested on a slightly larger group consisting of actual patients (typically a few hundred at most). In this phase, drug efficacy is assessed as well as continuing to monitor for serious adverse effects. Finally, phase III trials involve a much larger patient group to confirm efficacy and safety in a broader population. In all stages, the subjects are usually divided into a control (placebo) group and the drug test group, with neither group knowing what they are receiving. AIDS activists were incensed with this slow pace and the fact that half of the trial participants were only receiving placebos so had no chance for improvement. Activists argued that once a drug showed any promise at all that it should be made available to anyone who wanted to try it since the victims were all dying and had no other hope. The downside of the "right to try" demand is that without carefully controlled trials it is difficult to prove scientifically that a drug works or doesn't work against HIV (or any other disease). This conflict continued throughout the early years of the AIDS epidemic until effective drug therapies emerged and became available for everyone. Even as HIV drugs proliferated, the "right to try" concept spread and was adopted by patient groups with other serious diseases. Some states have even passed "right to try" laws, so this conflict between the scientific/medical community and patient groups about drug access has become part of our social structure and continues to this day.

As further studies showed that HIV could be spread via oral, genital, and rectal sexual activity (both heterosexual and homosexual), via blood exchange, and through pregnancy and breastfeeding, a new safety concept arose: "universal precautions". Essentially this concept taught that all blood and bodily fluids must be considered infectious and that direct contact with these substances should be assiduously avoided through the use of physical barriers. This concept had a major impact on three groups: healthcare workers/first responders, anyone sexually active and not in a monogamous relationship, and IV drug abusers. Among the first group, the use of face masks and protective gloves became an essential part of the job. Doctors, nurses, dentists, paramedics, policemen, and firemen had to assume that anyone they dealt with might be HIV positive. This concern about avoiding bodily fluids even led to new devices, such as the now-standard mouth shield to use in mouth-to-mouth resuscitation situations. In the second group, the need to protect young people from this sexually transmitted pathogen led to the breakdown of many old taboos. It was no longer sufficient to just hope that adolescents remained chaste and ignorant of sex when a lack of knowledge could lead to a fatal infection. Sex education, frank discussion of all types of sexual practices, and widespread promotion of condom use became commonplace. The collective advice for avoiding sexually transmitted diseases gave rise to the new catchphrase, "safe sex", which became a permanent part of our lexicon. Likewise, the stigma of drug abuse lessened as many organizations and communities began to focus on protecting IV drug abusers from HIV rather than ostracizing them. Education programs and needle exchange programs multiplied with the hope of stopping HIV spread. These significant social changes became a standard facet of American life in the twenty-first century and combined with the drug advances have greatly reduced HIV incidence and deaths.

In parallel with the social changes it inspired, there was an intense desire by both scientists and the public to understand where HIV originated and how it was disseminated throughout the world. Since HIV's discovery in 1983, there have been numerous and extensive investigations into the nature of this virus and its evolution. HIV is a virus in the family *Retroviridae*. Retroviruses are fascinating and unique viruses that all have a small RNA genome encoding four genes: Gag, Pol, Pro, and Env. Gag and Env each express a large primary protein that is subsequently cleaved into smaller proteins used to assemble the virion structure. Pol encodes a multifunctional protein that produces the viral replication enzyme (known as the reverse transcriptase or RT) and Pro encodes the protease enzyme (PR) that cuts the Gag and Env primary proteins. The infectious HIV virion contains the RT protein along with the viral RNA genome. Once the virus is taken up and uncoats in the cytoplasm of the infected cell, the RT enzyme converts the viral RNA into a double-stranded DNA molecule. This conversion of RNA to DNA was a remarkable and unexpected biochemical process whose discovery won the 1975 Nobel Prize in Physiology or Medicine for David Baltimore, Renato Dulbecco, and Howard Temin. The DNA form of the viral genome transports to the cell nucleus and uses another function of the RT protein to integrate the viral DNA into one of the cell's chromosomes. Since the viral DNA is biochemically identical to the cellular DNA, the integrated viral DNA, called a provirus, is now a permanent resident of that cell's genome. Neither the virus nor the cell has any mechanism to remove this proviral sequence, and the proviral DNA is passed forever to the daughter cells at each cell division. From its integrated location, the proviral DNA will transcribe into mRNA, translate its mRNAs into new viral proteins, and assemble new progeny virions as long as the cell lives. Because of the unique ability of these viruses to become a lifelong component of the genome of each cell they infect, the only way to remove them is for all the infected cells to die off or be killed by the immune system, often a difficult to impossible task. As noted in Chap. 1, our genomes are full of viral remnants, and many of these are remains of retroviral infections during our evolutionary history (see Chap. 9 for a full discussion of viruses in our genome).

Sequencing of the HIV genome revealed that it was not only a retrovirus but a subtype of retrovirus known as a lentivirus. Lentiviruses have the standard retrovirus genome with Gag, Pol, Pro, and Env along with two or more additional genes which make their biology even more complex and insidious. When HIV was discovered there were two known human lentivirus, HTLV-1 and HTLV-2, as well as several known lentiviruses among domestic animals, mostly associated with slow, chronic, wasting diseases. Analysis of the HIV genome quickly revealed that it was too distinct from HTLV-1 or 2 to derive from either of those viruses. Virologists recognize that newly appearing viral epidemics often arise when an animal virus is introduced into human populations such as occurred for the smallpox virus and the recent SARS-CoV-2. Comparison of known animal lentivirus genomes with that of HIV showed no close relationship, triggering searches for new animal viruses that could be the HIV forebear. Fairly quickly HIV-like viruses were found in captive macaque monkeys, and then in other wild simian species such as African green monkeys, sooty mangabeys, and mandrills. Collectively these primate retroviruses became known as simian immunodeficiency viruses (SIVs). Although none of these culprits were similar enough to HIV to be the direct precursor, these studies did reveal a rich diversity of SIVs and a significant frequency of infection among wild simian populations. The final breakthrough occurred in 1989 with the isolation of an SIV from chimpanzees (SIVcpz) that was highly related to HIV. Over the next 15 years, investigations of wild chimpanzee populations in Africa established that the human HIV strain was most highly related to a SIVcpz from the chimpanzee subspecies *Pan troglodytes troglodytes*, a species which is found in the central African countries of Cameroon and Gabon. The amazing close genetic relationship between HIV and SIVcpz leads to the inescapable conclusion that humans acquired this virus in Africa from this species of chimpanzee. As chimpanzees are our closest primate relative (98.8% DNA identical), a virus adapted to chimps would likely have only modest difficulty transferring to humans and becoming established. Parenthetically, there is a second much less prevalent form of HIV in Africa, designated HIV-2, that appears to have been acquired from sooty mangabeys.

There is now an excellent consensus in the scientific community about the major issues in HIV's genesis and dispersal to human populations around the globe. SIVcpz appears to have entered chimpanzee populations between 100,000 and 500,000 years ago, likely derived from an SIV of smaller primates that are hunted and consumed by chimps. This timeframe suggests that humans in Africa might have been exposed to SIVcpz multiple times through human hunting and consumption of chimps, a practice that continues to this day in parts of Africa. However, when human populations were small and isolated, any introduction of SIVcpz into humans would likely have quickly died out and not spread to become an established human pathogen. Molecular clock analysis based on genome sequence differences among numerous HIV isolates confirms a very recent origin for HIV, and various studies predicted that SIVcpz precursor to HIV became a human pathogen between 1896 and 1930. The most recent estimates refine this to an origin date at around 1910–1920, with biogeographical information pointing to the city of Kinshasa (formerly Leopoldville) in the Democratic Republic of Congo as the site of origination. Consistent with these molecular estimates, some stored serum and tissue samples from patients in central Africa as far back as the 1950s are HIV positive, confirming that the virus was well established and widespread by then. Remarkably, the virus appears to have entered the human population, established itself as endemic, and circulated for many decades before the infection was finally recognized in the 1980s. People in Africa certainly must have been suffering from AIDS-related illnesses during this period, but lack of medical care, poverty, malnutrition, and the prevalence of numerous other endemic diseases in Africa likely obscured the effects of HIV.

The obvious question is why then did SIVcpz become established as HIV only in the twentieth century? Most scholars attribute this event to two factors: the acquisition of mutations that made SIVcpz more adapted to humans and social conditions that facilitated virus transmission from person to person. Each time a human became infected with SIVcpz the simian virus could likely replicate to some degree. However, in most cases, the infection died out with the original victim or simply failed to spread to anyone else. Unfortunately, the viral reverse transcriptase enzyme (RT) has very low fidelity so that during even limited replication many mutations would arise in the viral genome. Selective pressure would favor mutant progeny viruses that were more biologically fit to replicate and transmit in humans. Early in the last century such a SIVcpz mutant likely arose in a single infected human and became the immediate forerunner of modern HIV. For example, we know that all HIV strains have either the amino acid arginine or lysine at position number 30 in the HIV Gag protein, while SIVcpz has a methionine at this position. A mutation changing this one amino acid from methionine to arginine or serine greatly improves the ability of HIV to replicate in human cells compared to SIVcpz and reflects an adaptation to humans. Another known adaptive change concerns the host restriction factor, tetherin, discussed in Chap. 5. SIVcpz uses a viral protein called Nef to counteract the antiviral effect of chimpanzee tetherin. However, human tetherin is different enough from chimpanzee tetherin that Nef is unable to disable human tetherin, so SIVcpz is severely restricted by tetherin in human cells. At some point HIV evolved to use a different viral protein, Vpu, to overcome the human tetherin restriction, and this adaptation was likely critical for effective viral replication and spread. Other such mutational changes undoubtedly occurred as parental SIVcpz transitioned to full-fledged human HIV, and because of the high mutation rate of SIV/HIV the transition may have happened within months to years, amazingly quickly in evolutionary terms.

In addition to adaptive mutational changes in the virus itself, most HIV historians agree that massive social changes in twentieth century Africa were critical for the establishment of HIV as an endemic human disease. Exposure to SIVcpz before the twentieth century likely failed to establish the virus as endemic because human groups in Africa were small and isolated, both culturally and geographically. Even if the virus established a chronic infection in some individuals, transmission to others would have been limited to a sexual partner and possibly other very close associates. Under these conditions, the virus was unlikely to spread widely and certainly not beyond the immediate village. As urbanization exploded in central Africa after World War I, Kinshasa became a large city that attracted a huge influx of citizens from rural areas. Along with this rapid urbanization came cultural disruption, family separation, poor sanitation, flourishing prostitution, and a high rate of sexually transmitted diseases (STDs), all conditions ripe for transmission of the newly adapted HIV. All it took was one individual with a well-adapted virus to introduce the pathogen into an environment where it could easily be transmitted. Since HIV causes only mild or no symptoms upon initial infection, this silent invader spread quickly and unnoticed for years among the young, sexually active population. There is also speculation that various vaccination campaigns and intravenous drug therapies for several tropical diseases in the first half of the 1900s contributed to the spread of HIV. Needles and syringes used for these injections were not typically sterilized between patients or were sterilized ineffectively, creating an inadvertent transmission mechanism that may have helped spread the virus far beyond the STD risk population. Since Kinshasa was a major transportation hub, infected individuals could easily carry the virus to other cities and villages to further the spread. Ultimately by the 1950s, the virus was endemic across central Africa, although no one yet recognized the disease consequences.

Once firmly ensconced in Africa, it was inevitable that the virus would spread to other regions of the world through commerce and visiting travelers. For historical reasons the primary language of the Democratic Republic of Congo (DRC) is French, making it an attractive destination for other French-speaking peoples, such as Haitians. It is believed that young Haitians working in the DRC during the 1960s became infected and brought the virus back to Haiti and then into North America. As previously discussed, the changing US sexual mores of the 1960s and 1970s, the gay liberation movement, and the growth of IV drug abuse facilitated the dissemination of HIV once it reached America. Similar transmission brought the virus to Europe and then eastward into Asia. For example, there is a documented case of a Norwegian sailor who died in the mid-1970s along with his wife and 7-year-old daughter, all of whom were posthumously shown to have HIV. He spent time in West Africa in the early 1960s and was known to have acquired gonorrhea there. Presumably, he also picked up HIV in Africa and helped introduce it into Europe, a story that was likely repeated by many other travelers over the years. Through this type of transmission, the deadly killer spread silently across the globe before exploding into public consciousness in the early 1980s.

HIV is now a permanent, endemic human virus, and until we find a vaccine it will likely remain a prevalent human pathogen. We are fortunate that this virus normally spreads via blood and unprotected sexual activity, both routes that are avoidable and somewhat slow for population spread. Imagine what might have happened to the human race if HIV was transmitted by aerosol like influenza or SARS-CoV-2? The virus might have spread around the world in months to years rather than decades and infected billions rather than millions. Under those conditions, scientists would likely have been unable to develop the lifesaving drugs that now make HIV a manageable chronic infection. So, for now, we just have to live with universal precautions, safe sex, and the awareness that there may be other more dangerous animal viruses lurking in the world just waiting to jump into human populations. If nothing else, the HIV epidemic helped raise our public and scientific consciousness about new and novel zoonotic viruses, revealed the global interconnectedness of all humans, and emphasized our precarious position in the viral world. Hopefully, we will all stay vigilant in the years to come.

Additional Reading

1. Scourge-The Once and Future Threat of Smallpox. Jonathon B. Tucker. Atlantic Monthly Press, New York, 2001.
2. Polio: An American Story. David M. Oshinsky, Oxford University Press, 2006.
3. The Origins of AIDS. Jacques Pepin, Cambridge University Press, 2011.

Definitions

Acquired immunodeficiency syndrome (AIDS) – An immunodeficiency syndrome caused by the human immunodeficiency virus (HIV).

Adult T-cell leukemia – A type of cancer involving T lymphocytes and caused by human T-cell leukemia virus type 1 (HTLV-1).

Attenuated vaccine – A vaccine that uses a live, weakened form of a pathogen as the material introduced into the vaccine recipient.

Attenuated virus – A vaccine virus that has been weakened so that it only causes mild or asymptomatic disease but still evokes a protective immunity.

Camelpox – A virus related to smallpox that infects and causes disease in camels.

CD155 – The cellular protein that serves as the receptor for poliovirus.

Central nervous system – The brain and the spinal cord.

Cowpox – A virus related to smallpox that infects and causes disease in cows.

Coxsackievirus – A virus of the picornavirus family with a linear, positive-sense, single-stranded RNA genome. Related to Echoviruses and poliovirus.

Echovirus – A virus of the picornavirus family with a linear, positive-sense, single-stranded RNA genome. Related to Coxsackieviruses and poliovirus.

Enterovirus – a subgroup of the picornavirus family containing poliovirus, Echoviruses, and Coxsackieviruses.

Env – The gene in retroviruses that encodes the envelope proteins.

Formalin – A solution of formaldehyde in water that is often used to inactivate viruses.

Gag – The gene in retroviruses that encodes capsid proteins.

Gastrointestinal – Relating to the stomach and the intestines.

Gonorrhea – A sexually transmitted bacterial disease caused by *Neisseria gonorrhoeae*.

Guillain-Barré syndrome – A paralytic condition caused by the immune system attacking the nerves that is often triggered by a viral infection.

Hemophiliac – An individual with a bleeding disorder caused by a defective gene for the blood-clotting protein known as factor VIII.

Horsepox – A virus related to smallpox that infects and causes disease in horses.

Human T-cell leukemia virus (HTLV) – A retrovirus of the lentivirus subtype that causes human T-cell leukemia.

(continued)

(continued)

Inactivated vaccine – A vaccine that uses a killed or inactivated pathogen as the material given to the vaccine recipient. Inactivated vaccines are incapable of causing the disease associated with the target pathogen.

Kaposi's sarcoma (KS) – A type of cancer, developing from cells that line the blood or lymph vessels, that is caused by Kaposi's sarcoma herpesvirus (KSHV) which is also known as human herpesvirus-8 (HHV8).

Killed-virus vaccine – Another term for an inactivated vaccine.

Lentivirus – A subtype of retroviruses with a genome that encodes additional proteins not found in leukemia viruses. HIV is a lentivirus.

Lymphadenopathy-Associated Virus (LAV) – The original designation of the virus that is now called HIV.

Molecular clock analysis – The process of using the mutation rate of biomolecules to determine evolutionary relationships and times of divergence.

Motor neurons – Nerve cells that conduct impulses from the brain to muscles to control movement.

Nef – A gene encoded by primate lentiviruses, including HIV. Nef is a virulence factor that is critical for viral replication and successful infection.

Opportunistic disease – An infection that occurs in immunodeficient individuals that is caused by organisms that are not generally pathogenic in healthy individuals with a normally functioning immune system.

Pan troglodytes troglodytes – A species of chimpanzees found in Central Africa.

Pneumocystis carinii (now called *Pneumocystis jirovecii*) – A type of fungus that is rarely harmful to healthy people but can cause life-threatening pneumonia in immunocompromised individuals.

Pol – The retroviral gene that encodes the viral reverse transcriptase polymerase.

PR – The retroviral protease protein.

Pro – The retroviral gene that encodes a protease.

Provirus – The DNA form of retroviruses that is integrated into the chromosome of host cells.

Recombination – The process by which DNA molecules break and rejoin pieces to create new genetic patterns.

Retroviridae – The family of single-stranded RNA viruses that replicate their genomes using reverse transcriptase.

Sequela – Any condition or problem that arises from a previous disease or injury.

Simian immunodeficiency virus (SIV) – A retrovirus related to HIV that infects primates and causes immunodeficiency in infected animals.

Syphilis – The sexually transmitted disease caused by the bacterium *Treponema pallidum*.

Taterapox – A member of the poxvirus family that infects the naked sole gerbil endemic to Africa.

Vaccine – The material used in vaccination to induce immunity in the recipient.

Variolation – The process of inducing anti-smallpox immunity by scratching smallpox infected pus into the skin.

Vpu – A gene encoded by lentiviruses that functions in the release of progeny virus from infected cells.

Abbreviations

ACT UP – AIDS Coalition to Unleash Power
AIDS – acquired immunodeficiency syndrome
CD155 – cluster of differentiation protein 155
DRC – Democratic Republic of Congo
Env – envelope protein
HTLV – human T-cell leukemia virus
KS – Kaposi's sarcoma
LAV – lymphadenopathy-associated virus
LGBT – lesbian, gay, bisexual, and transgender people
PR – protease
RT – reverse transcriptase
SIV – simian immunodeficiency virus
SIVcpz – simian immunodeficiency virus of chimpanzees
VAPP – vaccine-associated paralytic polio
VECTOR – State Research Center of Virology and Biotechnology
Vpu – viral protein U of HIV

7

SV40 – An Obscure Monkey Virus and the Golden Age of Molecular Biology

Keywords Polio • SV40 • Tumor virus • T antigen • p53 • Biosafety

The art and science of asking questions is the source of all knowledge.
Thomas Berger

Viruses, Viruses, and More Viruses

SV40, which stands for **s**imian **v**irus number **40**, is scientifically a highly influential virus yet it is largely unknown to the general public. From 1960 to 2020 this small virus appeared in over 20,000 publications. Many of these publications described seminal discoveries, yet this virus remains obscure outside the research community. As the name implies, there were 39 other simian viruses isolated before SV40. Additionally, many more simian viruses were identified in the years after SV40's discovery. Out of all these primate viruses, why was SV40 special and why did it become the subject of a major research effort for decades? How did the SV40 virus become one of the most important model systems at the beginning of the age of molecular biology? Before answering these questions, we need to go even further back in time. The history of simian viruses starts in the early decades of animal virology prior to World War II. During this era animals and animal tissues started being used for research on human viruses and diseases. Macaque primates from Asia, such as rhesus (*Macaca mulatta*) and cynomolgus (*Macaca fascicularis*) monkeys, became research mainstays. Both these species are genetically close to humans (>90% DNA identity) and share similarities with us in anatomy and

V. G. Wilson, *Viruses: Intimate Invaders*, https://doi.org/10.1007/978-3-030-85487-4_7

physiology. Importantly, it was often possible to propagate human viruses in these monkeys or their tissues, making the Macques excellent surrogates to study and produce human viruses. Not surprisingly though, researchers quickly found that wild primates carried their own indigenous viruses about which little or nothing was known. As early as 1934, Albert Sabin, of later polio vaccine fame, reported the first isolation of a novel virus from rhesus monkeys. He dubbed this new isolate the B virus and it was subsequently classified as a member of the herpesvirus family. Over the ensuing decades, more and more simian viruses were discovered as legions of researchers used these species for various purposes. Since biological characterization and classification were difficult for new isolates in that era, eventually the SV nomenclature was adopted starting with SV1. However, this sequential naming convention indicated nothing about the physical or biological properties, the disease potential, or the respective viral families of the isolates. For example, later studies showed that among the first 39 SVs isolated many were adenoviruses or picornaviruses, while some were myxoviruses or reoviruses. Instead of providing meaningful scientific information, the SV naming was merely a convenient method for coding these isolates and preventing them from being muddled in the literature. The SV nomenclature system was still in place when Maurice Hillman isolated SV40 in 1960.

Polio Vaccines and A Stealth Simian Virus

Chapter 6 covered the polio epidemics of the early twentieth century and the subsequent defeat of poliovirus by the Sabin and Salk vaccines. It might have been possible to use cultured human cells to grow the poliovirus stocks, but there was concern that human cells might harbor dangerous human viruses that would contaminate the vaccines, so monkey cell cultures seemed a safer choice. Therefore, to grow and produce the viral stocks for both vaccines, cultured kidney cells were used from rhesus and cynomolgus monkeys. While the possibility of monkey viruses in these two primate species was well known, great lengths were taken to ensure that these known viral agents were not present in the cells used for vaccine production. Typically, virus infection will cause damage (known as cytopathic effects) or death to cells, and these effects are detectable by microscopy. Cell cultures for poliovirus production were thoroughly vetted and deemed virus-free. Production of Salk's inactivated vaccine began in 1953, and the vaccine was widely used in the United States and globally thereafter. Sabin's attenuated vaccine went into large-scale production around 1957 and was licensed for US manufacture and distribution in 1961. Little did either of these remarkable scientists or the general public imagine that both vaccines were harboring a silent passenger that would be named SV40.

By the late 1950s, the search was on for different monkey species to use in research. The standard Macaques had many indigenous viruses that spread easily between captive monkeys during shipping and handling, making virus-free monkeys difficult to obtain. Maurice Hilleman, a Ph.D. microbiologist working at Merck & Company, recognized this problem and began hunting for alternative species for his vaccine work. On the advice of Dr. William Mann from the National Zoological Park, Dr. Hilleman tried importing West African Green Monkeys (*Cercopithecus aethiops*) rather than the Asian Macaques. By 1959 Dr. Hilleman had kidney cell cultures from these monkeys growing readily and they appeared virus-free and ideal for subsequent vaccine work. Dr. Hilleman then tried a fateful experiment whose result would ripple outward in the virology community for decades. He took the liquid media from cultures of supposedly virus-free rhesus and cynomolgus kidney cells and transferred that liquid onto cultures of Green Monkey kidney cells. Within 3–4 days after adding the exogenous media, the Green Monkey cells began losing their normal spindle shape and became rounded or shrunken. They also began to develop large vacuoles in their cytoplasm giving the cells a swiss-cheese appearance. These changes are classic cytopathic effects that are typical of virus infection. This finding indicated that the Macaque cells were actually harboring some type of virus that was being released into the media. When that virus-laden media was added to the Green Monkey cells, this unknown virus was able to infect those cells to cause the cytopathic effect. This was a stunning and remarkable discovery as the existence of a virus that could persist and reproduce in cultured cells without killing the cells or causing any detectable cytopathic effect was unknown at that time. It appeared that this mystery virus was highly adapted to Macaque kidney cells and underwent its life cycle in a manner that was completely innocuous in this cell type. The virus could infect Macaque cells, reproduce, and release new virions with no visible ill effects on the cells. Only when this unknown virus infected kidney cells from other monkey species did it exhibit a typical viral infection resulting in visible cell damage leading to cell death. Dr. Hilleman originally named this new virus the simian vacuolating virus, but in keeping with the nomenclature of that era, it was quickly designated SV40.

The discovery of SV40 immediately raised concerns about any research conducted using Macaque kidney cells, including vaccine production. Kidney cell cultures were usually created by mixing kidneys from 2–3 monkeys and then preparing a cell suspension to put into culture. As the Hilleman group tested independent kidney cultures derived from different rhesus and cyno-molgus monkeys, they determined that most cultures were SV40 infected. Even more troubling, when they tested batches of live vaccines made in Macaque kidney cultures, they were uniformly positive for SV40, including the new Sabin oral polio vaccine that was already being administered in some nations. Because of Hilleman's work the United States took strong measures to eliminate SV40 from future batches of the oral vaccine. However, batches already in existence were allowed to be used, so contaminated oral vaccine was administered in the United States until sometime in 1963. The situation was less satisfactory in other nations as their attempts to remove SV40 from their oral vaccines were not always successful. One retrospective study found that the live polio vaccine produced by an Eastern European manufacturer con-tained active SV40 until at least 1969. While later batches were not tested, the manufacturer didn't change practices until 1978 so all their polio vaccines until that time may have been contaminated with SV40. Similar studies on the more prevalent Salk-inactivated polio vaccine initially revealed a more favorable finding. To make the vaccine, the poliovirus grown in Macaque cells was treated with formalin to inactivate the virus, and it appeared that forma-lin treatment would also inactivate SV40. Unfortunately, this relief was short-lived. Subsequent more sensitive assays showed that SV40 was not as effectively inactivated by the formalin treatment as was poliovirus, resulting in some vaccine batches still containing viable SV40. Between the live and inactivated vaccines, estimates are that 10–30% of all polio vaccine doses administered in the United States between 1955 and 1963 contained live SV40, resulting in over 100 million Americans possibly being infected, most of them children. Since the Salk vaccine was globally used, the number of potentially infected individuals worldwide was hundreds of millions before SV40 contamination was finally eliminated. All the efforts to develop a safe polio vaccine free from monkey virus contamination had failed in the worst possible way, and a new, unknown virus had been inadvertently introduced into the human popula-tion. Additionally, several adenovirus vaccines used by the U.S. military from 1961–1965 also had SV40 contamination, contributing to the further intro-duction of this virus into humans.

Soon after the realization that tens of millions of children were inoculated with a live virus of unknown disease properties an even more disturbing observation was published. SV40 turned out to be a small DNA virus of the polyomavirus family, a family with known oncogenic capability. By 1962 the Hilleman group published evidence that SV40 caused tumors in newborn hamsters, and within a year several groups showed that SV40 could transform various types of normal human cells into malignant cells. Now there was not only a possible risk of some acute or chronic disease but also the potential that SV40 would cause cancer in vaccine recipients. It was this huge fear of a public health disaster that spurred decades of work to understand the biology of SV40 and the risk, if any, to the inadvertent recipients of this virus.

To Be or Not To Be – Is SV40 a Human Tumor Virus?

With millions of people worldwide infected with SV40, scientists in the 1960s realized that it was imperative to monitor people for any potential adverse effects of SV40 exposure. Adverse effects could range from acute illness to chronic infections to increased risk of cancer. While significant acute illness was never observed among vaccine recipients, the potential for chronic infection and/or cancer risk was an alarming possibility that has been studied ever since the 1960s. One of the first important questions asked was whether or not the vaccine-delivered SV40 could truly cause an active infection in humans, a necessary initial step for a virus to cause a clinical effect. Early studies of infants receiving contaminated oral poliovirus vaccine found that 15–20% of these children shed the SV40 virus in their stools for up to five weeks post-vaccination, strongly suggestive of an active infection. However, since shedding studies can only be done for a short interval immediately after vaccination, scientists turned to serology to detect evidence of SV40 infection in previously vaccinated individuals. If SV40 did replicate and reproduce in vaccine recipients, then these individuals should produce antibodies against SV40 proteins and become seropositive for SV40. In contrast, individuals with no infection would not have antibodies and would remain seronegative for SV40; such serological detection of specific antibodies is widely used as a marker of infection for many viral and bacterial diseases. Dozens of studies from the 1960s to the early 2000s used this approach with varying results. While some studies found no evidence of SV40 antibodies, most studies found SV40 antibodies in 2–50% of tested populations, with considerable variation in countries around the world. Surprisingly, there was little correlation between positivity for SV40 antibodies and exposure to contaminated poliovirus vaccine usage. Even people born long after the use of the contaminated vaccines ended were often found to have anti-SV40 antibodies. Some researchers equated SV40 seropositivity in people without exposure to contaminated vaccines as evidence that SV40 was now a circulating human virus. They reasoned that the vaccine introduced the virus into humans where it replicated, shed, and infected other people to become an established human virus that is now in permanent circulation around the world. Consistent with this idea, laboratory studies showed that SV40 could reproduce in various types of human cells, so productive infection and spread from the original vaccine contamination seemed plausible.

An important caveat to the circulating SV40 theory is the problem of specificity with the serological assays. We now know that poor reagents that lacked sufficient sensitivity and specificity confounded many early studies. As presented in Chap. 4, humans are naturally infected with human polyomaviruses (hPyVs) such as BKPyV and JCPyV. Both these viruses have proteins that are similar enough to SV40 to produce antibodies that can cross-react with SV40 proteins. These two hPyVs are quite ubiquitous so many people persistently carry these relatively silent viruses and have the anti-polyomavirus antibodies. Consequently, many studies to measure SV40 incidence in humans before around the year 2000 may have mistakenly detected human polyomavirus antibodies rather than true anti-SV40 antibodies. Such false positives likely led to an overestimation of SV40's prevalence in humans. Once this problem was recognized in the early 2000s, technical advances in serological assays were made to eliminate the cross-reactivity to the major hPyVs. These more specific assays still detect significant levels of SV40 seropositivity, from 5% to 20%, in tested adult populations. Some studies have also shown that seropositivity is first detectable in childhood and that the prevalence of seropositive individuals increases in older age groups, a pattern consistent with a pervasive infectious agent spreading naturally in the community. These more reliable observations provide strong evidence that SV40 is now an established human virus that spreads silently and effectively between humans. The data also suggest that the virus is widespread enough that it can easily be acquired during childhood, though the longer we live the more likely we are to encounter the virus, become infected, and develop antibodies. One remaining concern with this conclusion still involves the specificity of the serology. Since 2006 twelve new hPyVs have been discovered, and it is unknown if more exist. If any of these more recently discovered or unknown hPyVs are detected by the assay for SV40 antibodies then the problem of false positives becomes an issue again. Still, for now, the medical and scientific communities largely agree that humans accidentally introduced SV40 into human populations via the poliovirus vaccines where it established itself as a new, endemic human infection. Like the hPyVs, SV40 likely establishes persistent infections in humans with low-level or intermittent shedding. The anatomical location of the persistent virus is uncertain and it may occur in multiple tissues. However, the kidneys seem a prime candidate as this is where SV40 persists in rhesus monkeys and SV40 has been found in human urine.

If the status of SV40 as an established virus in humans is reasonably well accepted, the role of SV40 in human disease, particularly cancer, is still an unsettled issue. While there is a large body of research on this subject, much of it is conflicting and contentious, with no clear consensus on whether or not SV40 has been harmful to human populations. Proving that any infectious agent causes human cancer is a long and difficult process that requires multiple, corroborating pieces of evidence. Such evidence usually involves epidemiological data linking the agent to cancer incidence, finding the agent in tumor tissues and cells, demonstrating that the agent has oncogenic ability in model animals and/or human cell culture, and showing that molecular events needed for cancer formation in animals or cultured cells also occur in human tumors. Fulfilling all these requirements may take decades of work and can have many confounding problems. For SV40, some of these conditions are satisfied while others are still being investigated and debated in the scientific community, even after nearly 60 years of study.

As already noted, SV40 has oncogenic activity as it causes tumors in newborn hamsters and certain types of mice. It also readily transforms many types of cultured human cells into tumor cells, so it could potentially transform cells in people leading to tumor formation. However, we humans have sophisticated anti-tumor defenses that can detect and destroy many transformed cells long before they can develop into tumors, so possibly our bodies control SV40 very effectively and eliminate these aberrant cells before they cause cancer. One method to address this issue is epidemiological studies to determine if SV40-infected individuals have higher amounts of any type of cancer than people without SV40 infection. Such studies began in the early 1960s to look at the contaminated vaccine recipient populations, and similar studies were conducted in different population groups around the world for the next 40 years.

In 2003 the National Academies of Science published a safety review about SV40 and the poliovirus vaccines. This lengthy and detailed examination of the many epidemiological studies concluded that the data were too inadequate to make any firm conclusion about SV40 and human cancer. The earliest epidemiological studies simply tried to compare cancer incidence in groups who had received the poliovirus vaccines versus matched groups of unvaccinated individuals. Most such studies over the years failed to find any increased cancer rate in the vaccinated population. However, this approach has complications that make definite conclusions difficult. First, not every vaccinated person received a contaminated dose, and even for those people who did receive SV40, there was no way to tell how much virus any individual received. If SV40-induced cancer is rare or if a certain threshold of SV40 exposure is needed to initiate cancer, then many people in the vaccine group may have no higher risk than unvaccinated people. This situation could dilute any SV40-induced cancer incidence in the vaccine group to below statistical significance. Second, it is also possible that SV40-induced cancers take many decades to develop. If this is true then the negative results of the original epidemiological reports might simply reflect studies that were too short in duration to detect the viral effect.

To avoid the assumption that all vaccinated people received SV40, later epidemiological work used seropositivity to identify individuals who actually had an SV40 infection. While some of these studies provided evidence of SV40 association with cancers, other research groups found no such association and it was difficult to reconcile the often widely disparate results from these independent studies. Sadly, studies based on serological data suffered from the same confounding issues discussed above relating to a lack of specificity for the detection of SV40 antibodies. Many, if not most, of these studies were likely unreliable due to false positives from serological cross-reaction with the hPyVs. Given the uncertainty of who actually had a legitimate SV40 infection, any observed correlation between SV40 seropositivity and cancer incidence from these early studies must be regarded skeptically.

Because of the pitfalls associated with testing for antibodies against SV40, more recent studies turned to molecular diagnostics, particularly the polymerase chain reaction (PCR). PCR is a procedure that can amplify minute quantities of nucleic acid from clinical samples to produce sufficient amounts for detection and analysis. PCR is widely used in forensics to identify specific DNA in blood or cells collected from victims, and it is also used clinically to diagnosis a variety of infectious diseases. Similarly, PCR is applied to tumor tissues to detect viral nucleic acids such as SV40 DNA. Beginning in the 1990s many groups used this technology to assess a variety of different human tumors for the presence of SV40 DNA. SV40 DNA was found associated with cancers of the brain and bone, with malignant mesothelioma, and with non-Hodgkin's lymphoma, the same types of cancers caused by SV40 in animal models. Unfortunately, as with serology, there are technical issues that render much of this information suspect. Because of its extreme sensitivity, PCR is notorious for false positives that arise from inadvertent contamination of the samples being tested. Many of the labs doing the cancer testing were labs that worked on SV40 and/or used SV40 DNA sequences in their research. Such labs have significant environmental contamination with SV40 DNA that can easily be picked up and transferred to the tumor samples during handling and processing. If even a few molecules of extraneous SV40 DNA contaminated a sample it would give a false-positive result, possibly causing the prevalence of SV40 in the tumors to be greatly over-reported in some studies. In support of this concern, some samples reported as SV40 positive were actually negative when retested in other laboratories under more stringent conditions. Another technical issue that was discovered was the procedure for isolating DNA from tumor samples for use in the PCR assay. Some protocols were poor at recovering the very small SV40 DNA (~5200 base pairs), producing false negatives if SV40 was not successfully extracted from the tumor tissue. These technical issues likely contributed to the large disparity in SV40 prevalence in tumors from different studies, even when the same type of cancer was investigated. Because of all these uncertainties, there has never been concrete proof or general acceptance that SV40 causes human cancers. Whether or not the poliovirus vaccines had a long-term negative impact on our collective health due to the permanent introduction of SV40 into human populations continues to be debated. Some scientists are strong advocates that SV40 is oncogenic in humans, but others are equally sure that SV40 is just a harmless passenger in tumors and not a contributing factor in tumor formation. After 60 years of study, this remains a worrisome unsolved problem, and we may never have a definitive answer.

The SV40 Revolution

Hilleman's discovery of SV40 created enormous concern and anxiety about the impact of SV40 on human health but also helped jump-start the era of eukaryotic (organisms with a nucleus) molecular biology. While many researchers were focusing on the human health impact of SV40, others were trying to understand the biological and molecular features of this virus that allowed it to reproduce and transform normal cells into cancerous ones. As interest in and study of SV40 spread in the scientific community during the 1960s, another notion began to form about using this virus as a new model system. The first half of the twentieth century saw bacteria, particularly E. coli and its bacteriophages (see Chap. 12), as the major model system for studying genetics and molecular biology. By the 1960s, scientists were eager to apply the findings and concepts from bacterial cells (prokaryotes) to human cells (eukaryotes), but this was problematic. The human genome is 6.4 billion base pairs long, which is over a thousand times longer than the E. coli genome at 4.6 million base pairs. The huge size of the human genome was daunting and made it intractable to study with the tools available at that time. However, many researchers began to speculate that human DNA viruses could be used as surrogates for studying human DNA and cells. Viruses such as SV40 had tiny DNA genomes (5200 base pairs) that seemed much less formidable to investigate with available methodologies. Equally important, viruses had to reproduce inside eukaryotic cells and must be highly adapted to this environment. Consequently, they likely used similar, if not identical, processes for producing their messenger RNA (mRNA), translating their mRNAs into proteins, and replicating their DNA genomes. Since little was known about how these processes occurred in human cells, scientists hoped that viral studies would provide key insights into fundamental aspects of eukaryotic molecular biology and genetics. This hope was wildly successful with SV40 producing many of the earliest and most informative findings about the workings of our cells.

Coupled with the burgeoning interest in SV40 as a vehicle to explore eukaryotic cells, powerful advances in biological methodologies exploded in the 1970s and were instrumental in SV40 research. In particular, the discovery of restriction enzymes and the invention of DNA sequencing techniques provided completely new approaches that finally unlocked many of the mysteries of how eukaryotic genes are organized and regulated. Restriction enzymes (REs) are bacterial proteins that cleave double-stranded DNA at specific sequences. For example, *E. coli* produces a RE known as EcoRI that cleaves the sequence G/AATTC (the / indicates the site of cleavage between the G and the first A of the sequence). Every type of bacteria produces at least one restriction enzyme, and each RE has a unique target sequence that it recognizes and cleaves. Bacteria use these enzymes to attack and destroy invading foreign DNA such as bacteriophages that infect bacteria. The REs don't cleave target sequences that occur in their parental bacterial DNA because the bacteria modify the target sequences in their own genomes to make the sequences resistant. Thus, this restriction-modification system acts like an immune defense in bacteria and cleaves non-self DNAs but not self DNA. Hamilton Smith and co-workers purified the first RE in 1970, using the bacterium *Hemophilus influenzae* as the source. The Smith group subsequently characterized this RE and demonstrated its sequence-specific cleavage activity, work for which Smith ultimately shared the 1978 Nobel Prize in Physiology or Medicine. The power of different REs to make specific cuts in DNA was quickly developed into a methodology for cutting large DNAs into overlapping smaller fragments that were used to create physical maps of the original DNA. These physical maps could then be correlated with other known features to create functional maps of where genetic elements fell on the genome. The initial demonstration of this concept was by Daniel Nathans and colleagues who created the first restriction map using SV40 DNA; for his ingenious applications of REs Nathans shared the 1978 Nobel Prize with Hamilton Smith.

From this very basic research into how bacteria protect themselves from foreign DNA, the discovery of REs quickly led to the invention of cloning. Many bacteria, in addition to their DNA genomes, carry small, circular, independent DNAs known as plasmids. These plasmids often exist at tens to hundreds of copies per bacterium, so it is easy to produce large quantities of plasmid DNA from bacterial cultures. In contrast, it was much more difficult to produce large quantities of viral or human cellular DNAs, often making it impossible to get enough of these DNAs for convenient study. To circumvent this difficulty, Paul Berg and others reasoned that a feasible approach would be to insert the desired eukaryotic DNA into a bacterial plasmid. Then the plasmid could be introduced into bacteria where it would replicate to produce copious plasmid copies per cell, thus increasing the amount of the cloned eukaryotic DNA as well. Subsequently, the plasmid DNA could be harvested from the bacteria and the amplified eukaryotic DNA excised for study. All they needed were enzymes to cut the eukaryotic DNA and the plasmid DNA and then some way to glue the pieces back together. REs proved to be the perfect cutting tool, and another bacterial enzyme called ligase efficiently rejoined the cut DNAs. Berg quickly realized the power of REs and used them in 1972 to clone the first eukaryotic DNA, the entire 5200 base pair SV40 genome. This first recombinant DNA was a momentous confirmation of the concept of cloning, and it ushered in the golden era of molecular biology where exciting discoveries flowed in torrents. Now any piece of DNA from any source could be cloned, amplified, and studied in minute detail. Within three years of this first recombinant, two separate methodologies to sequence these cloned DNAs were independently introduced, one by Walter Gilbert and one by Frederick Sanger. Using their new sequencing technology, the Sanger group published the complete SV40 sequence in 1977, the first genome of a eukaryotic organism that was entirely sequenced. For these landmark achievements Drs. Berg, Gilbert, and Sanger shared the 1980 Nobel Prize in Physiology or Medicine.

Now that it was cloned and sequenced, SV40 became an even more attractive model for deciphering cellular processes. Mutations could be made anywhere in the SV40 genome and either the entire genome or selected fragments could be introduced into cultured cells to check for function. This combination of genetic and physical manipulation of the SV40 genome made it ideal for identifying novel features embedded in the virus's DNA and proteins. Other important discoveries made with SV40 include creation of the first transgenic mouse, isolation of the first eukaryotic transcription factor (named

Sp1), discovery of enhancer sequences that activate gene transcription, identification of the polyadenylation signal in mRNAs, discovery of nuclear localization sequences that direct certain proteins to the cell nucleus, and creation of the first in vitro DNA replication systems (along with Adenovirus). Collectively these seminal observations revealed fundamental mechanisms by which eukaryotic cells produce and process mRNAs, control the intracellular location of proteins, and replicate their genomes. By studying this one small DNA virus and how it reproduces in cells we gained profound insight into our own biology at a biochemical and molecular level. Ultimately, because of their useful functional properties, many pieces of the SV40 genome were cloned into commonly used research plasmids that were sold and distributed to laboratories throughout the world. Thus, the feared contaminant of the poliovirus vaccines became one of the most important research tools in the 1970s and 1980s and impacted nearly all research areas of human biology. Unfortunately, this widespread dissemination and use of SV40 sequences in research and clinical labs also contributed to the contamination issues discussed in the previous section.

p53 – The Guardian of the Genome

In addition to its invaluable contributions to our general understanding of the molecular biology of normal eukaryotic cells, SV40 has greatly impacted cancer research. At the time of SV40's discovery, cancer was viewed as a highly complex, multifactorial disease that likely required mistakes in many biochemical processes for a normal cell to convert to a cancerous one. One approach to understanding cancer initiation was to study cancerous cells from patient tumors and compare them to their normal cell counterparts in hopes of identifying key differences that triggered the transition from normalcy to malignancy. This avenue of research turned out to be frustrating as the observed differences were numerous and failed to reveal the critical initiation steps since the tumor cells were long past the transition point. To circumvent this problem, many researchers were turning to viral systems where tumor formation could be followed after viral infection of appropriate animals. For example, Hilleman's work in the early 1960s clearly showed that SV40 had cancer-causing ability in certain animals such as hamsters, and similar work with the related murine polyomavirus showed that it could cause tumors in mice. Yet these virus-animal studies were also limited as there was a significant time lag, often weeks to months, between infection and tumor formation. Furthermore, early cellular events still couldn't be followed after viral infection as the process was undetectable in the animals until a tumor finally appeared. Much simpler systems were needed where components could be manipulated, viewed, and analyzed more quickly. Fortunately, results from several fields converged to provide the necessary tools for attacking the tumor-igenesis problem.

One key advance was Wilton Earle's demonstration of chemical transfor-mation in 1943. Using a known carcinogen called methylcholanthrene he treated normal mouse cells in culture and observed that some cells "trans-formed" their appearance and properties, changing from elongated cells into a more rounded shape. More importantly, when Earle injected these trans-formed cells back into mice the cells grew into a tumor. These seminal experi-ments proved that normal cells could become cancerous in culture and established that malignancy was an independent cellular phenomenon that didn't require the cells to be part of whole organs or a living animal. Now the very early stages of the transition from normal to malignant cells were acces-sible to observation and study in cell culture. Additionally, this in vitro trans-formation typically took days to weeks and not weeks to months, so experiments could be done more quickly and easily than using animals. Other groups extended these observations and defined several properties of transformed cell growth that differed from normal cells. Perhaps most useful experimentally was the discovery that transformed cells typically lose contact inhibition. Contact inhibition is the property where normal cells stop growing when they become closely contacted by other cells. This is an innate ability of normal cells that controls their growth and density so that our organs and tissues stop growing when they reach the appropriate size. Because of contact inhibition, normal cells placed in cell culture will grow and divide until they completely cover the surface of the culture dish with a single layer of cells referred to as a monolayer. At this point, normal cells stop dividing and simply stay as this monolayer indefinitely. Transformed cells are like tumors in that they lose contact inhibition and don't stop growing in culture. The result is that the transformed cells continue to grow after the normal cells have stopped, and the dividing transformed cells pile up to form colonies (called foci) that are easily visible and recoverable for study. Each individual colony, or focus, results from a single transformed cell that continued to divide, so the entire colony of cells is genetically homogenous which facilitates further analysis.

A second important advance was the creation of stable lines of normal cells that could be propagated indefinitely. Previous work used primary cells that were taken directly from animal tissues and then placed in culture. These pri-mary cultures typically had mixtures of different cell types, and this heteroge-neity of cell types made the subsequent analysis more complicated. For example, different cell types in the primary cultures might exhibit different sensitivities to transformation, so the transformation studies were often highly variable in their results. Even more troubling was that all the cells in these primary cultures would eventually senesce (grow old) and die off, so new cultures had to be continually remade from new animals, adding more heterogeneity to the studies. The breakthrough that eliminated these issues was the establishment of the first stable cell line of normal cells named the 3T3 line. Published in 1963, the 3T3 cells were derived from mouse embryonic fibroblast cells. These cells retained their normal phenotype and growth properties yet somehow became immortalized and never senesced and died. Consequently, 3T3 cells were completely homogeneous and could be passed in culture indefinitely. These properties allowed these cells to be sent to labs around the world so that researchers could all be using a standard cell background. Using a common and consistent cell system allowed a reliable comparison of results across many labs and different types of experiments.

By the early 1960s, both mouse polyomavirus and SV40 were shown to transform cells in culture, including both rodent and human cells for SV40. Moving to the newly established 3T3 cells produced highly consistent transformation results that finally allowed a detailed analysis of the oncogenic mechanisms of SV40 and other viruses. Using a combination of genetics (i.e. viral mutants) and RE-generated fragments of the SV40 genome it was eventually shown that the viral transforming activity resided in a single gene named the A gene. For example, a restriction fragment of the SV40 genome containing only the A gene was sufficient to transform cells while no other region of the viral genome had any transforming activity. Similarly, mutations in the A gene eliminated transforming activity while mutations elsewhere in the genome did not affect transformation. Parallel studies found that the A gene expressed a single protein that reacted with antibodies from animals with SV40-induced tumors, so this protein became known as the SV40 tumor antigen or T antigen for short. Remarkably, this result indicated that for all its complexity and diversity, cancer could be initiated by the introduction of a single protein that somehow disrupted the normal cellular environment. Concurrent work with multiple retroviruses, a type of RNA virus, also showed that each different retrovirus carried a single unique gene that had transforming activity, and collectively these viral transforming genes became known as oncogenes. That single viral genes could cause cancer was a remarkable, unexpected, and stunning observation suggesting that there must be some very fundamental and critical cellular components that could be targeted to initiate the oncogenic transformation. While the SV40 A gene was eventually shown to express two other proteins that also contributed to oncogenesis, designated the small T antigen and the 17K T protein, the large T antigen became one of the most highly studied proteins and soon provided one of the seminal insights into human cancers.

If a single viral protein such as the large T antigen can trigger the transformation of a cell from its normal state to a malignant one, what is that protein acting on in the cell to elicit such a dramatic and pleiotropic change? The complete answer to this question is multifaceted and beyond the scope of this chapter, but one T antigen function revealed a central feature shared by a majority of cancers. Large T antigen is a 708 amino acid protein that is wonderfully complex and multifunctional. Some of its functions are mediated through direct binding with host proteins, and as of the end of 2020, there were 17 human proteins identified as binding partners of large T antigen. By binding to this multitude of cellular proteins the viral T antigen can alter each of their functions. Through this binding mechanism, the virus manipulates many biochemical pathways and fine-tunes the intracellular environment to promote successful viral reproduction. Among the 17 known partners, arguably the most important partner is the very first one discovered, a protein that we now call p53. By the late 1970s, the transforming ability of T antigen was well established and various investigators were using specific antibodies against T antigen to pull the protein out of cell extracts for further study. In 1979, several groups independently published reports that T antigen captured in this fashion brought with it another protein of cellular origin. This cellular protein bound to the large T antigen became known as the p53 protein based on its molecular weight of 53,000 daltons.

It would take another decade of work to fully appreciate the role of p53, but eventually, its function as the guardian of the genome was established, and in 1993 *Science* magazine named it the Molecule of the Year. P53 is a critical protein that helps determine the cell's fate following damage to the cell's DNA. DNA damage is a dangerous event for a cell as changes in the DNA sequence caused by the damage can lead to cancer formation. When cells detect DNA damage, p53 is activated and stops the cell's replicative cycle. Stopping the cell cycle allows repair systems time to try to correct the damage and restore the normal genome. In some cases, the damage is too severe for repair, and p53 then triggers a process known as apoptosis which kills the cell. Harsh as this seems, it is far better to kill off the occasionally damaged cells than allow them to propagate and potentially become cancerous, a scenario analogous to the immune defenses killing off infected cells to prevent further viral spread. The central and critical role of p53 in protecting us from cancer was best illustrated by a 1992 publication that deleted the p53 genes in mice. These "knockout" mice without any p53 developed perfectly in utero, were born healthy, and matured completely normally, so p53 seemed completely dispensable and its absence had no negative consequence on growth and development. However, these mice all developed multiple cancers starting at a young age, illustrating their inability to defend themselves from DNA damage and prevent oncogenic cells from arising. Analysis of human cancers confirms p53's importance as over 50% of all human cancers can be shown to have defects in p53. Most of these defects likely arise by spontaneous mutations that damage the p53 gene in a cell, making that cell unable to protect itself from further DNA damage. This defect leads to an accumulation of genomic mutations that eventually initiate cancer formation. These and other studies collectively proved that p53 is truly the guardian of our genomes and is one of our strongest and most critical protectors from cancer, a seminal finding that all began with SV40.

You may be wondering what the connection is between SV40 and p53 if p53's role is to protect us from cancer. It turns out that viral infection is another kind of damage to the cell that activates p53. As for DNA damage, activation of p53 by viral infection stops the cell cycle and can lead to cell death via apoptosis. Stopping the cell cycle limits viral replication by shutting down the production of many of the components that viruses need to synthesize their genomes, and cell death via apoptosis would completely prevent viral reproduction. Thus, p53 has a dual role in protecting us from cancer and contributing to our intrinsic antiviral defense. Because of p53's antiviral function, many viruses had to evolve mechanisms to block p53 function so that viral replication could proceed. SV40 overcomes p53 by having large T antigen bind p53 tightly. The binding of p53 to T antigen blocks p53 function, so even though p53 is still present in SV40 infected cells, it is not active and can't stop the cell cycle or induce apoptosis to limit viral reproduction. Many other viruses also have proteins that bind p53 and either block its activity and/or cause its degradation to remove it from the infected cell. This discovery of the central roles of p53 in anticancer and antiviral defense all stems from SV40, a small virus whose impact on twentieth-century biology was enormous.

Asilomar and the Birth of Biosafety

Along with its powerful impact on science, SV40 also had a profound effect on the realm of biosafety. In the highly regulated twenty-first century it may seem surprising, but the early years of molecular biology had essentially no regulations. As the technology for recombining and cloning DNA exploded in the 1970s, there were no regulatory limits or restrictions on what could be done with this recombinant DNA technology. More and more novel experiments were envisioned using restriction enzymes to recombine DNAs from different sources, leading to growing apprehension in the scientific community. Never before did we have this ability to create novel recombinant DNAs and organisms, and some researchers raised concerns that this work might produce unforeseen dangers and have unintended consequences. As the debate about the safety and ethics of this new science simmered, once again SV40 was at the center of the field. Paul Berg had begun to clone pieces of the SV40 genome and proposed putting the entire SV40 DNA into *E. coli*. This experiment alarmed many of his colleagues because *E. coli* is a common intestinal bacterium carried by most people. The idea of putting viral DNA with potential human cancer-causing ability into bacteria that live in the human gut seemed like a dangerous idea. Even though the laboratory strain of *E. coli* being used was supposedly unable to survive outside the lab, what if the recombinant lab strain transiently infected lab workers and transferred the SV40 DNA to the wild-type human *E. coli*? Would this wild-type *E. coli* that harbored SV40 DNA cause cancer in these lab workers? Moreover, would it spread to their close contacts and then out to the general population as intestinal bacteria are prone to do? Many argued that the risks were too great and the knowledge about how these new recombinant DNAs would work was too limited to even consider such an experiment. Similarly, Andrew Lewis at the National Institutes of Health had recently created an SV40-adenovirus recombinant. Adenoviruses transmit via the respiratory route, so could laboratory personnel become accidentally infected with the hybrid virus and spread it to the public? And again, would the easily spread hybrid virus have cancer-causing capacity? While Dr. Lewis's laboratory took great precautions with this hybrid, he worried that other labs might be less capable or less concerned and could accidentally release similar hybrids.

Considering the concerns and issues raised over these SV40 recombinants, Dr. Berg agreed to halt his experiment and convene a meeting of scientists to discuss how to proceed, not only with his own studies but with the emerging technology of genetic engineering in general. This first meeting in 1973 was held at the Asilomar Conference Center in Pacific Groove, California, and brought together nearly 100 scientists who were active in this new technology. The meeting had no formal outcome, though there was consensus that much more information was needed about the biological impact of recombinant DNAs. The group also proposed a temporary halt to the potentially dangerous work with recombinant SV40 and suggested that other research should proceed with great caution. Later that year a similar meeting was held on the East coast at a Gordon Conference bringing together U.S. academics and academics from across the world, as well as scientists from the government and private industries. This second group drafted a joint letter to the National Academy of Sciences and the National Academy of Medicine requesting the establishment of a formal expert panel to review the biohazards associated with recombinant DNA work. Following an additional informal study during 1974, feelings among scientists were still divided. The early 1970s were a turbulent time in America with the Vietnam War, Watergate, and President Nixon's resignation. The public was anxious and suspicious, and scientists were not immune to the social implications and consequences of their work. One faction of researchers demanded unrestricted experimentation, another faction lobbied for a complete cessation of recombinant DNA work, and a third group of moderates felt that there must be some oversight to forestall a public outcry. This latter group argued that scientists must help shape the rules with public transparency to prevent the government from imposing harsh restrictions that would hamstring the entire field. Ultimately the moderate position prevailed and the National Institutes of Health (NIH) was requested to establish an advisory committee to review and oversee recombinant DNA experiments.

The NIH agreed with the need for further review and appointed Paul Berg and several colleagues as the organizing committee for a group called the Assembly of Life Sciences on Recombinant DNA Molecules. The organizers sought the brightest and most knowledgeable molecular biologists from both the U.S. and the world for what would be a historic meeting. In early 1975 this group of 133 scientists reconvened at the Asilomar Conference Center along with 16 journalists and 4 lawyers. In three and a half short days this eminent group defined all the major principles that became the NIH Guidelines on Recombinant DNA, a document that would govern all future NIH-funded research. One of the key guiding tenets presented by the group was the concept of risk groups. Not all recombinant DNA experiments are equal based on the nature of the DNA being cloned, the type of cloning vector used, and the organism into which the recombinant molecules would be placed. The group proposed a series of four risk groups (what would become biosafety levels 1–4) from minimal to very high with different levels of laboratory practice and physical containment required for each level. For example, cloning a small piece of innocuous DNA from one bacteria and putting it into a related bacteria should have little or no risk and could be done in standard, low-containment labs (BSL-1). In contrast, cloning virulence genes from an untreatable pathogen such as Ebola would be considered a very high-risk experiment. Such high-risk experiments necessitated specialized containment facilities (BSL-4) to prevent the possible escape of the recombinant organisms. To further provide for experimental safety, the Asilomar group recommended the development of safer vectors and the creation of intentionally crippled recipient organisms that could not survive outside laboratory conditions. Lastly, they recommended that specific training requirements be mandated for all personnel who worked with recombinant DNA. This last point was to ensure that all workers had proper knowledge and techniques for the safe handling of microbial organisms. Since so many molecular biologists came out of chemistry and biochemistry backgrounds with little experience in the basic principles of microbiology, it was agreed that fundamental training was essential. Collectively, this set of principles and practices was devised to ensure appropriate safety for lab personnel and the surrounding community while still providing enough flexibility so that creative scientific ideas could be pursued.

The Asilomar report went to the National Academy of Sciences who then transmitted it to the NIH. The NIH converted the Asilomar organizing committee into the first Recombinant DNA Advisory Committee (RAC) and instructed them to expand the recommendations into formal and detailed guidelines. The RAC became a permanent committee at NIH and the first draft guidelines were published in June of 1976. After public comment and review, the final draft of the initial guidelines appeared in the Federal Register of October 1977. One of the requirements added was that each institution that performed recombinant DNA work must establish its own Institutional Biosafety Committee (IBC). The IBC is required to review and approve proposals for recombinant DNA work and to ensure that research labs at their institution practice appropriate biosafety measures. Importantly, each IBC must include at least two community members on the committee to ensure that the public is aware of the university research and can express concerns. The RAC and the IBCs continue to this day and provide invaluable guidance and protection to the researchers, the greater academic community, and the general public. The Guidelines have been modified many times since 1977 as new information and new technologies have emerged, but the basic principle of data-based biosafety to ensure adequate protection for all while not impeding important scientific advances remains unchanged. The development of recombinant biosafety as an obligation of both individual researchers and their institutions is another advance that owes a great debt to SV40. Because of SV40, biosafety guidelines developed early in the timeline of recombinant DNA applications prevented any significant bio-disasters from this type of research. The last 50 years have seen dramatic advances in biology and medicine from this technology, all done safely and as transparently as possible. This one small virus, discovered as an unfortunate contaminant of the poliovirus vaccines, dramatically changed twentieth-century science and became one of the premier eukaryotic model systems in the golden age of molecular biology. While most people have never heard of SV40, perhaps the readers of this chapter will now remember this name and appreciate its manyfold contributions to science and society.

Additional Reading

1. The Virus and the Vaccine. B. Bookchin and J. Schumacher. St. Martin's Press, 2004.
2. Immunization Safety Review: SV40 Contamination of Polio Vaccine and Cancer. The National Academies Press. 2003.
3. DNA Tumor Viruses and Their Contributions to Molecular Biology. J.M. Pipas. J. of Virology Vol. 93, 2019.
4. Asilomar Conference on Laboratory Precautions When Conducting Recombinant DNA Research – Case Study. M.J. Peterson. International Dimensions of Ethics Education in Science and Engineering. U. of Massachusetts, 2010.

Definitions

17K T protein – One of three proteins expressed from the SV40 A gene that is involved in viral reproduction and cell transformation.
3T3 – An immortalized cell line derived from mouse embryonic fibroblast cells.
A gene – The SV40 gene encoding the T antigens.
B virus – A primate virus of the herpesvirus family.
BK polyomavirus (BKPyV) – A human polyomavirus.
Cell cycle – The series of stages that a cell progresses through as it grows and divides.
Cell lines – Cultures of immortalized cells that do not senesce and die. Cell lines can be propagated indefinitely.
Cercopithecus aethiops – West African Green Monkeys.
Contact inhibition – The property where normal cells stop growing when they become closely contacted by other cells. Tumor cells usually lose contact inhibition.
Cytopathic effect – Any observable damage to cells caused by viral infection.
Dalton – A measure of the molecular weight of a biomolecule.
DNA sequencing – The process of determining the nucleotide order (primary sequence) of DNA.
EcoRI – A restriction enzyme that specifically recognizes and cuts DNA between the G and the A nucleotides in the sequence GAATTC.
Enhancer sequence – A DNA sequence element that stimulates transcription of a nearby gene.

(continued)

(continued)

Fibroblast – A type of cell found in connective tissue (e.g. skin and tendons) that secretes collagen.

Focus (plural foci) – The outgrowth of transformed cells into a colony that forms on a monolayer of normal cells.

Hemophilus influenzae – A bacterial pathogen that can cause bronchitis and pneumonia.

In vitro – Occurring outside of an organism, typically referring to studies done with cells, tissues, cell extracts, or purified biomolecules.

JC polyomavirus (JCPyV) – A human polyomavirus.

Knockout mice – Genetically engineered mice where a specific gene has been damaged or deleted so that the gene product can no longer be produced.

Ligase – A bacterial enzyme that catalyzes the joining of nucleic acid fragments.

Macaca fascicularis – Cynomolgus monkeys

Macaca mulatta – Rhesus monkeys

Macaque – A group of Asian primates that include rhesus and cynomolgus monkeys.

Mesothelioma – A cancer of the tissue that lines the lungs, abdomen, or heart.

Methylcholanthrene – A highly carcinogenic biomolecule.

Monolayer – A single layer of cells completely covering the surface of the culture dish.

Myxovirus – An RNA virus of the paramyxovirus or orthomyxovirus families.

Non-Hodgin's lymphoma – A cancer of the lymphocytes.

Nuclear localization sequences – An amino acid sequence in a protein that directs the protein to the cell nucleus.

Oncogene – Any gene whose protein product is capable of inducing cancer.

P53 – A cellular protein that protects cells from cancer and viral infection by inducing either cell cycle arrest or apoptosis.

Phenotype – The observable physical properties of an organism.

Plasmids – Circular DNA elements in bacteria (and some other organisms) that are independent of the chromosome. Plasmids often exist in high numbers of copies within a bacterial cell.

Polyadenylation signal – A specific sequence of nucleotides in an mRNA that directs the addition of multiple A nucleotides to the 3 prime end of the mRNA molecule.

Polymerase chain reaction (PCR) – A process used to vastly amplify DNA and RNA molecules.

Primary cells – Cells that were taken directly from animal tissues and then placed in culture. Primary cells typically have a finite lifespan and will eventually senesce and die.

Restriction enzymes (RE) – Bacterial enzymes that recognize and cleave specific DNA sequences.

Restriction-modification system – A protective system in bacteria that uses restriction enzymes to cleave foreign DNAs (e.g. bacteriophage DNA). Bacteria modify their own DNAs (chromosomal and plasmid) so that they are protected from cleavage by the restriction enzymes.

Serology – Any test or assay using serum. Typically serology refers to testing for specific antibodies in the serum or using the serum to detect specific antigens.

(continued)

(continued)

Seronegative – Lacking antibodies against a specific antigen, e.g. a virus.

Seropositive – Possessing antibodies against a specific antigen, e.g. a virus.

Small T antigen – One of three proteins expressed from the SV40 A gene that is involved in viral reproduction and cell transformation.

Sp1 – A cellular protein that regulates transcription from certain genes.

T antigen – One of three proteins expressed from the SV40 A gene that is involved in viral reproduction and cell transformation.

Transgenic mouse – A mouse that has been genetically engineered to contain foreign DNA in its genome.

Vacuoles – A membrane-enclosed space or vesicle that develops within the cytoplasm of a cell.

Vector – In the cloning process, a vector is a plasmid or virus that is used to carry a piece of foreign DNA. The vector is used to propagate and amplify the foreign DNA.

Abbreviations

3T3 – an immortalized mouse cell line

BSL – Biosafety level (there are 4 levels, BSL1-4)

EcoRI – *E. coli* R strain enzyme 1

IBC – Institutional Biosafety Committee

NIH – National Institutes of Health

P53 – protein with molecular weight of 53,000 daltons

PCR – polymerase chain reaction

RAC – Recombinant DNA Advisory Committee

RE – restriction enzyme

Sp1 – specificity protein 1

SV – simian virus

8

Viral Cancer

Keywords Retroviruses • Hepatitis C virus • Human papillomavirus • Hepatitis B virus • Epstein-Barr virus • Kaposi's sarcoma herpesvirus • Merkel polyomavirus

> *Most of the infections linked to human cancers are common in human populations; they are ubiquitous. They were present during the whole human evolution process.*
> Harald zur Hausen, 2008 Nobel Prize in Physiology or Medicine

The Infectious Beginnings of Cancer Research

Roughly 15–20% of human cancers are associated with viral infections, so yes you can catch cancer. Yet cancer is typically considered a genetic disease. Random mutations in our DNA cause genes to go awry, and when certain critical genes are affected their aberrant expression or function triggers the onset of cancer. These random mutations are caused either by faulty DNA replication or by environmental exposures such as radiation or the chemicals that we touch, breathe, and ingest. Mutations can occur in any of our body's cells (so-called somatic mutations) meaning that cancer can strike anywhere. Usually, cancers appear later in adulthood due to the slow accumulation of mutations in our genome throughout our lives. Given this genetic mechanism, how and why do certain viruses predispose us to cancer? The answers to this question go back over a hundred years to seminal work in the late 1800s and early 1900s.

The concept that microorganisms are responsible for certain diseases only became widely accepted in the late 1800s through the work of scientists such as Louis Pasteur and Robert Koch. Pasteur's work with sterilization ultimately convinced the world that infections were the result of contamination by pre-existing microorganisms present in the environment. Prior to his work, many believed that microorganisms arose out of nothingness by "spontaneous generation". Once Pasteur disproved this flawed belief, the field of microbiology exploded as researchers began to look for potential infectious agents as the underlying basis for many diseases. Building on this concept, a German physician, Robert Koch, developed a set of principles for proving that an organism caused a specific disease. Now known as Koch's postulates, a key element of his approach is to take an organism from a diseased animal or person and to reproduce the same disease after transferring the organism to a healthy recipient. In the decades that followed, this transfer of disease concept became a widely used tool for determining an infectious basis for many illnesses of unknown origin.

The first known application of the transfer concept to cancer was reported in 1908 by two Danish scientists, Drs. Vilhelm Ellerman and Olaf Bang. While studying leukemias in chickens they made an extract from leukemic cells and passed that extract through a filter to remove debris, remaining cells, and any bacteria. When the final extract was injected into healthy chickens many developed leukemia identical to the donor animal, suggesting that a very small infectious agent was responsible for causing the leukemia. In subsequent studies, Ellerman repeated this observation for different types of leukemias (e.g. erythroid, lymphoid, myeloid) and ultimately identified 8 different "strains" of his infectious agent. Unfortunately, this work went largely unnoticed because the science of virology was still in its infancy and leukemia was not even recognized as a true cancer until decades later. It would be years before these filterable agents were proven to be avian leukosis viruses of the retrovirus family.

The next advance came shortly afterward by the American veterinarian, Peyton Rous. In 1909 Dr. Rous took a faculty position at the Rockefeller Institute for Medical Research in New York City. While having no specific background in oncology, Rous was hired to take over the cancer research program of the Institute's director, Simon Flexner. Flexner wanted to change his focus to polio research as the disease was becoming a serious health issue in New York. Nonetheless, he wanted the cancer studies to continue and Rous seemed like a suitable young researcher to lead the project. In 1911 Rous performed transfer experiments using a chicken sarcoma, a solid tumor that was undisputedly a true cancer. Like Ellerman and Bang, Rous successfully transferred the sarcoma to healthy chickens using a cell-free filtrate, implying a potential viral agent. However, this work was again largely dismissed by the contemporary scientific community. Doubters argued that there might still be small leukemic cells in his extracts, and even if there were no contaminating cells, critics maintained that chicken sarcomas had no relevance to human cancers. Faced with little acceptance for his work with chickens Rous tried to find a filtrable agent in a mammalian tumor, but lack of success led him to abandon cancer research for 20 years.

Rous's interest in transmissible cancer was rekindled in 1934 by the work of a Rockefeller colleague, Richard Shope. Shope had recently shown that warts on jackrabbits could be transferred via cell-free filtrates leading to the discovery of a DNA virus that eventually became known as the Shope papillomavirus (related to the human papillomaviruses). Working with Shope, Rous determined that warts are generally benign tumors, though a subset of these lesions can progress to malignancy. With this demonstration that a mammalian virus could contribute to cancer development, research in tumor virology suddenly became an accepted, mainstream field. Rous continued his research in this area, and his original retroviral sarcoma agent proved to be the first discovered tumor virus. Many years later his eponymous Rous Sarcoma Virus (RSV) also proved instrumental in finally unlocking the molecular secrets of retroviral cancer. In recognition of his contributions, at the age of 87, Dr. Rous was awarded the 1966 Nobel Prize in Physiology or Medicine for launching the science of tumor virology.

Once tumor virology was no longer a fringe area, there was an influx of talented researchers leading to many new developments over the ensuing decades. Additional oncogenic avian retroviruses continued to be found, but perhaps a more important advance was the discovery of retroviruses in mice. The first observation was made by John Bittner in 1936 when he noticed that some mouse strains had a high incidence of mammary tumors whereas these tumors were absent in other strains. He then showed that the tumor-free mice would develop mammary tumors if they suckled from tumor-prone females as newborn mice. He reasoned from these studies that a tumor-causing agent was present in the breast milk of the tumor-prone mice. Filtering the breast milk and feeding the filtrate to newborn mice from the tumor-free strain was still able to induce mammary tumors in the recipient mice. These careful filtration studies established that there must be a viral agent in the milk, and subsequent work showed the agent to be the first discovered mammalian retrovirus, now called the mouse mammary tumor virus (MMTV). After the fortuitous discovery of MMTV, progress was slow on murine retroviruses until the 1950s when inbred mouse strains became more common. These inbred lines had more constant physical and genetic properties that gave greater consistency to experimental studies. Using an inbred line known as

AKR, Ludwik Gross, a Polish-born American physician, showed transmission of leukemia to suckling mice via mouse breastmilk. The agent (ultimately called the Gross leukemia virus) was also a retrovirus, so now retroviruses were capable of causing both leukemias and solid tumors in mice just as they were in chickens. Gross's work inspired the search for other murine retroviruses, and by 1970 roughly 10 additional mouse retroviruses were found; most retroviruses were named after their discoverers such as the Friend murine leukemia virus and the Moloney murine sarcoma virus. The existence of both leukemia and sarcoma-inducing viruses in mice strengthened the belief that many animal tumors might have a viral etiology. In the 1970s the tools and technologies existed to explore other animal species, and oncogenic retroviruses were soon found in other common animals such as cows, horses, sheep, cats, and goats. The further discovery of primate retroviruses in the early 1970s foreshadowed a role for these viruses in human cancers, though human retroviruses remained elusive for another decade. Collectively, the demonstration that the retroviral family played a broad and pervasive role in many animal cancers removed any lingering doubt about the oncogenic ability of at least some viruses.

The Oncogene Revelation

The burgeoning cadre of oncogenic retroviruses available by 1970 inarguably proved the innate ability of retroviruses to induce cancers and provided multiple model systems for exploring the molecular basis for this ability. While the transmissibility of both leukemias and solid tumors was indisputable, the underlying mechanism by which a retrovirus could initiate these cancers was still baffling. How could these relatively small viruses with linear RNA genomes elicit such dramatic transformation of normal cells into cancerous ones? Key answers would come from exploring the structure of the viral genomes and deciphering the retroviral replication process, work that led to two Nobel prizes. The replication studies were pioneered by Howard Temin at the University of Wisconsin's McArdle Laboratory for Cancer Research using Rous's sarcoma virus (RSV). Early in the 1960s, Dr. Temin observed that RSV replication required DNA synthesis and host RNA synthesis (transcription). These were puzzling and controversial results for several reasons. First, no known RNA virus made DNA during its genome replication process. Instead, RNA viruses replicate by copying their RNA genomes directly into new copies of RNA, so why would RSV require DNA synthesis? Second, RNA viruses did not use the host transcriptional machinery for viral mRNA production.

RNA viruses utilize their only viral enzymes to make their viral mRNAs, so why the requirement for host RNA synthesis? Additionally, both host DNA replication and transcription occur in the cell nucleus while RNA viruses replicate in the cytoplasm and wouldn't have access to these nuclear processes. These results implied that retroviruses must be distinct from all other RNA viruses and must have a very different reproductive process. It was also known at that time the retroviruses could be maintained in cells in some mysterious, non-replicating form that allowed the virus to persist indefinitely. It was Temin's brilliance that took all this novel and conflicting information and formulated a revolutionary new model to explain retroviral reproduction.

The so-called "central dogma" of molecular biology stipulated that in all life forms there was a unidirectional information flow from DNA to RNA (in the form of mRNA) to protein, a belief championed by James Watson of the DNA double-helix fame. The process was unidirectional because there was no known mechanism to convert protein sequence information back into RNA or RNA sequence information back into DNA. For RNA viruses, the prevailing belief was that they simply skipped the DNA stage and that RNA information went directly into proteins with no need for DNA. Nonetheless, to resolve his experimental data, Temin reasoned that retroviruses must violate the central dogma and go backward from RNA to DNA, a heretical model that ultimately proved correct. He proposed that the incoming viral genomic RNA was first converted to DNA, hence the requirement for DNA synthesis by an RNA virus (Fig. 8.1). This newly made viral DNA copy then somehow traveled to the nucleus where it became integrated into the host chromosome. The integrated viral DNA copy, which Temin called the proviral form, would then be a permanent resident of the host's genome, accounting for the stable persistence of the virus in cells. Finally, from the integrated proviral DNA, the host's transcriptional machinery could synthesize an RNA copy to regenerate the viral RNA genome, thus explaining the necessity for host RNA synthesis. Transcription of the proviral DNA could occur over and over to generate thousands of copies of the RNA genome for packaging into new virions. While this elegant model explained all the experimental observations, the one caveat was the lack of a mechanism to covert the initial viral RNA to the DNA form. However, Temin simply postulated that a viral enzyme must exist that could use RNA as a template to make a DNA copy, an enzyme he dubbed "reverse transcriptase". Subsequently, Temin and David Baltimore of MIT independently discovered the reverse transcriptase enzyme which validated Temin's model. Because retroviruses encode and express this unique reverse transcriptase, they have a distinctive life cycle. Unlike all other viruses, retroviruses not only infect our cells but as part of their normal viral reproductive

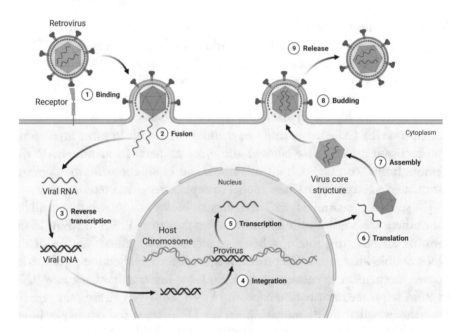

Fig. 8.1   The retroviral life cycle. Retroviruses have a single-stranded RNA genome and their virions are unusual in that each virion contains two copies of the genomic RNA. In addition to the RNA, the virion also carries the viral reverse transcriptase (RT) enzyme. After virion binding to the cell receptor (step 1) and fusion of the viral and cell membranes (step 2), the viral RNA is released into the cytoplasm along with the RT. The RT uses the genomic RNA as a template to produce a double-stranded DNA copy (step 3) while simultaneously degrading the RNA. The double-stranded DNA copy is transported to the nucleus and randomly integrates into a host chromosome (step 4) to generate the proviral form. Once integrated, the provirus is transcribed into RNA using the host cell RNA polymerase (step 5). The newly transcribed RNAs are both translated into new viral protein (step 6) and used as new genomic RNAs for packaging into progeny virions (step 7). The progeny virions bud out of the cell and acquire a membrane envelope from the cell membrane (step 8) and are ultimately released (step 9) to infect other cells. (Created with BioRender.com)

scheme, they also invade our DNA to become permanent passengers in our genomes (see Chap. 9 for more details). Temin and Baltimore shared the 1975 Nobel Prize in Physiology or Medicine for this paradigm-shifting work that became fundamental to understanding retroviral oncogenesis.

While Temin was deciphering the retroviral replication scheme, other researchers were directly tackling the oncogenic mechanism of the virus. Remember, DNA and RNA sequencing hadn't been invented yet and early retrovirologists had to use more circuitous approaches for interrogating viral genomes. A key advance occurred in 1970 when Peter Duesberg and Peter Vogt compared the genome structure of oncogenic RSV with that of a variant

RSV that had lost transforming ability. The non-oncogenic RSV variant was replication-competent and seemed normal in every way except for lacking the ability to cause sarcomas. When the linear genomes were compared, they found that the non-oncogenic variant genome was smaller because it was missing about 20% of the genome from one end, implicating this region as containing the transforming activity. Just as SV40 and other polyomaviruses had a viral gene (encoding T antigen) with transforming ability, it appeared that this sarcoma virus also carried a transforming gene. Eventually, this pre-dicted transforming region from RSV was cloned and its intrinsic transform-ing activity was confirmed; simply introducing this single gene into susceptible cells could induce transformation without any other viral contribution. This first retroviral transforming gene became known as the Src gene (pronounced "sark") in honor of its sarcoma virus origin. As other known sarcoma viruses were examined, they each were found to carry a transforming gene, and the term oncogene was coined to refer to any transforming gene carried by this type of virus (Fig. 8.2). Unexpectedly, the oncogenes from different sarcoma viruses were each unique and were not related. This diversity of oncogenes indicated that there must be many ways to drive cells into the oncogenic state and not just a single pathway.

With the discovery of oncogenes, Howard Temin and others quickly pro-posed an explanation for the aggressive tumorigenicity of sarcoma viruses based on the retroviral life cycle. During infection, the DNA form of the retroviral genome randomly integrates into the DNA of the host cell. Since this integrated proviral DNA contains the oncogene, every infected cell will be producing the oncogene protein, potentially turning it into a cancerous cell. When an animal is infected by one of these sarcoma viruses there are tens of thousands of cells infected within days, and these infected cells are all at risk for transformation. Consequently, transformation should start quickly upon infection, occur independently in many cells, and rapidly give rise to multiple tumors. These predicted events are observed during experimental infections, thus the clinical picture of sarcoma virus cancer is elegantly explained by the viral life cycle and the presence of oncogenes in the genomes of this group of viruses. Still, what the model failed to explain was the actual function of the oncogenes. What are all these diverse oncogenes, what do they do for the virus, where did they come from, and how do they cause transfor-mation? Figuring that out would be the basis for another round of Nobel Prizes.

The answer to the question about the role of oncogenes in the viral life cycle was already implied by the initial RSV studies. Duesberg and Vogt observed that the non-transforming RSV variant without an oncogene was completely functional for infection and reproduction. This implied that the

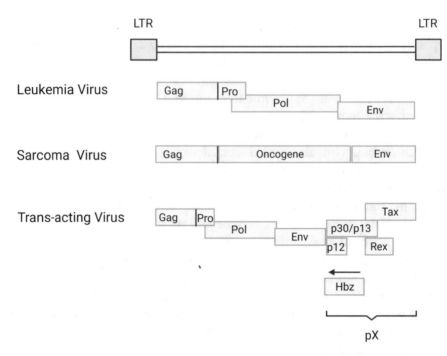

Fig. 8.2 Organization of retroviral genomes. Depicted are the proviral forms that consist of double-stranded DNA encoding the viral genes flanked on both ends with the long terminal repeat (LTR) sequences. The LTRs contain the transcriptional regulatory elements (promoters and enhancers) that control the transcription of the viral genes. Leukemia viruses are wild-type viruses fully capable of replication and reproduction to generate progeny virions. These viral genomes contain the four basic retroviral genes: Gag (capsid proteins), Pro (a protease enzyme), Pol (the reverse transcriptase and integrase), and Env (surface proteins). Sarcoma viruses typically have lost some of the standard four genes and have replaced the missing viral sequences with an oncogene derived from a host proto-oncogene. Because they lack a complete complement of the four essential leukemia virus genes, sarcoma viruses are defective and are unable to reproduce by themselves. Instead, sarcoma viruses only reproduce during a co-infection with a complementary leukemia virus that supplies the missing protein products. The trans-acting viruses are a subfamily of retroviruses known as lentiviruses. The hallmark of lentiviruses is the presence of other genes (indicated in the pX region) in addition to the four standard leukemia virus genes. Shown is the genome organization for a representative lentivirus, HTLV-1. Tax and Hbz are the genes whose protein products are implicated in the transforming activity of this virus. The arrow above the Hbz gene denotes that it is transcribed from the right-end LTR in contrast to all the other viral genes that are transcribed from the left-end LTR. (Created with BioRender.com)

oncogene had no real function for the virus, a conclusion borne out for every other sarcoma virus. It was the answer to the second question about the origin of the oncogenes that was a Nobel Prize-winning revelation. In the 1970s, the evolutionary history of viral genes was unknown, but most scientists assumed that viral genes and host genes were derived from different ancestors and shared little or no relatedness. Thus, it was shocking when Michael Bishop and Harold Varmus showed in 1976 that the *src* gene from RSV was highly homologous with a sequence in the chicken genome. As more and more sarcoma viruses were tested, the oncogene for each virus had a counterpart in the genome of its host animal. This led to the inescapable conclusion that retroviral oncogenes derive from host genes, another dramatic paradigm shift that earned Bishop and Varmus a Nobel Prize in 1989. It turns out that sarcoma viruses are a bit of an aberration. Each sarcoma virus is derived from a wild-type parental retrovirus without an oncogene. On rare occasions during infection, the viral genome will accidentally capture a host gene and incorporate this gene into the viral genome, usually at the cost of losing part of the authentic viral RNA. If this hybrid genome gets packaged into a virion and the captured gene has oncogenic capability then a sarcoma virus is born. Consequently, not only did retroviruses steal host genes, but some of these host genes must have an intrinsic transforming capability, and such genes in the host are now called proto-oncogenes. This realization that animal genomes contain potential transforming genes that are used against the host by the sarcoma virus was a stunning disclosure. To borrow Walt Kelly's famous phrase, "We have met the enemy and he is us".

Within a few years, the oncogenic potential of proto-oncogenes was confirmed with a key twist. By this time a retroviral oncogene called *ras* was identified in the Harvey murine sarcoma virus. Surprisingly, when the *ras* proto-oncogene from normal human cells was tested it did not have transforming activity, while this same gene isolated from human bladder cancer cells was a potent transformer. The important difference was a mutation in the cancer cell version of *ras* that conferred oncogenic activity. As more viral oncogenes and their corresponding cellular proto-oncogenes were compared it became clear that the viral versions all had significant differences from the original cellular genes. The viral versions all either had mutations or lacked critical regions of the wild-type gene. These various changes in the viral oncogenes make their protein products highly active and unregulated compared to the wild-type protein in the cell, which creates a biologically dangerous situation. For example, imagine a water pipe with a faucet on the end. A "wild-type" faucet can be turned to adjust the flow velocity with great precision from completely off to wide-open as needed. But a broken or "mutant" faucet may leak continually at a high rate of flow without any way to shut off or regulate the output. Like the faucet, the proteins made by cellular proto-oncogenes perform needed functions, but these proteins are carefully regulated so that they are only active at the times and in the amounts needed. If they lose regulation through somatic mutation then their constant activity can lead to cellular transformation. Similarly, the oncogenes carried by retroviruses produce unregulated proteins whose constant expression from the proviral DNA drives the cell into the oncogenic state. So, it just comes down to genetics. Intact genes produce healthy proteins whose functions are carefully integrated with the cell's needs for normal maintenance and growth. Mutant genes, whether derived from mutation of our own genes or by the introduction of mutant proto-oncogenes via retroviruses, can disrupt cellular homeostasis and promote uncontrolled growth leading to cancer. Over 100 oncogenes have been identified, confirming that there are many and diverse ways in which cells can go awry.

Understanding how retroviruses carried and delivered oncogenes was an important part of the puzzle but still didn't address the function of each different oncogene or its proto-oncogene counterpart. Sorting out the biology and biochemistry of each oncogene was a long and tedious process, but the collective results were invaluable in advancing our knowledge of cellular function and the oncogenic processes. Our bodies are made of organ systems that are each composed of multiple types of cells. These organs and cells must act in a highly integrated and coordinated fashion to perform their separate functions that collectively keep us alive and healthy. Individual cells are constantly dying through normal senescence and turnover, as well as by injury and illness (viral-mediated cell killing for example). New cells are produced by growth and division to replenish lost cells, and this growth must be carefully regulated to prevent accidental overgrowth or unchecked cancerous hyperproliferation. This implies that organs and cells must communicate to convey the proper information about when and where to grow, but in the 1970s little was known about the molecular or biochemical nature of the intercellular communication process. Oncogenes provided the window into this vast and complex network that we call cell-to-cell signaling. As each unique oncogene was discovered, its cellular homolog was eventually identified and characterized. Different subsets of proto-oncogene proteins were found to act together in functional networks that became known as signaling pathways or cascades. Elucidating the physical and biochemical properties of each proto-oncogene protein revealed its place in the cascade and its functional role in signal transmission. While many discrete signaling pathways exist, these collective studies revealed that the basic conceptual features of signaling are conserved across all of the pathways. These studies also finally demonstrated how retroviral oncogenes usurp these pathways to invoke out-of-control growth.

For each growth signaling pathway, there is a cellular receptor protein that is embedded in the cell membrane with portions of the protein extending both outside and inside the cell. The external portion is the functional receptor that sits poised to receive a signal known as a ligand. Growth ligands are molecules, usually themselves proteins, that are excreted by cells when growth is needed. When excreted, the ligands will diffuse until encountering an appropriate receptor on target cells. In the absence of a ligand, the receptor is inactive and the target cells do not grow and divide. However, once a receptor binds its cognate ligand, the growth signaling pathway is activated to stimulate the reproduction of the target cell. Ligand binding changes the shape of the receptor which induces the segment of the receptor inside the cell to become biochemically active. The active intracellular domain of the receptor interacts with a second protein (or protein complex) inside the cell causing activation of the second protein. The second protein activates a third protein, the third activates a fourth, and so on to transmit the signal along a series of intracellular protein signaling transducers. Eventually, this signal enters the nucleus and activates transcription factors that induce the genes responsible for growth and cell division, causing the cell to reproduce. To prevent runaway growth there are mechanisms built in to stop the cascade and return the cell to normal stasis when reproduction is no longer needed.

We now know that cellular proto-oncogenes encode the components of these signaling pathways. Proto-oncogene proteins can be ligands, receptors, signal transducers, or the final transcription factors that turn on gene expression to reprogram the cell from static mode to growth mode. When a retrovirus introduces its oncogene into a susceptible cell, the oncogene protein functions in the same signaling pathway as its original proto-oncogene progenitor. The critical difference is that the viral oncogene protein is constantly expressed from the proviral DNA. This expression is unregulated and is not deactivated by the normal cell control mechanisms. Consequently, whatever signaling pathway the oncogene is in will become permanently activated. Thus, by acquiring critical growth pathway genes and aberrantly expressing them in infected cells these viruses disrupt the cell's normal growth control. This disruption drives the cell into runaway growth and eventual transformation into a cancer cell that can multiply into a tumor. Unfortunately, this is a powerful though completely unintended mechanism for producing cancer in the host animal. Keep in mind that this transformation activity and resultant cancer production are of no actual importance to the virus. Acquisition of the oncogene is purely inadvertent and its addition to the viral genome adds no advantage to the virus in terms of its reproduction, survival, or transmission. The generation of these sarcoma viruses is just a genetic accident that happens by chance, just as random somatic mutations in critical genes can also convert our cells into cancer. The only positive note is that these viruses gave us the tools to discover and understand growth signaling pathways. This exploration greatly expanded our understanding of both normal cell growth and how disruption of normal growth can be oncogenic.

The final remaining mystery concerned the numerous leukemia viruses. Unlike the sarcoma viruses, the examination of leukemia virus genomes revealed no oncogenes (see Fig. 8.2). Instead, these viral genomes just contained the standard four genes needed for retroviral viability and reproduction called Gag, Pro, Pol, and Env. This finding was not completely unexpected as the cancer presentation with leukemia virus infection is drastically different from that of sarcoma viruses. After animal infection with leukemia viruses, the cancers are slow to develop with all the cancer cells deriving from a single oncogenic transformation. Since every transformed cell in the resultant cancer is derived from that single initial transforming event this is known as a monoclonal tumor. Furthermore, only some infected animals ever develop leukemia, and the majority of animals remain cancer-free. In contrast, sarcomas develop very quickly after infection, multiple tumors arise from independent transformations (polyclonal tumors), and almost every animal infected with a sarcoma virus develops tumors. Furthermore, sarcoma viruses can efficiently transform susceptible cells in culture, while leukemia viruses have little or no transforming activity in cultured cells even though they can cause cancer in the same cells in an animal. Once the oncogene mechanism was established for sarcoma viruses, the more indolent cancer development from the oncogene-free leukemia viruses initially seemed even more puzzling. Remarkably, proto-oncogenes again were found to be at the heart of the cancer-inducing mechanism for leukemia viruses.

The solution to the leukemia virus mystery was solved after examining the sites of integration of the proviral DNA in tumor cells. The proviral DNA integration is a relatively random event meaning that it can insert itself anywhere throughout the host genome. In normal cells infected with retroviruses, the proviral DNA is found in many different genomic locations, but never adjacent to any cancer-associated host genes. In leukemic cells, we observe a completely different finding, and the integration site always places the viral DNA near a cellular proto-oncogene. It was known that proviral sequences contain elements (promoters and enhancers) that drive high levels of viral gene transcription, and it was discovered that these viral transcriptional elements can also inadvertently drive expression from nearby cellular genes. This transcriptional property led to the promoter insertion model to explain leukemia virus transformation. The model stipulates that viral infection leads to random proviral integration in the host cells. The host genome is vast in size compared to the viral genome, and in most host cells the integration event doesn't place the viral transcription elements near any proto-oncogene. In this situation, nothing happens to those cells and they retain their normal phenotype as they go on producing new virions. Alternatively, the proviral DNA may randomly integrate close enough to a proto-oncogene to permanently activate its expression. This event is functionally the same as introducing that active oncogene via a sarcoma virus, and this rare cell with the turned-on proto-oncogene becomes transformed (Fig. 8.3).

The random nature of the proviral integration event that leads to cancer with leukemia viruses explains the cancer presentation in infected animals. Integration next to a proto-oncogene is a very low probability event that never happens in most infected animals so they never develop leukemia; it's like getting hit by lightning, many people are out in storms but few ever get hit. Even in the animals that do eventually develop cancer, it takes a long time to occur

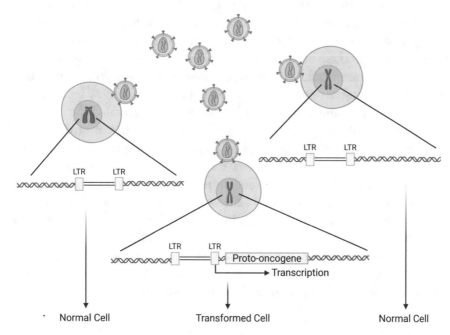

Fig. 8.3 The promoter insertion model of Leukemia virus transformation. When leukemia viruses infect cells they randomly integrate the double-stranded DNA form of their genomes into host chromosomes. When the integration occurs in a harmless region of the chromosomal DNA then there is no damage from the proviral integration and the cells remain normal. However, when integration occurs proximal to a proto-oncogene then the proto-oncogene can be constitutively transcribed into mRNA by the adjacent viral long terminal repeat (LTR). The continuous translation of the proto-oncogene mRNA into the protein product can drive the transformation of this cell. (Created with BioRender.com)

because typically millions of cells must become infected to have any chance of the oncogenic integration event happening. But if that oncogenic integration event does eventually happen in a single cell of the animal then that cell can transform and reproduce to form the leukemia. Since the entire leukemia is derived from that single transformed cell, it is a monoclonal tumor, unlike the polyclonal tumors with sarcoma viruses. Lastly, transformation in culture rarely occurs with leukemia viruses because there are simply too few cells in a culture dish to have much probability of hitting a proto-oncogene with the proviral DNA. Therefore, while sarcoma and leukemia viruses both transform using host genes, their tumor characteristics are quite different. Sarcoma viruses are quite potent carcinogens because their genomes carry the activated oncogenes, while leukemia viruses are much weaker carcinogens because they only transform if they accidentally trigger a host proto-oncogene through nearby insertion.

Unraveling these two mechanisms of retroviral transformation was a major scientific achievement in the twentieth century that extended the field of cancer biology down to the molecular level of individual proteins and their functional pathways. These once-obscure animal retroviruses proved to be the key that unlocked many cellular mysteries, and together the sarcoma and leukemia viruses completely changed our concept of cancer and how it could initiate. The discovery and characterization of oncogenes revealed the molecular underpinnings of cell growth regulation and how disruption of the normal signaling pathways led to uncontrolled proliferation and transformation. We now know that many spontaneous cancers have mutations in a proto-oncogene that constitutively activate the protein, therefore, proto-oncogenes are not just involved in retroviral cancers, but are widely involved in many non-viral cancers. In addition to its importance for cancer, the discovery of the signaling pathways was also critical for understanding many other areas of normal biology and human health. Still, all the work with retroviruses was done in animal models, and the search for elusive human cancer viruses was racing in parallel with the avian and murine retroviral studies.

Human Retroviruses at Last

By the 1970s there was a widespread and diverse array of retroviruses identified in chickens, mice, and other animals although human retroviruses were still unknown. The inability to find comparable retroviruses in humans was a frustrating scientific problem that confounded many researchers. The breakthrough occurred in 1980 when an American scientist, Dr. Robert Gallo, was able to isolate a retrovirus from a human T-cell lymphoma cell line. The virus was much more fastidious for growth in culture compared to the common avian and murine retroviruses which accounted for the previous failures to find human retroviruses. However, Gallo persevered for many years in the quest for human retroviruses and ultimately developed the cell culture conditions that led to the detection of a new virus. The isolate became Human T-Cell Leukemia Virus 1 (HTLV-1) and was eventually accepted as the first authentic human retrovirus discovered.

Over the next several years evidence accumulated that HTLV-1 was truly an oncogenic virus and not simply a passenger in the lymphoma line. Before Gallo's work, a rare, aggressive adult T-cell leukemia (ATL) was described in Japan in the late 1970s. At that time, the limited geographic distribution of the disease suggested a possible indigenous infectious agent, though none could be isolated. Once Gallo isolated HTLV-1, other investigators used his techniques to examine ATL cells and found a virus that ultimately proved to be the same as HTLV-1. Epidemiological work eventually provided convincing data that the HTLV-1 virus was the causative agent of ATL, making it not only the first human retrovirus discovered, but the first identified human oncogenic retrovirus. The virus spreads via breast milk, sexual intercourse, and contaminated blood, leading to a life-long, chronic infection. In the endemic area, breastfeeding appears to be the primary mechanism of the generational persistence of the virus. Once the virus became established in the population, infected mothers could pass it on to their offspring in breast milk with the risk of transmission increasing the longer the infant is breastfed. Once grown, females who were chronically infected as infants again passed the virus to the next generation through breastfeeding to perpetuate the cycle of infection and transmission. Interestingly, as viral reagents became available the distribution of HTLV-1 was found to be much greater than just in Japan. Other clusters of infection occur in central coastal Africa, many countries of the Caribbean basin, Taiwan, and Papua New Guinea, all places where ATL also occurs. In these indigenous areas, the overall population infection rate for HTLV-1 hovers around 5% though many think this is an underestimate due to limited surveillance of the general population in those countries. In regions where this virus is not endemic, like the United States, the prevalence is only 0.018% (about 2 infections per 10,000 people).

As data about this infection and disease accumulated, ATL had a low incidence even among the chronically infected individuals, and when it did occur it was typically seen many decades after the initial infection. This presentation is reminiscent of leukemia viruses, not a sarcoma-type virus with a genomic oncogene, and analysis of the HTLV-1 genome quickly confirmed that an oncogene was not present. Surprisingly, the examination of HTLV-1 proviral DNA in tumor cells failed to find consistent integration near proto-oncogenes. The lack of physical juxtaposition of the viral integrant with a proto-oncogene indicated that the promoter insertion mechanism of leukemia viruses could not be the basis for ATL, implying that there must be a third oncogenic mechanism. Further analysis of the virus revealed a novel gene region termed pX (see Fig. 8.2). The pX region is absent in typical leukemia and sarcoma viruses, and its presence defines a subset of retroviruses known as lentiviruses (HIV also falls into this lentivirus category). pX encodes several small proteins that perform critical functions in the viral reproduction cycle. All but two of the encoded proteins, Tax and HBZ, are also important for the oncogenic activity of the virus. Unlike typical sarcoma virus oncogenes, Tax and HBZ are purely viral genes with no cellular counterparts. Tax and HBZ function by interacting with more than 100 cellular proteins and these interactions alter numerous cellular processes to shape the internal environment to better support viral reproduction. The current model of HTLV-1 transformation postulates that the combined effects of Tax/HBZ on cellular proteins promote genetic damage in the cell. Over decades of chronic infection, this random DNA damage accumulates and mutations with oncogenic capacity may occur giving rise to ATL. Therefore, ATL is a coincidental event in HTLV-1 infected individuals which explains the low incidence and the long lag period between infection and disease. In this model, the virus isn't directly inducing cancer, but instead is creating intracellular conditions that favor the chance of oncogenic mutations arising in an infected cell. Because of its dependence on these purely viral proteins, the novel HTLV-1 oncogenic process was designated as the trans-activating mechanism. Variations of this indirect model where virus-induced genetic damage promotes cancer will prove to be quite common among oncogenic DNA viruses.

In the years following HTLV-1's discovery, three additional related HTLVs were found (HTLV-2, -3, and -4). HTLV-2 spreads via the same routes as HTLV-1 but only causes asymptomatic infections with no specific disease attributed to this virus. HTLV-3 and HTLV-4 are extremely rare and have only been isolated from a few individuals in parts of Africa. Like HTLV-2, neither HTLV-3 nor -4 has been associated with any disease, though both viruses cause persistent infections in humans. All four HTLVs are highly related to simian T-cell leukemia viruses (STLVs) and are believed to have entered human populations from contact with infected monkeys as we saw for HIV. Due to their extreme rarity, some researchers speculate that the occasional HTLV-3 and -4 cases may just represent recent exposures by those individuals to the corresponding simian viruses rather than true human retroviruses. Consequently, the current cadre of true human transmittable retroviruses is four, HTLV-1 and -2 along with the recently acquired HIV-1 and -2 (see Chap. 6). So unlike chickens and mice, humans have a paucity of retroviruses, and neither a sarcoma nor a leukemia-type retrovirus has ever been found in humans, just the trans-activating HTLV-1. While not a huge contributor to the human cancer burden, HTLV-1 remains unique as the only human oncogenic retrovirus. This rarity is fortunate as there is no cure for HTLV-1 infections and no treatment for ATL.

Herpesviruses Turn Oncogenic

The history of retroviruses extends back for decades, and these viruses were critically important to our understanding of cellular growth, cell-to-cell signaling, and oncogenesis. However, long before the discovery of HTLV-1, a common DNA virus claimed the title as the first human oncovirus. The eponymous Epstein-Barr virus, named after Drs. Epstein and Barr, is a widely ubiquitous herpesvirus that is associated with several malignancies, most commonly Burkitt's lymphoma. Burkitt's lymphoma was first identified and characterized by Dr. Denis Burkitt, a British physician living and working in Uganda post-World War II. Burkitt noted that many children in the malarial belt of Africa developed aggressive lymphatic tumors of the head and neck. Reporting his findings in 1958 he speculated that these tumors might be of viral origin and spread via the same mosquitos that carried malaria. Although mosquito-borne transmission of the virus proved incorrect, the tumors eventually bore his name and were found to occur worldwide.

Several years after his initial report on the lymphatic tumors, Dr. Burkitt was lecturing about his findings on a trip to London. His talk was particularly exciting for Anthony Epstein who was working on the Rous sarcoma virus and found the possibility of a human tumor virus intriguing. After arranging for Burkitt to send him tissue samples, Epstein applied a new technology, electron microscopy, to look for viruses in the samples. Electron microscopy was a powerful new tool that allowed the visualization of objects roughly a thousand times smaller than what could be seen with standard light microscopy. Now the formerly invisible world of viral particles could finally be revealed, and the shape and size of individual virions could be observed at last. The disadvantage of this technique is that the field size, the area of the sample in view at any moment, is exquisitely small. Consequently, unless the number of viral particles in a clinical sample is large, finding the virions is a little like searching for a needle in the proverbial haystack. Epstein persevered for several years without success until he managed to grow tumor cells in culture. These cultured cells now produced sufficient quantities of viruses for easy detection by electron microscopy. Epstein and his collaborator, Yvonne Barr, finally visualized the viral culprit in 1964. This Epstein-Barr virus (EBV) was quickly shown to be a new human virus that was eventually classified as a herpesvirus (now also called HHV4 for human herpesvirus 4).

With the virus in hand, reagents were developed to track infection and two surprising results were quickly obtained. First, most people had antibodies against EBV indicating that these infections were extremely common; we now know that around 90% of adults globally have been infected. Second, EBV was proven to be the causative agent for a common malady, infectious mononucleosis, often called mono for short. Mono presents with fever, fatigue, head and body aches, a sore throat, and swollen lymph nodes in the neck and armpits. This constellation of symptoms was first described in the medical literature in the late 1800s, and its current name of infectious mononucleosis was proposed in 1920, yet by the 1960s the cause of this disease had not been identified. The breakthrough that made the EBV-mono connection was an unfortunate though fortuitous laboratory mishap. In 1967, a research technician working with EBV developed mono. Prior to the disease, he was negative for antibodies to EBV, but during the disease he seroconverted and developed anti-EBV antibodies, strongly suggesting that his disease was related to EBV. Once this connection was made, epidemiological studies soon confirmed that infectious mononucleosis was indeed the result of EBV infection. The virus is shed in saliva where it is often transmitted and acquired by teenagers and young adults through kissing or by sharing food or drinks. Interestingly, most EBV infections are asymptomatic or very mild, especially when the virus is contracted in infancy or early childhood from infected mothers. It is only when the virus is contracted during adolescence or later that the infection sometimes manifests as mono. The absence of disease following most infections explains why 90% of adults have been infected but only a much smaller number ever presented with mono.

EBV, like all the other herpesviruses, is a large DNA virus encoding between 50 and 100 proteins. This large number of proteins makes deciphering the virus's functions and abilities a complex enterprise, and we still have only a partial understanding of how it causes disease. We do know that the primary target cells for this virus are B lymphocytes, the immune cells that produce antibodies. In most infected B cells, EBV undergoes a typical productive replication cycle with the generation of new virions. These productively infected cells multiply and are responsible for causing disease. However, in a small number of infected cells, the virus establishes a permanent, latent infection with the viral DNA persisting in the cell nucleus. These latently infected B cells become immortalized and can remain present for the life of the infected individual. During acute mono, there is a transient increase in productively infected B cells as the virus induces these cells to proliferate. The B cell expansion is followed by a large increase in T cells that try to kill off these proliferating B cells. Eventually, the immune system shuts down viral reproduction and kills off most infected B cells. However, a few latently infected B cells escape immune destruction and remain circulating forever. In immunocompetent people, this small number of infected B cells simply remains with no adverse consequences, but with immunosuppression these EBV-B cells may expand and give rise to Burkitt's lymphoma or related B cell cancers.

Given this normal presentation of EBV infection, why then was Burkitt's lymphoma more common in EBV-infected youngsters in the malarial regions of Africa? While still lacking in some details, the explanation appears to be that EBV transforms indirectly as we saw for HTLV-1. Children infected with EBV have circulating B cells containing the viral DNA. If these individuals contract malaria there is a massive immune response to the malarial parasite. This response stimulates B cell proliferation, including the B cells containing the EBV DNA, giving rise to many more EBV-infected B cells than are normally present. Like the HTLV-1 Tax protein, certain EBV proteins promote genomic instability of the host cell DNA, including events that alter large regions of chromosomes. So, as this large cadre of EBV-B cells proliferate they accumulated random DNA damage as do HTLV-1 infected cells. In most cells nothing harmful transpires, but if one specific event occurs then that cell becomes transformed and proliferates into Burkitt's lymphoma. We know the actual event, a chromosomal translocation that switches segments of DNA between two different chromosomes. The specific translocation giving rise to Burkitt's lymphoma exchanges one end of chromosome number 8 with the corresponding end of chromosome number 14. The importance of this translocation is due to a proto-oncogene named *c-myc* that is present on the transferred region of chromosome 8. Moving the *c-myc* gene to chromosome 14 places it under the control of a strongly active transcriptional unit where *c-myc* expression is not regulated appropriately and the protein is constitutively produced in this cell. As for other proto-oncogenes, permanent expression of *c-myc* protein drives the cells further into transformation and cancer development. Since this specific translocation event is random, it may or may not happen in any one individual, so many children with EBV and malaria never develop lymphoma. Nonetheless, because of the DNA-damage promoting effect of EBV and the proliferative effect of malaria on B cells, the odds of this carcinogenic event are much higher in the dually infected population. Sadly, many children do develop a malignancy. In this model, EBV is not directly oncogenic, but like HTLV-1 simply promotes a cellular environment where greater than normal amounts of DNA damage accumulate. A similar scenario likely occurs in anyone carrying EBV who becomes immunosuppressed. As their EBV-B cell population expands, the odds of the *c-myc* translocation happening increases, giving rise to Burkitt's lymphoma in some unfortunate individuals. Even in immunocompetent people, with or without EBV, the Burkitt-inducing translocation can randomly happen at a low frequency in any B cell to give rise to the spontaneous Burkitt's lymphomas that occur anywhere in the world. Although this cancer isn't confined to EBV-infected people, its development is greatly facilitated by the presence of EBV along with conditions that promote B cell proliferation. Because of this important role in Burkitt lymphoma, EBV clearly deserves its status as the first discovered human oncovirus.

EBV remained the only human oncogenic herpesvirus until the AIDs epidemic exploded in the 1980s. Among the many ailments arising in patients with AIDS was a formerly rare skin cancer called Kaposi's sarcoma (KS). Before HIV, KS was seen primarily in older men of Mediterranean, Middle Eastern, or Eastern European descent where it was an indolent cancer that was rarely life-threatening. Surprisingly, even a hundred years after its original description in1872 there was nothing known about its etiology. Suddenly, many of these HIV-infected young men were developing this skin cancer in a much more serious and aggressive form, stimulating intense study into possible causes. Still, it took nearly 10 years before researchers finally identified a herpesvirus in tissues from a KS lesion. The new isolate shared some distant relatedness to EBV but was clearly distinct. Multiple studies quickly confirmed that this virus was present in the KS tumors from patients with AIDS and played a causal role in the disease development, and accordingly, it was named the Kaposi's sarcoma herpesvirus (KSHV). As the eighth human herpesvirus discovered, it is also called HHV8. Like its distant relative, EBV, KSHV is primarily spread in saliva, though sexual activity and contaminated blood are also sources of infection. The initial infection may be asymptomatic, but can also have associated fever, rash, and fatigue, though nothing so specific, alarming, or prolonged as to raise much concern in most individuals.

Another similarity between KSHV and EBV is the discordance between the incidence of infection and the prevalence of KS. Estimates place the average worldwide incidence of infection between 5% and 20% of the population, with certain geographic hotspots approaching 50%, yet KS remains a very rare cancer in immunocompetent individuals. As with EBV, the low cancer rate with KSHV reflects the very indirect and poorly understood mechanisms of cancer initiation for both viruses. KSHV expresses at least a dozen viral proteins that disrupt the normal biology of infected cells by stimulating growth signaling pathways, suppressing apoptosis, evading the immune system, fostering cellular immortalization, and promoting genetic instability. In immunocompetent people, the viral proteins suffice to keep KSHV persistent at a low level throughout the person's lifetime but are rarely enough to cause cancer development. In the immunosuppressed AIDS patients, there is a higher KSHV load and more KSHV protein expression that seems to allow a greater chance for viral-transformed cells to arise and become tumors. As with EBV, transformation by KSHV is a chance event and not driven by a strong oncogene. Since the transforming event only occurs with a low frequency, not even every AIDS patient with KSHV develops KS. Similarly, anyone who is KHSV positive and immunosuppressed because of organ transplants or other medical conditions is also at risk for KS. Even the situation in elderly individuals in endemic regions may involve immunosuppression. As we age there is generally a decline in immune function. In KSHV positive individuals, if their immune function falls beneath a protective threshold it may allow increased viral replication. The more active virus leads to a greater risk of transformation similar to that seen in the frankly immunocompromised patients. Whether or not there are other genetic or environmental factors associated with KS cases has not been clearly defined in these endemic populations.

Based on the evidence to date, cancer induction for both EBV and KSHV is a rare byproduct of each virus's manipulation of the cellular environment. Both are persistent viruses that promote cell proliferation and impair the DNA damage responses in the host cells, a combination that leads to a gradual accumulation of DNA damage and mutation to a greater degree than in uninfected individuals. Nonetheless, since specific mutations that invoke transformation occur randomly, such mutations may or may not ever occur in the lifetime of any given infected individual, allowing most infected people to remain cancer-free. Even if transforming events do occur, many transformed cells are simply destroyed by a healthy immune system. However, if immune function dwindles due to disease, medication, or advancing age then the risk of transformed cells surviving to form tumors increases. Unfortunately, since much of the world's population is chronically infected with one or both viruses, the risk for these two cancers will remain until we find cures or preventative vaccines for these herpesviruses.

Hepatitis Viruses and Hepatocellular Carcinoma

There are five well-documented human hepatitis viruses, A-E. Each one causes or contributes to liver infections that manifest as clinical hepatitis characterized by fatigue, jaundice, and abdominal pain among other symptoms. Surprisingly, although these viruses all infect the same organ and cause diseases with similar clinical symptoms, none of these five viruses are related and each belongs to a different viral family. Hepatitis B virus (HBV) and hepatitis C virus (HCV) can both establish chronic liver infections and are the two of the five hepatitis viruses with oncogenic potential. HBV is a DNA virus with a small, circular, partially double-stranded genome (Family *Hepadnaviridae*) while HCV is a linear, single-stranded, positive-sense RNA virus (Family *Flaviviridae*).

The origins of HBV are hazy and still controversial among scientists. For many years the accepted dogma was that HBV entered human populations relatively recently, on the order of hundreds to a few thousand years ago. This position is challenged by sophisticated phylogenetic sequence analysis of human HBVs and related viruses in animals. HBV-like viruses are ancient and have a genetic history that extends back tens of millions of years. Primate HBV-like viruses likely arose around 78 million years ago, although the exact time when the progenitor virus entered the modern human lineage and became HBV is still uncertain. The recovery of HBV DNA in 7000-year-old human remains is consistent with this much older introduction into humans, but still leaves a huge gap in the evolutionary history of the HBV-human relationship. Regardless of when this virus first entered humans, our discovery of HBV only occurred after a long and frustrating search for the agent. To even further complicate matters, by the late 1940s it was apparent that at least two agents, presumably viruses, caused hepatitis. One agent was associated with fecal contamination of food or water and was designated Hepatitis A, while the other agent, which became known as Hepatitis B, was associated with contaminated blood and was referred to as "serum hepatitis". Another key difference between these agents was that the B agent could cause chronic liver abnormalities after recovery from acute disease and the A agent did not.

Hepatitis B virus was finally identified after a series of serendipitous events beginning in 1964. In that year a geneticist, Dr. Barry Blumberg, discovered a new blood protein in a sample from an Australian aborigine. Blumberg and collaborators went on to show that this blood protein, dubbed the Australia antigen (Au), was present to different degrees in human populations from different geographical locations. As they surveyed different population groups, they made a strange finding that infants often had a much lower incidence of the Au than did adults. If Au was a human blood protein then it should be present at birth and the incidence shouldn't change with age. The fact that adults more often displayed Au was the first inkling that Au might not be an innate human protein but instead was something acquired during life. Ultimately it was shown that Au wasn't a human protein at all and instead was the surface protein from HBV. The original aboriginal sample was likely from a chronically infected individual with virions in their blood. Other Au positive blood samples from around the world reflected the incidence of HBV infection in different population groups, not genetic variations in the prevalence of a human antigen.

The next fortuitous event was a laboratory accident reminiscent of the discovery of EBV. One of Blumberg's technicians who worked with Au developed hepatitis symptoms and seroconverted to become Au positive – so much for laboratory biosafety in the era before the SV40 concerns. This observation inspired numerous studies that included transmission experiments with human subjects. The transmission work involved the administration of sera or filtrates from Au positive individuals into healthy, Au negative recipients to see if the recipients became ill and Au positive. Horrifyingly, most of these experiments were conducted with soldiers, prison inmates, or institutionalized patients with mental illness or with intellectual disabilities. Such work would not be allowable under today's more stringent protection for human subject rights and safety but was instrumental in proving that Au was truly associated with serum hepatitis B. Other studies used electron microscopy to reveal virions in samples purified from Au positive serum, finally confirming that a viral agent existed. Collectively this work indisputably established that Hepatitis B was caused by a new virus, and HBV entered the realm of human viruses. Dr. Blumberg was a co-recipient of the 1976 Nobel Prize in Physiology or Medicine for his seminal work that lead to the discovery of HBV and its role in hepatitis B.

With the virus and reagents in hand, a clearer picture of HBV infections soon emerged. Besides blood, the virus is found in many bodily fluids including saliva, tears, semen, and vaginal secretions. Infected mothers can easily pass it to their infants during birth while adults typically acquire it through sexual activity or exposure to contaminated blood (for example drug users sharing needles). Once contracted, HBV can cause either acute or chronic infections depending on the age at which infection occurs. Initial infection as an adult typically results in acute hepatitis with complete recovery and elimination of the virus in 95% of patients, with only around 5% of adults developing a chronic, life-long HBV infection. In contrast, infection in early infancy leads to chronic infections in nearly 90% of the cases, while exposure in later childhood yields a rate of chronic infections around 20%. Acute cases have no long-term issues, but chronic infection predisposes to liver disease and a type of liver cancer known as hepatocellular carcinoma (HCC). The mechanism of oncogenesis by HBV is again indirect and depends on the long-term effect of viral protein expression. The primary culprits are a viral protein known as HBx and the virion surface protein that was formerly called Au and is now known as the S protein. Both HBx and S interact with cellular proteins to drive cellular proliferation, disrupt growth control mechanisms, and promote genetic damage to accumulate in the host cell DNA. Additionally, chronic HBV infection produces continuous long-term inflammation in the liver that also promotes tumorigenesis. As for the oncogenic herpesviruses, these HBV effects create an environment where oncogenic mutations are more likely to accumulate over time, thus these chronically infected people have a 100 to 300-fold higher risk for liver cancer than does a comparable uninfected person. While there is still no broadly effective treatment for chronic HBV, there is a very safe and effective vaccine to prevent these infections. Since the vaccine was introduced in 1981, the incidence of chronic HBV infections and HBV-associated HCC cancer is drastically reduced in vaccinated populations around the world. The good news is that since there are no non-human reservoirs of HBV, extensive vaccination could someday purge this virus from the world as was done for smallpox and soon for poliovirus. Eradication of HBV would eliminate all the HBV-associated liver disease and cancer, saving nearly 900,000 lives per year.

After HBV and the non-oncogenic HAV were discovered it became clear that some cases of viral hepatitis were caused by neither of these viruses, leading to the adoption of the term non-A, non-B (NANB) hepatitis. Eventually, NANB hepatitis was ascribed to a third virus designated hepatitis C virus (HCV). As detailed in Chap. 4, the discovery of HCV occurred in 1989 through sophisticated molecular biology. HCV is an RNA virus and is completely distinct from HBV, yet there are many similarities between the two viruses. Transmission is via the same routes, both viruses can cause acute hepatitis, and each can establish chronic infections in the liver. Unlike HBV, HCV infections are more likely to become chronic in adults than in infants or children. Many adults who contracted HCV have such mild symptoms that they are not diagnosed and do not realize that they have become chronically infected. And while HCV infections are less prevalent than HBV infections, the resulting HCV chronic infections also predispose patients to liver disease and HCC. Once again, cancer initiation involves multiple viral proteins (NS3, NS5A, and C) that combine to disrupt cellular homeostasis by stimulating proliferation, promoting DNA damage, and restricting DNA damage repair systems. Chronic inflammation in response to persistent HCV infection also occurs. The role of the inflammatory response in HCV liver cancer is likely even more important than it is for HBV as there is a direct connection between the degree of inflammatory damage and the risk for cancer in patients with HCV. Overall, the oncogenic mechanisms of HBV and HCV are conceptually similar even though the specific details are much different. For both viruses, there are viral proteins that bind to numerous cellular proteins to alter the function of these cellular proteins. Coupled with the negative effects of virus-induced inflammation, the result is increased DNA damage. Over time, the accumulation of small mutations and chromosomal aberrations increases the risk of oncogenic transformation in the cells of these chronically infected persons, putting them at a much higher risk for liver cancer.

A preventative vaccine for HCV remains elusive, so reducing infections and corresponding liver cancer through this approach is not yet possible. Fortunately, a large group of effective anti-HCV drugs was developed in the last 10 years. Cure rates for chronic infections now approach 90%, a much better success rate than with the older interferon-based therapies. While drug therapy is more difficult and costly to implement than a vaccine, the availability of these drugs should impact HCV liver disease and cancer in countries where drug therapy is possible. Like HBV liver cancer, the next decades should see a greatly reduced incidence of HCV-associated hepatocellular carcinoma as our assault on viral-related cancers continues.

Human Papillomaviruses – Cancer as an STD

As presented in Chap. 3, human papillomaviruses (HPVs) are an ancient infection that has co-evolved with humans since the early primate times. These small DNA viruses are mostly innocuous and usually cause inapparent infections in our skin. They spread by direct person-to-person skin contact and start being acquired at birth from our mothers. Occasionally they manifest as benign warts and more rarely can develop into cancers of the epithelium (the outer skin and the cells in our oral and genital regions). Cervical cancer is the most common HPV-associated cancer, but anal cancer, some oral cancers, and certain skin cancers are also the result of these infections. Collectively, HPV-associated cancers represent the largest group of all virus-related cancers worldwide and are estimated to cause 4–5% of total cancer cases.

The speculation that cervical cancer had an infectious basis emerged well over 100 years ago. By the late 1800s and early 1900s, the association of cervical cancer with sexual activity was becoming well established. Early epidemiological investigations revealed that cervical cancer rates were very high in sex workers while the disease was almost nonexistent in nuns. Subsequent studies noted that women with multiple sexual partners or who were married to men with previous multiple partners had greater than average rates of cervical cancer. These correlative studies suggested a possible sexually transmitted agent, but identifying the agent was beyond the capabilities of science and medicine in that era. Over the first half of the twentieth century, as various bacterial agents were isolated for sexually transmitted diseases (STDs), many of these bacteria were examined for a role in cervical cancer without success. Later in the 1960s and 1970s, there was a flurry of studies looking at the herpes simplex virus (HSV), a known viral STD that caused oral and genital lesions. While initially promising, HSV eventually proved not to be the culprit in cervical cancer. As the possible HSV connection with cervical cancer floundered, Harald zur Hausen, a young German virologist, was proposing a bold new hypothesis that an unknown papillomavirus might be the causative agent of cervical cancer.

Dr. zur Hausen trained in both medicine and virology. As part of his training, he spent time in the United States where he participated in studies showing that the Epstein-Barr virus (EBV) had transforming ability. This EBV work influenced his thinking, and Dr. zur Hausen returned to Germany with a distinct appreciation that viruses could have oncogenic activity. Once in Germany, he encountered reports of women with genital warts who subsequently developed cervical cancer. Being familiar with Shope's work on rabbit papillomaviruses and their oncogenic potential, zur Hausen reasoned that an undiscovered human papillomavirus might play a role in cervical cancer. Eventually, his research group was able to isolate HPVs 16 and 18 that together are responsible for 75% of cervical cancer cases. The remaining 25% of cases are caused by several additional oncogenic HPVs, thus cervical cancer is almost exclusively the result of infection by this viral family. Interestingly, over 200 HPVs have been found but only 15 are highly oncogenic. The remaining HPVs just cause benign warts, although the oncogenic potential of a few of these is still uncertain. Exactly what distinguishes the oncogenic types from the non-oncogenic types is still unclear as all of these viruses have similar genomes and express similar proteins. It is likely that the proteins from the oncogenic HPVs differ in quality or activity from their non-oncogenic counterparts, and that these subtle differences determine whether a particular HPV has oncogenic potential.

For HPV to cause cervical cancer or other types of cancer, the viral infection must be persistent. Many infected individuals eventually resolve the infections and become virus-free, but for unknown reasons, the virus can establish chronic infections that persist for decades in some individuals. Even in these persistent infections, transformation does not regularly occur without a second event where the viral DNA accidentally integrates into the cellular DNA. For the oncogenic HPVs, three of the early viral proteins, E5, E6, and E7 are important in transformation, with E6 and E7 being the primary viral oncogenes. As with the hepatitis B virus HBx and S proteins, each of the HPV oncoproteins makes interactions with numerous cellular proteins that drive the cells to proliferate. E6 and E7 can both induce host genomic instability and increased mutation rates that can ultimately result in transformation. However, during normal persistence, the virus regulates the E6 and E7 protein expression to keep their amounts in the cell at low quantities that are insufficient to drive significant DNA damage and transformation. Transformation only occurs when the virus DNA inadvertently integrates into a chromosome in such a way that disrupts the regulation of E6 and E7. Without regulation, the amounts of the E6 and E7 proteins rise intracellularly and reach levels that promote sufficient DNA damage to facilitate transformation. Because both the viral DNA integration and the transforming mutations in the cell DNA are random events, only a subset of persistently infected women will ever develop cervical cancer. As with the other DNA viruses in this chapter, oncogenic HPVs are creating cellular conditions that can lead to transformation and malignant tumors; without persistent HPV the odds of women ever developing cervical cancer approach zero. Fortuitously, it proved relatively easy to develop a protective vaccine against the major oncogenic HPVs. First introduced in 2006 for protection against HPVs 16 and 18, the current HPV vaccine now protects against the nine most common oncogenic HPVs. In only 32 years, Dr. zur Hausen's prophetic hypothesis was proven correct, the specific oncogenic papillomaviruses were identified, and a protective vaccine was developed. For starting this research field, finding the most important oncogenic HPVs, and shepherding the entire field through its modest beginnings, Dr. zur Hausen shared the 2008 Nobel Prize in Physiology or Medicine.

Merkel Polyomavirus – Finally an Oncogenic Human Polyomavirus

Although it was not discovered until 2008, in many ways the history of the Merkel cell polyomavirus started in the 1950s with a persistent scientist. Dr. Sarah Stewart was a pioneering woman in science who earned her Ph.D. in microbiology from the University of Chicago in 1939 and in 1949 became the first woman to earn an M.D. from Georgetown University Medical

School. Fascinated with the idea of cancer-causing viruses she began working with the mouse leukemia virus in 1951 after she took a position at the National Cancer Institute in Baltimore. Ludwick Gross in New York had recently identified a mouse leukemia virus, one that eventually bore his name, by doing transfer experiments. Stewart and her collaborator, Beatrice Eddy, tried to repeat Gross's leukemia experiments using filtrates made from ground-up organs from leukemic mice. To their surprise, when injected into recipient mice the mice developed solid tumors of the parotid glands rather than leukemias. By passaging these mouse extracts in cultured monkey cells and then back through mouse embryos they eventually obtained a cell-free extract containing a high transforming activity that was presumably a new virus. Since this new virus could cause over 20 types of tumors in mice it was dubbed the polyomavirus, polyoma meaning "many tumors". In only a few short years the SV40 virus was found in the contaminated polio vaccines. SV40 was also shown to be a polyomavirus with transforming ability although its role in human cancers remains disputed (see Chap. 7). With both a murine and primate polyomavirus in hand, the race was on to understand their transforming mechanisms and to find other polyomaviruses.

It would take another decade, but in 1971 the first two human polyomaviruses were identified. The JCPyV and BKPyV viruses were both isolated from immunocompromised individuals and were named using the initials of these respective patients. We now know that both viruses are extremely widespread with 70–90% of adults showing infection. Both viruses can establish permanent persistent infections, but fortunately they rarely cause disease in immunocompetent individuals and are not considered cancer-causing. Work on human polyomaviruses (hPyVs) languished for 30 years as no new human members of this viral family were discovered, and the field only began to reinvigorate in the early 2000s. Since then 12 more hPyVs have been identified, although only one is considered oncogenic. The fifth overall hPyV was identified by complex molecular analysis of cells from Merkel cell carcinomas (MCC). MCC is a rare and highly aggressive form of skin cancer with a mortality rate of around 30%. Because MCC was more common in the elderly and the immunosuppressed there was suspicion that it might have a viral etiology like KSHV. The suspicion was correct and it turned out that MCC cells harbor a novel hPyV that is now called Merkel cell polyomavirus (MCPyV).

Like the JC and BC viruses, MCPyV is quite common with incidence rates of 40–80% on the skin from healthy individuals. By age five, 20–40% of children are seropositive indicating that they have been infected, and the percentage of seropositive individuals increases with age. Transmission is likely by direct skin-to-skin contact, though other forms of spread have not been ruled out. As for several of the other cancer viruses, the oncogenic disease with

MCPyV is a rare occurrence and is usually associated with some level of immunosuppression from age, disease, or medications.

Mechanistically, MCPyV is a small DNA virus that depends on its T antigens for transformation as do all the polyomaviruses. For MCPyV both the large and small T antigens appear to contribute to transformation though many details about the process remain unknown. One salient observation is that in MCC tumor cells the viral DNA is integrated into the host genome and the integrated viral genome is mutated and only expresses a truncated large T protein. There is no clarity still on why a truncated T antigen is critical to tumor formation, but the apparent need for these two events, viral DNA integration and large T truncation, may explain the rarity of MCC. In most MCPyV-positive people, integration and/or T gene mutation never occur so they remain healthy and unaffected by the MCPyV. Transformation only initiates in unfortunate individuals who acquire both random events in the same cell. The risk of the double event is small, but it does manifest in an unlucky few during decades of persistent infection. Thus, while not a common human cancer that contributes large numbers of cases each year, MCPyV still counts as one of the seven deadly human cancer viruses.

The Seven Deadly Cancer Viruses

The seven cancer viruses presented in this chapter remain major contributors to the human cancer burden worldwide. These seven viruses, and perhaps others yet to be discovered, are the causes of much misery, pain, and cancer death. Nonetheless, the lessons learned from these oncoviruses have had great scientific and medical impacts. The study of these viruses expanded our understanding of the fundamental events that disrupt normal cell growth and lead to transformation with subsequent tumor formation. A common theme emerging from this work is that viruses seek to modify the cellular environment to promote conditions that favor viral replication and persistence. Most of these alterations are mediated through viral proteins that interact with cellular proteins to subvert the host's normal protein function. While beneficial to the virus, there can be significant collateral damage caused by these virus-induced changes. The infected cells typically have increased proliferation along with more DNA damage and/or less ability to repair normally acquired DNA damage. Over long periods of viral persistence, even small effects lead to increased accumulation of mutations and chromosomal aberrations in infected individuals compared to uninfected people. This increase in DNA abnormalities puts these virus-positive individuals at greater risk for the development of transforming events that initiate a cancer. Of course, other environmental exposures and personal genetics both make contributions to the overall risk, but the virus is the critical promoting factor. So even if the effects of these seven human oncoviruses are mostly indirect, the bad news is that you can literally "catch" cancer through contracting one of these viruses. Conversely, if we can someday cure each of these infections with antiviral drugs or prevent these infections with vaccines, then all the virus-associated human cancers will disappear, and a substantial portion of the total human cancer burden will be eliminated.

Additional Reading

1. The History of Tumor Virology. R.T. Javier and J.S. Butel. Cancer Research 68:7693–7706. 2008.
2. Viruses and Human Cancers: a Long Road of Discovery of Molecular Paradigms. M.K. White, J.S. Pagano, and K. Khalili. Clinical Microbiology Reviews 27:463–481. 2014.
3. Human Oncogenic Viruses: Nature and Discovery. Y. Chang, P.S. Moore, and R.A. Weiss. Philosophical Transactions 372: 20160264. 2017.

Definitions

AKR – An inbred mouse strain.

Australia antigen (AU) – Originally believed to be a human blood group antigen that turned out to be the hepatitis B virus S protein.

Avian leukosis virus – A retrovirus that causes a B-cell lymphoma in chickens.

Burkitt's lymphoma – A B-cell lymphoma in humans that can be caused by the Epstein-Barr virus.

Cervical cancer – A cancer of the cervix associated with human papillomavirus infection.

C-myc – A cellular proto-oncogene located on human chromosome 8.

Cognate ligand – The specific ligand that matches with a particular receptor.

Epithelium – A type of tissue the covers our body surface as well as lining many internal structures.

Erythroid – Relating to red blood cells or the precursors to these cells.

Friend murine leukemia virus – A type of retrovirus causing leukemia in mice that was discovered by Charlotte Friend.

Gross leukemia virus – A type of retrovirus causing leukemia in mice that was discovered by Ludwik Gross.

HBx – The hepatitis B virus X protein.

HBZ – A regulatory protein expressed by HTLV-1.

Hepadnaviridae – The taxonomic family of double-stranded DNA viruses that infect the liver and replicate using reverse transcriptase.

(continued)

(continued)

Hepatocellular carcinoma (HCC) – The most common type of liver cancer that arises from transformed hepatocyte cells.

Herpes simplex virus (HSV) – A member of the herpesvirus family that causes oral and genital lesions.

Human herpesvirus 4 (HHV4) – A double-stranded DNA virus of the herpesvirus family that is also known as Epstein-Barr virus.

Human herpesvirus 8 (HHV8) – A double-stranded DNA virus of the herpesvirus family that is also known as Kaposi's sarcoma herpesvirus.

Infectious mononucleosis – A disease, typically seen in teenagers, that is characterized by an increase in white blood cells and is caused by Epstein-Barr virus infection.

Kaposi's sarcoma herpesvirus (KSHV) – A double-stranded DNA virus of the herpesvirus family that causes Kaposi's sarcoma.

Leukemia – A cancer of the blood-forming cells. There are many types of leukemia depending on exactly which type of blood cell is transformed.

Long terminal repeats (LTRs) – Transcriptional regulatory sequences (promoter/ enhancer) that are present on both ends of the proviral form of retroviruses.

Lymphoid – Relating to lymphocytes or the tissues that produce lymphocytes.

Malaria – A febrile illness caused by the plasmodium parasite that is transmitted to humans via mosquitoes.

Merkel cell polyomavirus (MCPyV) – A double-stranded, circular, DNA virus of the polyomavirus family that is associated with the development of Merkel cell carcinoma.

Moloney murine sarcoma virus (MoMLV) – A type of retrovirus causing sarcomas in mice that was discovered by John Moloney.

Monoclonal tumor – A tumor that develops from a single transformed cell.

Mouse mammary tumor virus (MMTV) – A type of retrovirus transmitted in breast milk that causes breast cancer.

Myeloid – Relating to the bone marrow or blood-forming cells present in the bone marrow.

NANB – Non-A, non-B hepatitis, that is now known to be caused by hepatitis C virus.

Oncovirus – Any virus with transforming or tumorigenic activity.

Parotid glands – The salivary glands that reside in front of the ears.

Polyclonal tumor – A tumor derived from multiple transformed cells.

Promoter – A sequence element in DNA that is located adjacent to genes and which is involved in regulating the transcription of genes.

Proto-oncogene – A cellular gene that encodes a protein that can have oncogenic potential when inappropriately expressed or regulated.

pX – A gene region present in HTLV-1 that is involved in the oncogenic ability of this virus.

Ras – The oncogene carried by the Harvey murine sarcoma virus.

Rous Sarcoma Virus (RSV) – A type of retrovirus causing sarcomas in chickens that was discovered by Peyton Rous.

S protein – The surface protein of the hepatitis B virus.

Sarcoma – A solid tumor that develops in any tissue other than the epithelium.

Sera – The plural of serum, a yellowish fluid that remains after blood coagulates.

(continued)

(continued)

Shope papillomavirus – A double-stranded, circular DNA virus of the papilloma-virus family that causes skin lesions in rabbits and which was discovered by Richard Shope.

Simian T-cell leukemia virus (STLV) – A retrovirus of monkeys that is related to the human T-cell leukemia virus.

Somatic mutations – A change in the DNA sequence that occurs in any cell of the body except the germ cells (eggs or sperm).

Tax – A transcriptional regulatory protein expressed by HIV-1.

Abbreviations

ATL – adult T-cell leukemia
Au – Australia antigen
EBV – Epstein-Barr virus
HBx – hepatitis B virus X protein
HBZ – HTLV-1 bZIP protein
HCC – hepatocellular carcinoma
HHV – human herpesvirus
hPyV – human polyomavirus
HSV – herpes simplex virus
KSHV – Kaposi's sarcoma herpesvirus
LTR – long terminal repeat
MCC – Merkel cell carcinoma
MCPyV – Merkel cell polyomavirus
MMTV – mouse mammary tumor virus
NS3 – nonstructural protein 3
NS5A – nonstructural protein 5A
RSV – Rous sarcoma virus
S – The hepatitis B virus spike protein
Src – Sarcoma
STLV – simian T-cell leukemia virus
Tax – transactivator of the X gene region

9

The Viruses Within – How Human Are We?

Keywords EVEs • HERVs • Syncytin • Arc • Cancer • Autoimmune disease

*If Charles Darwin reappeared today, he might be surprised to learn that humans
are descended from viruses as well as from apes.*
Robin Weiss

Adam and EVEs

Viruses are the bane of bacteria, plants, and animals, including humans, and
have caused disease and death for eons. Yet for all their horrendous effects on
cellular life forms, viruses have been incredibly important to cellular evolution
and have contributed to making humans what we are today. This remarkable
and unexpected finding only became fully appreciated once we could sequence
the whole genome of any organism from bacteria to humans. Sequencing
revealed a stunning result – we are all part virus. When genomic sequences
from any organism are examined, there are many stretches interspersed in
each genome where the DNA sequence is related to known viruses. Collectively
these viral-related sequences are known as EVEs for endogenous viral elements. This discovery was a little like finding mysterious foreign words in a
book printed in English; why were these viral "words" present in our genomic
"books", how did they get there, and did they have any meaning in the story
of our lives?

By comparing the EVE sequences within and between species it is possible
to derive their evolutionary relationships and construct phylogenetic trees.

From these studies, it is clear that EVEs are the remnants of ancient viral infections that have persisted in the genomes of the hosts. Interestingly, because they reflect ancient events, EVEs offer a unique perspective on viral evolutionary history. Once they become part of the host genome, the integrated viral DNA evolves along with the host DNA. Host cells have many repair mechanisms that correct mutations while viruses lack such systems. Consequently, exogenous viral genomes accumulate mutations and evolve much faster than their related EVE sequences. Due to these different evolutionary rates, modern viruses have diverged more extensively over time than have the EVEs, resulting in sequences that are different though still related. Hence, EVEs are fossils, and like fossilized skeletons, they are a snapshot of these ancient progenitors of modern viruses and provide insight into what these ancient viruses were and what their genomes looked like. Paleovirologists study these prehistoric sequences in our genomes to understand how modern viruses arose and how they have changed over time.

Besides their value for understanding viral evolution, much recent attention has focused on where EVEs came from and what effect they might have on the host. An important and fundamental question is how did external viruses become EVEs? To transition from exogenous infectious invaders to endogenous integrated EVEs required several events to happen. First, a virus that infected a cell had to insert its viral DNA into the genetic material of the host cells so that it was permanently maintained. While most viruses don't normally integrate their DNA into the host genome, this insertion process is standard for retroviruses. Retroviruses use their viral reverse transcriptase to convert their genomic RNA to double-stranded DNA inside the host cell. After the conversion of viral RNA to DNA, another viral enzyme called integrase inserts the viral DNA into the host chromosomal DNA where the integrated viral sequence is called a provirus. Since retroviruses always integrate their viral DNA form into the host DNA to form proviruses during their normal life cycle, EVEs derived from retroviruses are the most frequently found in nature. Usually, the integration event happened in one of the general cells of the body, the somatic cells, in which case there was no long-term consequence genetically. Somatic cells with integrated proviral DNA eventually senesce and die off or are lost from the gene pool when the host organism dies, preventing the provirus from passing on to the next generation. Consequently, the frequently occurring somatic integration events never create an EVE. The critical second step in becoming an EVE was for the integration event to occur in a germ cell, the progenitors of eggs or sperms. Next, a mature egg or sperm with the integrated proviral DNA had to be used for fertilization. If this happened, the proviral DNA sequence was now part of the embryo's genome and was present in the DNA of every cell of the resultant offspring, including each of their germ cells. Finally, if this individual survived and reproduced all of their subsequent offspring would also have the viral sequence in their genomes. If those offspring survived to reproduce and their descendants continued to survive then an EVE was established. As you might imagine, this sequence of events required to generate an EVE is very rare and highly unlikely, yet over millions of years of evolution it is has happened multiple times throughout plant, animal, and human lineages. Even if there is no single, primordial Adam in our history, there are certainly numerous EVEs that helped populate the human genome.

HERVs, HERVs, and More HERVs

In human genomes, these retroviral EVEs are called human endogenous retroviruses (HERVs). The accumulated HERVs now constitute a remarkable 200,000,000 plus base pairs or approximately 8% of the entire DNA in our genomes. Since retroviral genomes are no more than 10,000 bases long, at first thought it would appear that retroviruses must have invaded our genomes around 20,000 times to generate the current viral content in our DNA. However, it turns out that the gigantic quantity of HERV DNA in genomes originated from less than 40 separate viral invasions over the last 150 million years; that's how rare these events are and how difficult it is to establish a new HERV. But 40 events would only account for 400,000 base pairs of proviral DNA, so where did the rest come from? The answer is that each of the 40 retroviruses that entered our genomes likely continued to replicate and produce new viruses. As these progeny retroviruses reinfected the germ cells they introduced additional copies of the proviral DNA leading to the accumulation of copies in the germ cells. Alternatively, some scientists speculate that the proviral DNA may have been able to copy itself directly within the germ cell and reinsert the copy back into the host genome even in the absence of viral reproduction. In either case, if the resultant egg or sperm with the additional copy was used for fertilization then the succeeding generations inherited both the original viral element and the copy. One original integration would expand to two copies, then three, and on and on. Over millennia these copies accumulated and slowly spread throughout our genomes like invasive weeds, and there are now approximately 200,000 copies derived from these roughly 40 inserted retroviruses. Fortunately, this increase in copies did not go on indefinitely. Slowly the HERVs were inactivated so that today they no longer replicate and spread as they did after the initial invasion; otherwise, we would have even more HERVs cluttering up our genomes. For example, most of the existing copies have lost portions of the viral protein-coding sequences through genetic deletion. The remaining HERV sequences often represent only fragments of the original viral genome that are no longer functionally active for viral reproduction or protein expression. Additionally, so-called epigenetic modifications of the HERV sequences reduce or eliminate their expression, though these modifications can be reversed under certain conditions. Both the genetic and epigenetic mechanisms have combined to limit HERV spread, so at least for now, our repertoire of HERVs is fairly constant!

The most ancient of our HERVs, designated HERV-L, actually precedes humans as it arose 100–150 million years ago. Consistent with this very ancient viral invasion, HERV-L is found in all placental mammals. In contrast, our most recent HERV, a member of the HERV-K family, likely invaded the genome of our primate ancestors a mere 2–3 million years ago. After the human lineage diverged from primates, our HERV-K copies expanded to produce at least 29 distinct insertions that are unique to humans and are not found in our closest relative, the chimpanzee. As humans and chimps share an overall genome sequence identity of 98–99%, there is even speculation that the human-specific HERV-K elements may contribute to the distinct differences between our species. Furthermore, HERV-K elements are such recent acquisitions in evolutionary terms that there are still polymorphisms in human populations. Simply put this means that not every human has the same number and location of the HERV-K sequences. Just as differences in HERV-K elements could help explain differences between humans and chimpanzees, differences in HERV-K elements among the human population likely contributes to our genetic diversity and phenotypic variations.

Between the origination of the ancient HERV-L and the relatively more recent HERV-K, other retroviral invasions reflect the course of animal to human evolution. A few HERVs, such as HERV-F/H and HERV-FRD, occurred early in the primate lineage and are widely found in both Old World and New World primates, as well as humans. However, most HERVs originated after the split between Old World and New World primates and are found exclusively in humans and Old World apes and monkeys. And of course, many other retroviral EVEs are found throughout plants and animal lineages that are unrelated to humans and their HERVs. While it might seem that these viral invasions and the subsequent expansion of the viral sequence copy numbers would inevitably be harmful to the host, this is not the case. As presented in the following sections, there are several well-understood examples of how HERVs positively contributed to human evolution and development. It will be fascinating to see if proof develops to support any functional contributions for all the HERV elements in our genomes.

Beyond HERVs

While HERVs are the most common type of EVE, less common EVEs have been found that did not originate from retroviruses, so how did this happen? For any virus with a DNA genome, integration into host DNA can happen randomly at a low frequency via recombination. Recombination is a natural process in the cell nucleus where DNA is broken, pieces of DNA are exchanged, and the ends of the pieces are rejoined to seal the breaks. It can be used to move DNA segments around, delete DNA, or insert DNA. In the cell, recombination normally has several functions including repairing damaged chromosomes, increasing genetic diversity during meiosis in germ cells, and rearranging our immunoglobulin genes in B cells to increase antibody diversity. However, the enzymes that catalyze recombination can sometimes act on non-chromosomal DNAs in the nucleus, such as viral DNAs, resulting in viral DNA insertion into the chromosome. Again, for this integration to become an EVE it had to happen in a germ cell and get passed to subsequent generations. Perhaps even more surprisingly, some non-retroviral RNA viruses also became EVEs. This is unusual because no other type of RNA virus normally converts its RNA to DNA during its life cycle. The most likely possibility is that other RNA viruses that became EVEs had their RNA genomes inadvertently converted to DNA by reverse transcriptase supplied by retroviruses or other reverse-transcriptase-encoding elements already in the cell. Integrations by DNA virus genomes or non-retroviral RNA virus genomes are much rarer than integration by retroviruses, consequently, EVEs related to these other viruses occur much less frequently than HERVs. Yet for all their rarity, numerous examples exist of different RNA and DNA viruses that have become EVEs throughout the plant and animal kingdoms. Nearly every family of RNA and DNA viruses is found in the genomes of some plant or animal species, showing that given enough time viruses can likely invade any cellular genome.

Unlike many species, humans have very few non-HERVs in our genomes, including no evidence for endogenous DNA virus elements. In contrast, there are scattered reports of non-retroviral RNA sequences in our genome related to modern viruses such as Ebola, Marburg, and poliovirus, but very little is known of these potential EVEs. The one clear example of a non-retroviral

RNA virus-derived EVE in human genomes is related to the bornavirus group. Modern bornaviruses cause neurological diseases in several animal species, including common domestic animals such as horses, cattle, sheep, and dogs. There is currently no evidence for any human-specific bornaviruses although some humans have antibodies to bornaviruses indicating that they have been infected. Recent findings indicate that humans can develop encephalitis after exposure to animal bornaviruses, suggesting that inapparent exposures also likely occur from animal contact and account for antibody development in some people. Our bornavirus EVE entered the primate lineage more than 40 million years ago, so like many HERVs, it is an ancient component of the human genome. There are now 7 copies of the bornavirus insert in our genome, though all are partial versions with portions of the original viral genome deleted. Interestingly, each EVE fragment comes from the viral gene that encodes a protein called N; N stands for nucleocapsid and is the protein that comprises the virion capsid itself. As all 7 copies derive from the N gene they are designated EBLNs for endogenous bornavirus-like N gene. Some of the EBLNs still encode complete or partial versions of the N protein that have biological activity in our cells. While not fully characterized, there is tantalizing evidence that these EBLNs help to regulate certain cellular functions and may have anti-tumor activity. Using acquired viral sequences for functions beneficial to the host is an emerging paradigm that is more firmly documented for some HERV elements presented in the following sections.

Assailants or Allies?

As noted above, human genomes contain 200,000 HERV copies that collectively total 200,000,000 base pairs of viral DNA, roughly 8% of our total DNA, spread throughout our 23 pairs of chromosomes. It would seem like this amount of foreign DNA randomly inserted into our genome would disrupt our normal DNA function just as weeds in a field can damage crops. How is it possible for our genomes to tolerate this massive invasion without devastating consequences? How did all these viral sequence insertions not hit critical elements in our DNA and cause terrible genetic damage? The answer to these questions is two-fold. First, throughout our evolutionary history, some integration events likely did hit critical genes or regulatory elements in our genomes. Some of these negative events were fatal in utero or in early childhood and were never passed on to the next generation. Other insertions that weren't fatal may have seriously disadvantaged the individuals who inherited these events. Just like a harmful mutation in our DNA, a viral sequence insertion that made the recipient less competitive would be under negative selection pressure. Eventually, such lineages would have died out and we would no longer see these insertions in the current human population. Therefore, the HERVs that remain are ones that had little or no harmful effect on human survival and reproduction and were either genetically neutral or were beneficial to us. Still, you might ask how the 200,000,000 base pairs of foreign DNA sequences that are present today could penetrate our genome without hitting something important. The second part of the answer is that our genomes are filled with "empty space" that can easily tolerate the insertion or deletion of sequences. Our genomes contain about 25,000 genes that encode all the wonderful proteins that exist in our bodies. Yet the total amount of DNA in these 25,000 genes is less than 2% of our genome. We know some of the remaining DNA has structural and regulatory functions, but much of it seems to have little or no discernible function. It is still an unsettled debate, but some scientists argue that 70–90% of our DNA has no essential activity and is "junk DNA". Consequently, viral insertion into these large and seemingly functionless regions had no impact on the health of our genomes. That isn't to say that the remaining HERV insertions had no consequence, just that they didn't kill us outright or impair our reproductive fitness significantly – more on this later.

In contrast to the potentially harmful effects, there is accumulating acceptance in the scientific community that viral insertions played an important, beneficial role in our genetic evolution by providing raw materials for the development of new functions. Remember, there is more viral DNA sequence in our genomes (8%) than there are sequences for our own protein-encoding genes (2%). Viral genomes are compact and efficient with very little junk DNA. Almost every portion of a viral genome is either a protein-encoding gene or a regulatory sequence that controls viral replication and/or gene transcription into mRNA. For retroviruses that comprise our HERVs, the proviral genome consists of two identical transcriptional regulatory regions called LTRs (long terminal repeats) and 4 genes named Gag, Pro, Pol, and Env. Gag encodes the viral capsid proteins, Pol encodes the reverse transcriptase (and other activities), Env encodes the viral surface proteins that locate in the virion envelope, and Pro encodes a protease protein that cleaves both the initial Gag and Env proteins into smaller proteins. The four genes are arranged sequentially in a linear cluster, and the gene cluster is flanked by an LTR at each end. When a retrovirus inserts its proviral form into a germ cell and becomes an established HERV, individuals with this HERV now have four additional genes and two additional sets of regulatory elements in their genomes. As each initial HERV expanded into multiple copies there was a corresponding enlargement of this pool of protein-coding genes and regulatory elements. Additionally, as the number of copies of the HERV sequences increased this favored recombination between the copies, leading to random rearrangements and hybrids between the viral genes and regulatory sequences. Since neither the intact nor altered copies of these viral components were needed for cellular function, they were completely malleable for further mutation and selection. Stated another way, the viral sequences were fodder for genetic experimentation. If random changes in the viral DNA generated a protein or regulatory element with a new function that improved the survival of the host then individuals with that particular HERV eventually replaced individuals without the beneficial HERV. Over time the new function became part of the normal human gene repertoire. Using this abundant supply of viral raw materials, evolution was able to gradually adapt and improve the human species to our present form. It's a little like genetic roulette. Each organism has its own genome that is its "stake" in the genetic evolution game. However, there is only so much you can gamble with your own genome without risking extinction. Incursions of viral sequences were like an influx of free cash that could be gambled at will with no repercussions. In most cases, if acquired viral sequences changed through mutations or rearrangements there was no effect on the host. But occasionally you hit a jackpot and gained a valuable new activity that gave the host a selective advantage. Offspring that inherited the favorable changes had a survival advantage and the beneficial viral sequences eventually became fixed in our genomes. Therefore, viruses were not only deadly infections but have also been critical contributors to genomic evolution as the viral sequences were co-opted by the hosts for cellular use. Without viral integrations into host genomes, most species, including humans, wouldn't have evolved to what they are today.

The HERVs That Made Us Human

While the functional contribution of most HERV sequences remains unknown, there are now several well-documented examples of retroviral sequences that became critical additions to the human genome during our evolutionary past. Perhaps the best-studied case is a placental protein called syncytin-1. The placenta is a complex and multifunctional organ unique to mammals that enables them to give birth to live offspring. The placenta connects the mother to the embryo and fetus through the umbilical cord which allows entry of oxygen and nutrients along with the removal of waste from the developing offspring. Placentas also produce hormones necessary for pregnancy and help prevent pathogens from reaching the embryo and fetus. Exchange of nutrients and waste without actual mixing of the fetal and maternal blood supplies is accomplished by diffusion across a gap called the intervillous space. The intervillous space forms between the maternal uterine wall and branched projections from the placenta known as chorionic villi. The placental villi are composed of fetal cells called cytotrophoblasts. In humans, the outmost surface layer of the villi forms by fusing thousands of cytotrophoblast cells together to create a few giant, multinucleated cells known as syncytiotrophoblasts that are each 4–5 inches long. Functionally these gigantic syncytiotrophoblast cells help prevent maternal white blood cells from entering the fetal bloodstream where they could attack fetal cells. Normally, white blood cells enter tissues by squeezing through the junctions between individual cells. Since the syncytiotrophoblasts are single giant cells that have no junctions, they restrict the entry of maternal lymphocytes from the intervillous space into the villi and help prevent immune rejection of the fetus. The fusion of the cytotrophoblasts into syncytiotrophoblasts is mediated by the syncytin-1 protein which is expressed primarily in fetal tissue. It turns out that the critical syncytin-1 protein is not expressed from a uniquely human gene, but instead originated as a HERV protein encoded by a retroviral Env gene.

The story of syncytin-1 began around 25–30 million years ago, after the separation of Old World and New World primates. Old World primates suffered an integration by a retrovirus whose remnant progeny, known as the HERV-W family, are now seen in Old World monkeys, apes, and humans. Over the millennia the primordial ancestor of the HERV-W family amplified and degraded to generate over 200 copies in the human genome, each of them just partial fragments of the original provirus genome. Like all retroviruses, the HERV-W progenitor possessed an Env gene encoding the envelope protein that was present on the virion surface. A primary function of envelope proteins is to mediate the fusion of the virion lipid membrane with the host cell membrane, an action that releases the virion nucleocapsid into the host cell to initiate infection. Due to its innate membrane fusing capacity, the envelope protein can also fuse cells together as well as fusing virions with a host cell. Expression of the envelope protein from the proviral DNA in host cells likely caused cell-to-cell fusion and was selected against since most fusions would be harmful to the host. Slowly, the Env gene was deleted from many of the proviral copies, and the expression of the remaining copies was repressed in most types of cells due to its negative effects. In cytotrophoblast cells, however, fusion proved advantageous as it reduced cell junctions and helped protect the developing fetus from maternal white blood cells. Because of this beneficial effect, Env gene expression was maintained in cytotrophoblast cells throughout primate evolution. Primates became increasingly dependent on this viral envelope protein until it eventually became an integral and necessary part of our placental development. Now all human births rely on a retroviral protein that we assimilated into our gestational process.

Syncytin-1 isn't the only retroviral protein contribution to human gestation. Another evolutionary issue in the development of mammals was the fetal rejection problem. The placenta and the growing fetus are akin to transplanted organs in that they express many paternal proteins that are foreign to the mother and would stimulate immunological rejection. The early embryo, known as a blastocyst, is also a bit like a parasite or tumor in that its fetal cells invade the mother's uterine tissue during the formation of the placenta. For successful implantation and maintenance of the developing fetus and embryo, the mother's immune response needs to be suppressed otherwise a maternal immune attack would destroy the foreign cells. Part of the rejection solution is provided by a second retroviral envelope protein that has been named syncytin-2. Syncytin-2 arose from a different retroviral insertion than syncytin-1, and its lineage is referred to as HERV-FRD. Unlike syncytin-1, syncytin-2 is found in both Old World and New World primates, hence its origin is an even older retroviral insertion than for the HERV-W family. Like syncytin-1, syncytin-2 has a fusogenic ability and both syncytins contribute to cytotrophoblast fusion during placenta formation. Importantly though, syncytin-2 retains an envelope protein sequence known as the immunosuppressive domain (ISD). ISDs are a normal component of retroviral envelope proteins and are important for establishing viral infection. The ISD function blocks the host immune response immediately upon virion entry to the host cell and allows the incoming virus an opportunity to establish a foothold. Our primate ancestors co-opted the immunosuppressive function of the envelope protein to help with the fetal rejection problem and this primordial retroviral envelope protein evolved into what we today call syncytin-2. Interestingly, syncytin-2 not only acts locally in the uterus but also is excreted in small lipid vesicles that can circulate throughout the body and may have more systemic immunosuppressive effects. Syncytin-1 also has a functional ISD, and together the two syncytins are likely major contributors to the complex and changing immunosuppression that occurs throughout human pregnancy.

Because the HERV-W and HERV-FRD type retroviral invasions were in the primitive primate lineages, proteins directly related to syncytin-1 and -2 are only found in primates. Nonetheless, most but not all other non-primate mammals have proteins similar to syncytin-1 and -2 that perform the same functions. For example, mice have two proteins named syncytin A and syncytin B, but these mouse proteins did not develop from the same progenitor viruses that gave rise to syncytin-1 and -2. Instead, the ancestral mouse lineage was infected with two different retroviruses and the murine syncytins evolved from the envelope proteins of those retroviruses. Likewise, all carnivorous mammals share a third distinct syncytin designated syncytin-Car1 that arose from yet a different retroviral envelope protein. The conclusion is that this represents convergent evolution where independent events evolved towards the same functional endpoint. Retroviruses have assaulted every species and have integrated proviruses into the genomes of all animals. As placental animals arose, the utilization of proviral expressed envelope proteins happened independently in different species to facilitate trophoblast fusion and to help limit immune rejection. Apparently, the beneficial effects that the envelope proteins conferred in gestation were so powerful that their selection happened over and over again as different mammalian lineages evolved. This is a beautiful example of evolution utilizing virus-supplied genes as raw materials to craft new properties and features in cellular lifeforms.

If the role that retroviral envelope-derived proteins play in mammalian gestation was not remarkable enough, retroviral LTR elements may make an even broader contribution to fetal development. Most of the HERVs that remain in our genome are just orphan LTRs without any remaining viral genes adjacent. LTRs are retroviral DNA sequences that control viral gene expression through sequence elements known as promoters and enhancers. Promoters define the start sites for mRNA transcripts while enhancers control the amount of mRNA synthesized. LTRs function by binding multiple combinations of host cell proteins called transcriptional factors (TFs). Depending on the specific combination of TFs bound to the LTR its activity varies from fully off to maximally on. When these orphan LTRs integrated adjacent to human genes then the LTRs could direct mRNA production from our genes. Because of the large number of orphan LTRs in our genome, it is estimated that there is 10 times more transcription originating from these HERV LTRs than from our authentic human promoters. Nowhere is this more prominent than in the early fetus and placenta where LTRs from at least a half-dozen HERV families are highly active. Though the precise effect that each LTR has on fetal and placental development remains mostly obscure, several clear examples of human gene regulation by individual LTRs have accumulated in recent years. For example, two growth factors produced by trophoblasts, pleiotrophin and insulin-like growth factor, both have their mRNA transcription driven by HERV LTRs. Other studies have shown that there is a complex, orchestrated pattern of human gene expression driven by the HERV

LTRs that initiates at fertilization and progresses as the fetus grows. LTR expression is highest at the 1–2 cell stage and decreases as growth continues with the LTRs turning specific human genes on and off to regulate fetal development. Experimentally blocking LTR function leads to a cessation of fetal development, illustrating just how critical these viral elements are to gestation. Most researchers now believe that during evolution we co-opted numerous retroviral elements to fine-tune the developmental process, and eventually these viral sequence modules became indispensable components of human fetal development. Like it or not, the very earliest and most fundamental aspects of human gestation are completely dependent on viral elements that invaded our ancestral genomes and helped direct the growth and development of the human fetus. We don't just have viral sequences languishing in our genome, we are all functionally part virus.

That viral LTRs and proteins play an essential role in human gestation is remarkable, but what if viruses also helped shape brain evolution leading to the very consciousness that makes us self-aware? Emerging evidence supports a major and dramatic viral impact on our brains. Similar to the syncytin story, animal brains have co-opted a retroviral protein to fulfill a critical need for cell-to-cell communication within the brain. Neuroscientists call this brain protein Arc in humans and add a prefix to designate the comparable protein in animals, for example in rats it is rArc and in the fruit fly (*Drosophila melanogaster*) it is dArc. The Arc protein was first identified in 1995, and decades of subsequent study have revealed an essential role in learning and memory. Using genetic tools to remove the gene for Arc in mice produced animals that were physically normal, but that had impaired long-term memory without any deficit in short-term memory. Similarly, using inhibitors to prevent Arc production in normal animals caused long-term memory loss in both mice and rats without impairing short-term memory. Though comparable functional studies have not been done in humans, the presence of Arc in the human brain is consistent with a requirement for Arc in human long-term memory formation.

While still very mysterious, many neuroscientists believe that memory storage in the brain involves engrams which are specific assemblies of individual neurons that form relatively stable networks. Critical in the formation of these networks are the numerous branching projection from nerve cells called dendrites. Each dendritic branch can contact many other neurons to form vast, complex linkages of neurons. The formation, dissolution, and reformation of dendritic connections is essential for long-term memory and is known as neural plasticity; genetic and biochemical experiments establish that Arc is a major contributor to neural plasticity. Upon neural activation, the gene for

Arc is transcribed into mRNA that is rapidly transported to the dendrites where Arc protein is produced. Accumulating Arc regulates receptors on the surface of the dendrites and modulates neuron-to-neuron communication, presumably to help develop memories. In humans, deficits in Arc production and/or regulation are linked with several neurological conditions including Alzheimer's disease, Fragile X syndrome mental retardation, Angelman syndrome, autism spectrum disorder, and schizophrenia. This broad connection with neurological disabilities implies that Arc is not only required for long-term memory formation but also for normal cognition.

The origin of Arc as a retroviral protein was first noted in the mid-2000s when protein sequence analysis found a relationship between Arc and the Gag protein of retroviruses. While our syncytin protein evolved from an ancient retroviral Env protein, our Arc protein was co-opted from an ancient Gag protein. Interestingly, just as independent viral invasions resulted in syncytin-like proteins developing in different animal lineages, a similar situation exists for Arc. The Arc genes in tetrapods (reptiles, amphibians, and mammals) all derive from the same ancestral Gag gene indicating that the viral invasion event must have occurred before these lineages diverged. In contrast, insects also have an Arc gene that functions in their brains, but their gene is derived from a different retroviral Gag. For both syncytin and Arc, the fact that distinct retroviral genes independently evolved to become animal proteins with similar functions again highlights the concept of convergent evolution. The different retroviral genes provided important starting materials for the evolving organisms, but the final outcomes represent the most optimal solution to an organism's needs, and many organisms converged on comparable mechanisms.

Given that disparate Gag proteins evolved into Arc proteins for both tetrapods and insects, an important question is why did a retroviral Gag protein get co-opted for brain function? The Gag protein is the retrovirus capsid protein that assembles around the viral RNA to package the RNA into a protective nucleocapsid structure. Gag proteins spontaneously self-aggregate into a roughly spherical shape and no other viral product is necessary for the formation of the nucleocapsid. During virus production, the nucleocapsid subsequently acquires a membrane envelope from the host cells, with the envelope containing many copies of the viral Env protein. The completed virion is then the delivery vehicle that transmits the viral RNA information from the original infected cell to a new cell. Exciting work from 2018 revealed that Arc not only influences dendritic plasticity in the cell where it is expressed but also retains its viral functions of self-assembly and RNA packaging. In both mouse and *Drosophila* studies, Arc protein in the dendrites formed capsid-like structures that encapsulated Arc mRNA. While lacking the Env protein necessary for authentic virion formation, the Arc capsid is released from the producing cell inside small lipid vesicles called exosomes. Exosomes are normal cellular products that ferry both proteins and RNAs between cells. Circulating exosomes are taken up by recipient cells and release their cargo inside the cells. The Arc-containing exosomes likewise deliver Arc protein and mRNA to recipient neurons or neuron target cells such as muscles. In the *Drosophila* system, disrupting this transfer prevented plasticity, suggesting that this novel cell-to-cell communication through Arc-exosome pseudovirions is critical for neuronal function in animals from insects to humans. Just as syncytins retain the Env fusion function and repurpose it for cell-to-cell fusion, Arc proteins retain the Gag self-assembly and RNA transport function and redirect these features to transmit information between neurons and their target cells. Since there are more than 100 other human proteins derived from Gag proteins, most with unknown functions, it will be exciting to see what other adaptations have occurred. Some of these 100 proteins whose counterparts were tested in mice also affect cognition, therefore, the viral contributions to mammalian brain function may extend well beyond the Arc protein. Furthermore, while Arc is almost exclusively found in the brain, other Gag-derived proteins may act in other organs and in roles unrelated to brain function. As early life forms developed into multicellular organisms they would have needed mechanisms to transmit information signals between cells. Because of their capacity for self-assembly and RNA transfer, perhaps Gag proteins provided a malleable template that evolution could domesticate to develop different cell-to-cell communication networks.

A final example of how retroviral invasions have benefited humans is their contribution to the evolution and function of immune systems. When ancient retroviruses initially invaded early animal genomes, the resultant proviruses were replication-competent and continued to produce viruses in many cells. Such chronic infections likely carried some health risks for the affected individuals which would have provided strong selective pressure to develop measures that stopped viral production. Many researchers suspect that this kind of continuous infection provided much of the evolutionary impetus for the development and expansion of primitive early immune systems into the complex, tripartite system discussed in Chap. 5. Beyond this general influence, we have some good evidence and many hints for specific roles that HERV elements play in mammalian immune systems, though much remains to be discovered and understood. Not only did the presence of endogenous retroviruses promote immune system development, but retroviral elements were also incorporated as part of our antiviral defenses. Like the syncytin and Arc stories, our ancestors utilized both retroviral proteins and LTRs as raw materials for genetic experimentation, and the successful experiments remain part of our immune arsenal to this day as we'll see in two examples below.

One immune function with a clear impact from HERVs is the interferon response. As described in Chap. 5, interferon is an important antiviral protein that is rapidly synthesized after the innate immune system of an infected cell detects viral molecular signatures via the pattern recognition receptors (PRRs). The produced interferon is exported from the infected cells, bound by receptors on surrounding cells, and triggers a signaling cascade that results in hundreds of genes in the recipient cells being transcribed into mRNA and then translated into proteins. Expression of these many proteins from this network of interferon-stimulated genes (ISGs) creates a cellular environment poised to resist viral infection. By linking together the expression of many genes, the cell ensures that there are multiple and redundant defense mechanisms all simultaneously induced by interferon. Mechanistically, the coordinated transcription of these genes occurs because upstream of every interferon-responsive gene there are similar DNA sequences, known as interferon regulatory elements (IREs). IREs bind the specific transcription factors (TFs) that are activated by the interferon signaling cascade, and this binding induces mRNA production. Consequently, any gene that acquired appropriately located IRE sequences became part of the interferon-regulated network. For many of the known human ISGs, their IREs are derived from HERV LTRs. As ancient HERVs expanded and spread throughout genomes, random insertion placed LTRs in the vicinity of many of our genes. If a particular LTR could bind the interferon-induced TFs and was positioned correctly near a gene, then that gene became responsive to interferon. Over time, a variety of genes were brought into the interferon network by co-opting the viral LTRs to mediate interferon-induced transcription. Thus, using viral LTRs as functional building blocks facilitated the evolution of a large and complex array of genes inducible by interferon.

A second example of HERV contributions to anti-viral immunity relates to a principle known as resistance to superinfection. This principle refers to the observation that when a cell is infected by a virus it is often resistant to being infected again (i.e. superinfected) with the same virus or a closely related virus. In essence, the first virus infecting a cell is claiming that cell for its own reproduction and wants to prevent other viruses from entering and competing for the cellular raw materials. There are multiple ways in which this resistance can be produced, but all involve either protein or RNA from the first virus interfering with a step in the life cycle of the second virus. As some endogenous retroviruses still express proteins and RNA, they may make cells resistant to infections by related modern retroviruses. Studies in the last decade have confirmed this idea, and there is strong evidence in animals that endogenous retroviruses (ERVs) mediate resistance to exogenous retroviruses. For example, sheep are susceptible to an exogenous retrovirus named the Jaagsiekte sheep retrovirus (JSRV). However, sheep also harbor a related endogenous retrovirus (ERV) known as the endogenous Jaagsiekte sheep retrovirus (enJSRV). EnJSRV entered sheep genomes roughly 5–10 million years ago and expanded into multiple copies with the copy number varying in different domestic breeds. Two of the enJSRV copies continue to express the viral Gag protein that forms the viral capsid, and this endogenous Gag protein confers some protection against the exogenous JSRV. Both copies of the endogenous Gag protein have the same mutation that changes a single amino acid in the protein from arginine to tryptophan. Because of this mutation, the mutant Gag proteins are defective for normal capsid formation. When a cell is infected with exogenous JSRV, the exogenous virus makes wild-type Gag protein while the endogenous virus produces the mutant Gag protein. Within the cell when capsids are being assembled, both the normal and the flawed Gag proteins are used to build the capsids. Because every capsid has a mixture of wild-type and mutant Gag proteins, all these mixed capsids function poorly, and very few are ever released from the infected cells. This assembly interference by the mutant Gag protein greatly reduces viral spread within the animal and allows other host defenses more opportunity to clear the infection.

Another example of Gag-mediated superinfection resistance was described in certain mouse strains. Many strains harbor an ancient ERV known as muERV-L that likely entered the murine lineage over 40 million years ago. MuERV-L expresses a Gag protein referred to as Fv1 that confers resistance to exogenous murine leukemia viruses (MLVs). Surprisingly, the amino acid sequence of Fv1 is not highly related to the Gag protein sequence of MLVs, indicating that the mechanism of resistance is not through capsid function interference as it is for the enJSRV Gag protein. Consistent with a different mechanism, Fv1 has activity against a broad spectrum of MLVs and blocks their infection at a much earlier step in the life cycle long before capsid assembly. While the actual mechanism remains obscure, the block does occur before the proviral DNA from the exogenous MLV integrates into the host DNA. Without proviral integration, no new virions can be made and the infected cell cannot produce viral progeny that spread to other cells. Similar examples have been found in a variety of other animal species, so the ability of ERVs to help protect animals from exogenous retrovirus infection is widespread with a well-documented benefit against these external invaders.

Unlike its prevalence in animals, there is no evidence yet for HERV-mediated superinfection resistance against the two natural human retroviruses, HTLV-1 and -2. A possible explanation is that our many HERVs were so efficient at providing resistance against exogenous human retroviruses that these viruses have all become extinct during our evolutionary history with only the two HTLVs remaining. Still, our HERVs do retain the capacity to interfere with exogenous retroviruses as seen with HIV now that it has become a new human retrovirus. A human HERV called HERV-K expresses a Gag protein that provides some resistance against HIV through a mechanism similar to the enJSRV Gag protein. In HIV-infected cells, the HERV-K Gag proteins interact with the HIV Gag proteins and interfere with the proper assembly of the HIV capsids. The hybrid HIV/HERV-K Gag capsids that do form are released less efficiently from the cells and have reduced infectivity for new cells. While insufficient to offer full protection against HIV, this observation nonetheless illustrates that human HERVs can combat infections by exogenous retroviruses through mechanisms of superinfection resistance similar to those elucidated in animals. Maybe someday a chance mutation in the HERV-K Gag gene will increase its potency against HIV and offer better protection to the individuals that harbor this mutation.

The Dark Side

The above examples illustrate how retroviral genetic information was used for beneficial purposes during animal and human evolution, with other instances likely being found as we further explore the human genome and its viral origins. Still, most of our HERVs have not been thoroughly examined and their contributions to our genome or health are unknown. Some HERV elements could have harmful rather than helpful effects under certain conditions. Evolutionarily, as long as a HERV sequence did not decrease reproductive fitness, there was no selection against it even if it caused health issues later in life. Numerous examples now exist of HERVs linked to complex diseases from neurological conditions to cancer, though in most cases the evidence is merely correlative rather than causal, and molecular mechanisms are either lacking or are just speculative. Unfortunately, the large number of copies spread throughout our genomes makes the analysis and characterization of HERV's role in complex human disease problematic. Coupled with the myriad of other environmental and genetic factors that can influence disease manifestation in any individual, teasing out the impact of HERV elements is an arduous process that is not yet complete for any disease. Still, even if the HERVs are not actually causing diseases, it's important to investigate these associations as they may provide greater insight into disease risk factors, may identify HERV RNAs or proteins that could be used as disease markers for diagnostic purposes, or may even lead to new therapeutic strategies.

One intriguing and frightening possibility is that HERVs contribute to certain human cancers. Many types of cancers including breast, colorectal, lung, ovarian, bladder, and melanomas show elevated expression of HERV RNAs and proteins compared to their corresponding normal tissue. Whether the heightened expression of these HERV products preceded these cancers and contributed to cancer initiation is unknown, but some interesting possibilities have been proposed. For example, the Env protein's immunosuppressive property is important in embryogenesis, but the inappropriate expression of Env in adult tissues could reduce local immune detection and destruction of transformed cells, thus facilitating tumor formation. Additionally, some studies report that Env has inherent transforming activity under some conditions, thus it could be acting as an oncoprotein when turned on in adult cells. Similarly, some HERV-K sequences encode two small viral proteins named Rec and Np9 that both display transforming capacity. Normally, adult cells repress the production of the viral proteins due to epigenetic modifications that block the LTRs from producing RNA transcripts. However, many of the same environmental agents that promote cancers such as DNA-damaging chemicals, UV light, radiation, and certain viral infections can overcome the epigenetic modifications that normally suppress transcription from the HERV LTR elements. As a result, an LTR associated with the Env, Rec, or Np9 genes could become activated and produce the respective protein product. This raises the possibility that HERV protein products induced in cells by these environmental exposures could act as factors or co-factors to transform cells just as we saw for the oncoproteins of exogenous tumor viruses. Of course, many other factors such as personal genetics and immune status would also contribute to whether or not a transformed cell expanded into cancer in any particular individual.

In contrast to a role for HERV proteins, other models propose that cancer may be driven simply by activated HERV LTRs themselves. For example, there are numerous dormant LTRs in the vicinity of cellular proto-oncogenes. If a dormant LTR becomes activated this could turn on the adjacent cellular proto-oncogene inappropriately to drive cancer initiation similar to the mechanism of promoter insertion for exogenous leukemia viruses. Alternatively, activated LTRs may be more prone to inter-LTR recombination. Recombination between LTRs at different chromosomal locations would lead to chromosomal aberrations and rearrangements such as are commonly seen in tumors. Unfortunately, in looking at human cancer tissue we have no way to determine if the increased activity of the HERV LTRs preceded the cancer or came after the cancer developed. It is possible that once transformation occurred that transformation-related changes in the cell environment activated the LTRs. In this case, the increased HERV activity may not have any functional role in cancer and may simply be an unintended after effect. Consequently, whether HERV LTR activation is a cause of cancer or an outcome of cancer cannot be distinguished, and much work is still needed to tease out the role of HERVs, if any, in each individual type of cancer. The good news is that if HERVs are causal for some cancers, then drugs targeting HERV transcription or proteins could be developed as a new type of anti-cancer therapeutic.

A second area where there has been great interest and much research is the role of HERVs in autoimmune diseases such as rheumatoid arthritis (RA), multiple sclerosis (MS), and lupus. In general, autoimmune diseases arise from a failure to distinguish self from non-self. Normally, our immune system recognizes our own macromolecules (e.g. proteins and nucleic acids) and will not react to them, with immune attacks being reserved only for foreign molecules. In autoimmune diseases, our immune system becomes confused and begins to chronically attack self-molecules leading to tissue destruction and permanent inflammation. Depending on the anatomical location of the immune attack, the effects can range from the joint pain of arthritis to the neurological symptoms of MS. Like cancer, these autoimmune diseases are complex entities that likely have multiple causes or pathways that all lead to the same pathology and presentation in the patient. Consequently, HERVs may be important only in a subset of clinical cases. This potentially limited contribution confounds attempts to confirm that HERV elements are critical for specific diseases. To date, the data mostly show associations without proving causality, though there are fascinating possibilities that are being explored.

Two of the best-studied autoimmune diseases that have a possible role for HERVs are RA and MS, both of which have correlations with the HERV-K family. For RA, more people with the disease have antibodies against the HERV-K Env protein than do non-affected individuals. The antibody results indicate that HERV expression is higher in the RA patient population, but the antibodies against the Env protein may simply be a marker of HERV expression and not a causative factor for RA. Instead, the influence of HERV-K is more likely mediated through the HERV-K Gag protein expressed by one member of this family called HERV-K10. The mRNA that encodes this Gag protein is significantly higher in blood cells of patients with RA compared to healthy controls, and even higher in affected joints, an observation consistent with a connection between Gag and RA. Importantly, the HERV-K10 Gag protein shares regions of amino acid similarity with several human proteins, including type II collagen, a known target for autoimmune antibodies in RA. Because of this similarity to type II collagen, some individuals may make antibodies against the foreign HERV-K10 Gag protein that cross-react with and attack the type II collagen in their joints. Thus, the viral Gag protein may be a trigger that starts the production of the anti-type II collagen antibodies seen in patients with RA. Why this only happens in some individuals is unknown, but could be due to some environmental or genetic event that initiates Gag expression and subsequent antibody production. It is also possible that anti-type II collagen antibodies can be induced in other ways, so Gag is not necessarily the sole or primary trigger for RA, and may only constitute one possible pathway to this disease.

A role for HERV-K in MS is also suggested by studies that found an association between certain HERV-K variants and MS, as well as increased expression of Gag RNA in blood and brain cells from individuals with MS. However, a mechanism by which HERV-K expression could be contributing to the onset and progression of MS is much less clear than for RA. One interesting hypothesis relates to a potential interaction between the HERV-K Gag protein, Epstein-Barr virus (EBV – the agent of infectious mononucleosis), and a host antiviral restriction factor called TRIM5α. EBV has long been known as a risk factor for MS development, although the basis for this risk is still uncertain as 90% of adults have had EBV infections and yet only a small number of individuals develop MS. One possibility relates to the known ability of EBV to activate HERV LTRs, possibly indirectly through the immune response to EBV infection. Certain transcription factors that are induced as part of the immune response to EBV can bind to the HERV LTRs and turn on transcription from these LTR promoters leading to Gag protein production. It is known that certain HERV-K DNA sequence variants correlate with MS. There are speculations that these variants may allow overexpression of Gag protein from their activated LTRs or have Gag genes that produce Gag proteins with slightly different amino acid sequences. Overexpression of Gag or production of variant forms of Gag protein may facilitate the ability of Gag protein to bind to the cellular TRIM5α protein. In infected cells, TRIM5α normally binds to the capsids of exogenous retroviruses, and the binding of capsids signals the protective release of pro-inflammatory cytokines. TRIM5α binding to endogenous HERV-K Gag protein may result in a similar release of pro-inflammatory cytokines. Additionally, individuals with MS often have TRIM5α variants, and perhaps these TRIM5α variants are ones with enhanced or dysregulated cytokine signaling. Then in individuals harboring some of these rare variants, EBV infection could start a chronic inflammatory response through HERV-K Gag induction, binding of Gag to TRIM5α, and overabundant TRIM5α-mediated release of cytokines. In this model, one route to MS would require both a genetic risk (the HERV-K and/or TRIM5α DNA sequence variants) and an environmental exposure (EBV infection) to trigger the development of MS. The rarity of all these risk factors occurring in any given individual would account for the relatively low rates of MS even though EBV infection rates are quite high.

The models for how HERVs contribute to rheumatoid arthritis and multiple sclerosis are still very speculative, and we are a long way from any definitive explanation of how endogenous retroviruses are influencing these diseases or any other autoimmune disorders. Still, the correlations are fascinating and the models are plausible. Future studies will continue to test predictions of these models and will either confirm or refute specific features of the proposed molecular mechanisms. As presented in the previous sections, HERVs had a well-documented positive impact on animal and human evolution, but it is also highly likely that there has been a disease cost incurred by these genomic invaders. The aberrant expression of HERV proteins, HERV RNAs, and HERV LTRs driving cellular genes provides a myriad of opportunities for molecular mishaps that could trigger or contribute to the progression of many diseases. Again, as long as the negative impact of HERVs didn't reduce reproductive fitness, they wouldn't be selected against. Consequently, the types of diseases most likely related to HERVs are those that typically appear later in life such as cancer, autoimmune diseases, and neurological disorders. Only time and further study will disclose the full extent of the relationship between our endogenous viral passengers and our biological health. For now, the one certain thing is that we will undoubtedly be surprised and amazed as future research reveals new and novel connections between endogenous viruses and human health and disease.

Additional Reading

1. Symbiosis: Viruses as Intimate Partners. Marilyn J. Roossinck and Edelio R. Bazán. Annual Review of Virology 4:123–139. 2017.
2. Viruses and Cells Intertwined Since the Dawn of Evolution. Julia Durzyńska and Anna Goździcka-Józefiak. Virology Journal 12:169. 2015.
3. Co-option of Endogenous Viral Sequences for Host Cell Function. John A. Frank and Cédric Feschotte. Current Opinion in Virology 25:81–89. 2017.

Definition

Alzheimer's disease – A degenerative disease of the brain characterized by mental deterioration.

Angelman syndrome – A genetic disease characterized by mental disability and movement disorder.

Arc – A brain protein of viral origin that is important for neural plasticity.

Autism spectrum disorder – The range of neurodevelopment disorders characterized by difficulties with social interaction, speech, and nonverbal communication.

Blastocyst – An early stage of embryonic development where the developing cells begin to form a sphere with a central cavity.

Bornavirus – A family of single-stranded, negative-sense RNA viruses.

Chorionic villi – Tiny, finger-like projections from the placenta.

Collagen – A type of structural protein found in connective tissue such as skin.

Cytotrophoblast cells – A type of cell that covers the blastocyst surface and will eventually fuse to become syncytiotrophoblasts.

Dendrites – Branched extensions from nerve cells that function to transmit nerve impulses from cell to cell.

Drosophila melanogaster – The scientific name of the fruit fly.

Embryo – The developmental stage in humans from about the second to the eighth week after fertilization.

Endogenous Jaagsiekte sheep retrovirus (enJSRV) – A retrovirus that entered sheep genomes 5–10 million years ago.

Engrams – Stable networks of neurons that function to form memories.

Epigenetic – Heritable modifications to the DNA that influence gene expression. These are not mutations because these modifications do not change the nucleotide sequence of DNA.

Exosome – Extracellular lipid vesicles released by cells to ferry biomolecules between cells.

Fetus – The stage of human gestation from after week eight to birth.

Fragile X syndrome mental retardation – A genetic disease due to an abnormality with the X chromosome that causes intellectual disabilities.

Fv1 – A type of gag protein expressed from an endogenous mouse retrovirus.

Germ cell – The reproductive cells that will produce eggs or sperm.

Human endogenous retrovirus (HERV) – Retroviral DNA sequences that have become a permanent part of the human genome.

Immunosuppressive domain (ISD) – A region on retroviral envelope proteins that can suppress the host immune response.

Insulin-like growth factor – A small protein hormone that promotes the growth of many cell types.

Integrase – A retroviral enzyme that catalyzes the insertion of the DNA form of the retroviral genome into the host cell chromosome to form a provirus.

Interferon regulatory element (IRE) – DNA sequences that bind transcription factors induced by interferon. These elements coordinate the expression of many genes in response to interferon signaling.

Intervillous space – In the developing embryo this is the space between the chorionic villi and the syncytiotrophoblast layer.

(continued)

(continued)

Jaagsiekte sheep retrovirus (JSRV) – An exogenous retrovirus that causes lung cancer in sheep.

Lupus – An inflammatory autoimmune disease where the immune system can attack many tissues, including joints, skin, lungs, the brain, and others.

Macromolecule – Any large biomolecule, typically referring to proteins and nucleic acids.

Marburg virus – A single-stranded, negative-sense RNA virus of the filovirus family that causes a severe hemorrhagic fever.

Meiosis – A specialized type of cell division used to generate the haploid egg and sperm cells.

Melanoma – A type of skin cancer involving pigment-producing cells (melanocytes).

Neural plasticity – The ability of neural networks to change through growth and reorganization in response to what the individual experiences.

Np9 – A protein expressed from an endogenous retrovirus that has transforming activity.

Oncoprotein – Any protein with transforming or tumorigenic activity.

Paleovirologist – A scientist who studies ancient and extinct viruses.

Placenta – The structure that develops during pregnancy to nurture the embryo and fetus. The placenta provides oxygen and nutrients and also removes waste.

Pleiotrophin – A human protein, produced by trophoblasts, that has growth-promoting activity.

Polymorphisms – Genetic variations that exist within a population.

Pseudovirion – A viral particle that lacks replicative ability.

Rec – A protein expressed from an endogenous retrovirus that has transforming activity.

Rheumatoid arthritis (RA) – An inflammatory autoimmune disease that attacks the joints, particularly in the hands and feet.

Schizophrenia – A mental disorder characterized by thoughts, speech, and actions that are divorced from reality.

Superinfection – The second infection of an already infected cell.

Syncytin A and B – The syncytin 1 and 2 equivalents found in mice.

Syncytin-1 – A retroviral-derived protein expressed by the placenta that mediates the fusion of cytotrophoblast cells into syncytiotrophoblast cells.

Syncytin-2 – A retroviral-derived protein expressed by the placenta that helps prevent maternal rejection of the embryo and fetus.

Syncytin-Car1 – The syncytin found in carnivorous mammals.

Syncytiotrophoblast cells – Giant cells formed by the fusion of multiple cytotrophoblast cells.

Tetrapods – The group that includes reptiles, amphibians, and mammals.

Transcription factors (TF) – Proteins that bind to the DNA sequence in promoters and enhancers to regulate gene expression.

TRIM5α – A host protein that functions as a restriction factor for retroviruses.

Trophoblast – The cells that form the outer layer of the blastocyte.

Abbreviations

EBLN – endogenous bornavirus-like N gene
enJSRV – endogenous Jaagsiekte sheep retrovirus
ERV – endogenous retrovirus
HERV – human endogenous retrovirus
IRE – interferon regulatory element
ISD – immunosuppressive domain
JSRV – Jaagsiekte sheep retrovirus
MLV – murine leukemia virus
MS – multiple sclerosis
muERV – endogenous retroviruses of mice
N – nucleocapsid
Np9 – nuclear protein of 9000 daltons
RA – rheumatoid arthritis
TF – transcription factor

10

Vaccines and the Conquest of Viruses

Keywords Inactivated vaccines • Attenuated vaccines • Subunit vaccines • mRNA vaccines • Anti-vaccine movement

Throughout human history, viral diseases have had their way with us, and for just as long, we have hunted them down and done our best to wipe them out.
Jeffrey Kluger

The Era of Vaccines

The nineteenth and twentieth centuries saw many remarkable advances in health and medicine, including such life-saving discoveries as anesthetics, antibiotics, imaging technologies (e.g. X-rays, MRIs, PET, and CT scans), and organ transplants. Among this list of advances, vaccines certainly deserve a high place as collectively these disease-preventing prophylactics saved an enormous number of lives. The World Health Organization (WHO) conservatively estimates that vaccines currently save 2–3 million lives per year with more liberal estimates closer to 5 million lives saved. Since the first vaccine was administered in 1796 by Edward Jenner, the cumulative estimates indicate that vaccines prevented hundreds of millions of unnecessary deaths. No other single medical advance has saved as many lives and stopped so much disease and human misery. In addition, by preventing diseases from even occurring, vaccines save billions in medical treatment and recovery expenses. While there can be adverse effects associated with vaccines, these are mostly quite mild and manageable, and the overall benefit from vaccines for both individuals and society vastly outweighs the small risk.

© The Author(s), under exclusive license to Springer Nature Switzerland AG 2022
V. G. Wilson, *Viruses: Intimate Invaders*, https://doi.org/10.1007/978-3-030-85487-4_10

As detailed in Chap. 6, modern vaccination began with Edward Jenner's use of live cowpox virus as a surrogate for smallpox. Without any knowledge of viruses and no understanding of the immune system, he intuited a fundamental concept: infection with and recovery from a mild disease agent can confer resistance to a related severe disease agent. By intentionally infecting people with relatively harmless cowpox he was able to generate an immune response that recognized the virulent smallpox virus and effectively neutralized it to prevent disease. We now know that cowpox and smallpox viruses are closely related and share significant antigenic similarities which allow the development of a cross-reacting immune response. Even the minimal disease that humans developed from cowpox infection was sufficient to evoke a protective immunity against other poxviruses, making this approach feasible and effective. However, Jenner was fortunate that the human virus, smallpox, had close relatives in domestic animals (cows and horses) that provided a safe and accessible source of material for his vaccination process. A similar situation does not exist for most other significant human viral diseases, therefore Jenner's success with cowpox as a smallpox vaccine remained a unique and unduplicated feat for nearly a century.

The concept and application of vaccination finally expanded from the isolated success against smallpox to a more generally available practice in the late 1800s, thanks to the insightful efforts of Louis Pasteur and many of his contemporaries. Pasteur's seminal studies contributed to the proof that microorganisms caused disease and led to the first laboratory-created vaccine against chicken cholera (a bacterium), followed quickly by a vaccine against another bacterial disease, anthrax in cattle. In both cases, the vaccine was created by developing weakened or attenuated forms of the bacteria that caused little disease yet evoked protective immunity. The success of these vaccines against livestock diseases encouraged Pasteur to pursue a vaccine against rabies which was increasing in feral dogs in Paris. While he was not aware that rabies is a viral disease instead of a bacterial one, he utilized the same approach and eventually developed a procedure to attenuate the rabies agent. By 1884 he proved the efficacy of his vaccine by showing that vaccinated dogs were protected from disease when injected with virulent rabies. The next year, three individuals bitten by rabid dogs were given the vaccine in a humanitarian effort to save their lives, and none developed rabies. These human trials not only proved the general efficacy of Pasteur's vaccine approach but also demonstrated that vaccination can be successful after exposure to disease if the disease has a long incubation period (weeks to months) as does rabies. The next several decades brought similar success as many groups developed vaccines against prevalent diseases, including plague, human cholera, and typhus.

However, these subsequent vaccines were all against bacterial diseases as viruses were still poorly understood and mostly impossible to propagate outside of infected animals or humans.

Fortunately for the world, some progress with viral vaccines was made even with this limitation, for example, a vaccine for yellow fever was developed in the 1930s. Yellow fever is caused by a flavivirus that spreads via mosquitos in tropical regions of Africa. The virus causes a hemorrhagic infection characterized by fever, headache, nausea, vomiting, jaundice, muscle pain, and fatigue with a significant fatality rate. In the 1600s and 1700s, the African slave trade brought the disease to North and South America and even into Europe where it remained problematic throughout the 1800s and into the early 1900s. During the Spanish-American war in 1898, estimates suggest that over 10 times more American soldiers were killed by yellow fever than by combat during the invasion of Cuba. This devastating effect on the military, combined with the disease persistence in U.S. territories, stimulated an American effort to find the source of the disease and develop a vaccine. While it took another three decades, eventually Max Theiler of the Rockefeller Institute in New York City was able to grow and propagate the virus in chicken eggs. By passing the virus through successive rounds of growth in eggs he created a weakened strain of the yellow fever virus that he named 17D. The 17D strain proved to be an effective attenuated vaccine, a discovery that earned him the 1951 Nobel Prize in Physiology or Medicine. Theiler's 17D strain remains the basis for the yellow fever vaccine to this day, and the vaccine along with massive mosquito control programs reduced yellow fever to a manageable disease with a much more limited geographic distribution.

Both the U.S. military and chicken eggs also played critical roles in a second early twentieth-century vaccine, the influenza vaccine. The horrible influenza pandemic of 1918–1919 killed roughly 50 million people worldwide, with many U.S soldiers dying from influenza, both at home and in Europe. Some of the first influenza cases on American soil occurred on military bases, quickly sensitizing the military that they needed to protect troops. At this time the cause of influenza was still unknown though a few scientists were beginning to suspect a viral origin for this disease. However, the causative virus wasn't confirmed until a British research group published studies in 1933 showing the isolation and transmission of a virus that we now designate influenza A. It was soon discovered that this virus could be grown and propagated easily in fertilized chicken eggs, and this provided a large-scale source for viral production and vaccine development. Efforts in Great Britain and the United States quickly focused on finding a workable vaccine, with both countries eventually developing inactivated virus vaccines. In the United

States, Thomas Francis, head of the US Army Commission, led the work with colleagues that included Jonas Salk of subsequent poliovirus vaccine fame. They used the chemical formalin to inactivate the influenza A virus and tested this killed virus as a vaccine for the U.S. military. When a second influenza virus, type B, was discovered in 1940, the initial formalin vaccine was modified to include both the A and B type influenza viruses. This bivalent vaccine was widely administered to U.S. troops during World War II and by 1945 was considered safe and effective enough to be licensed for civilian use. Salk would utilize his experience with formalin and influenza virus as the basis for his future work creating the formaldehyde-inactivated poliovirus vaccine. While the yellow fever and influenza virus vaccines were great successes, there were many more common and medically problematic viral diseases rampant in the early twentieth century whose viral agents were not amenable to growth in chicken eggs, stimulating the search for new methods.

The technological advance needed for widespread viral vaccine development was the discovery of cell and tissue culture conditions that allowed the growth and propagation of viruses in the laboratory. Much credit goes to Ross Harrison, an embryologist at Yale University who developed a procedure for growing frog tissue in a solution of lymph fluid. Because his solutions quickly became contaminated with bacteria, he developed a series of practices for sterilizing all the required materials and equipment. These practices collectively became known as aseptic technique, and these sterile methods remain a cornerstone of cell culture work to this day. Ultimately he was able to keep his tissues viable and uncontaminated for weeks at a time which allowed him to observe the growth and development of nerve fibers in culture. Subsequently, Montrose Burrows and Alexis Carrel adopted Harrison's techniques to create cultures from tissues of chickens and several mammals including dogs, cats, and rats. By testing different salts and nutrients they created complex liquid media that could sustain these tissues for months, coining the term "tissue culture" to describe the procedure. Cells established directly from animal tissues are called primary cultures, and typically such cells will only persist for weeks to months before they senesce and die. Surprisingly, Burrows and Carrel also established the first permanent cell line in 1912 from tissue derived from a chicken embryo heart. Rather than senescing, this line of cells grew continuously for 34 years until it was finally destroyed after Carrel's death. However, their immortal chicken line remained an anomaly for over 30 years as attempts to create other continuous lines repeatedly failed due to cellular senescence.

From the early introduction of aseptic technique and the initial primitive culture media of Burrows and Carrel, the subsequent decades produced several other important scientific innovations that greatly advanced the cell culture field. Much work went into identifying the nutrients and compounds that cells need for proper survival and growth in culture. This work eventually led to the creation of defined media with consistent properties such as the eponymous Basal Eagle Medium (BME) and Dulbecco's Modified Eagle's Medium (DMEM), named after Drs. Eagle and Dulbecco, respectively. Another important advance was the use of an enzyme called trypsin to create single-cell suspensions. Introduced by Peyton Rous of Rous Sarcoma Virus (RSV) fame, tissues treated with trypsin dissociated into single cells that could be dispersed in liquid media and placed into a culture dish. The single cells would settle out of suspension, adhere to the surface of the dish, and begin to grow and divide. Once the dish was filled with cells, the cells could be retreated with trypsin to release them from the dish. The released cells were put back into suspension, divided into equal aliquots, and seeded into multiple new dishes. After seeding, the cells adhered and grew to fill each of these new dishes so that now the original one dish of cells had become many dishes. This splitting process could be repeated over and over to yield more and more plates of cells until the culture eventually senesced and died. As it was easy and efficient, trypsinization became another standard technique for cell culture that is still used today, and scientists call this process "splitting cells" or "passaging cells". One of the problems that arose from cell splitting was bacterial contamination introduced during the manipulations. The development of antibiotics in the 1930s and 40s was quite helpful as culture media could be supplemented with antibiotics to prevent this inadvertent bacterial contamination that ruined cell cultures. In addition to biological innovations, there were also equipment advances that greatly facilitated cell culture, for example, biological safety cabinets. These devices used filtered airflow to provide sterile working environments that protected both the cultures and the people working with the cultures. Scientists sat outside the cabinets with only their arms inside where they could manipulate the cell cultures with minimal risk of contamination from organisms in the outside air. Growth chambers specifically for cell cultures were also invented to maintain temperature, humidity, and CO_2 levels at the optimal value for cell cultures. Colloquially these devices have become known as CO_2 incubators, and they are found in virtually all of the thousands of labs today that work with cell cultures.

The biological and instrumentation advances in the first half of the twentieth century allowed cell culture to become a widespread and fundamental component of biological research touching every subfield from anatomy to virology. Still, the process of cell culture was frustrating due to the inability to keep cells alive permanently. Having to go back to a new animal or human tissue every time a cell culture died out added time, expense, and unwanted heterogeneity into every study. The breakthrough occurred in the 1940s when Wilton R. Earle at the National Cancer Institute used carcinogens to transform mouse fibroblasts. Unlike normal cells, these transformed mouse cells were immortal and could be passaged indefinitely without senescence. As all the descendent cells were identical to the original parental cells these transformed cultures were highly homogeneous. Cultures with this immortal property are called cell lines to distinguish them from transient primary cultures, and Earle's mouse cells became the L cell line, the first of many to come. One of the next, and now most famous cell lines, was the HeLa line. Derived from the cervical cancer tissue of a woman named **He**nrietta **La**cks, HeLa cells were immortal and prolific reproducers that quickly became a favorite of any scientist who wanted to work with human cells. Many years later it was determined that HeLa cells originated by transformation with the human papillomavirus type 18 (HPV 18) that is a frequent cause of cervical cancer. The creation of the HeLa line spurred attempts to establish other cell lines from different cancer types with great success. As various animal and human cell lines were developed, virologists now had a constant source of homogeneous, permanent cell lines in which to grow viruses for study and vaccine development.

Due to the technical advances in both cell culture and our basic understanding of virus biology, the roughly 15-year period from the mid-1950s to around 1970 was the "golden age" of viral vaccinology. Starting with the inactivated polio vaccine in 1955 (see Chap. 6), a succession of vaccines against common bacterial and viral diseases emerged. Scientific efforts focused first on the most serious diseases, and as these were conquered then less serious illnesses were addressed. Victories during this era included the oral poliovirus vaccine in 1962, the measles vaccine in 1963, the rubella ("German measles") vaccine in 1966, the mumps vaccine in 1967, and the trivalent measles/mumps/rubella (MMR) vaccine in 1971. The combined effect of these vaccines on viral disease morbidity and mortality was astounding. The formerly rampant childhood illnesses of polio, measles, mumps, and rubella all diminished drastically. For example, in the 5 years before the measles vaccine was introduced in 1963, the United States averaged a little over 500,000 cases per year. By 1968 the number of annual measles cases had declined to approximately 25,000 per year, and by the 1980s it was down to our current rate of a few hundred cases a year. There was a corresponding decrease in measles deaths from roughly 500 deaths per year to zero deaths in most years since 1980. Similar declines were seen for the other major childhood diseases after vaccine introduction, emphasizing just how effective and impactful vaccines have been in modern medicine. After the golden era, there have been continual advances with several new and important viral vaccines developed for hepatitis A, hepatitis B, rotavirus, human papillomaviruses, chickenpox, and SARS-CoV-2. Still, there are important viral diseases that we haven't conquered, for example, HIV and hepatitis C virus, so the quest for new, better, and more innovative vaccines continues.

Vaccinology 101

All vaccination is based on the concept of immune memory. When we encounter a new pathogen, our intrinsic and innate immune systems attack the invader and try to suppress the infection (see Chap. 5). If not completely successful, the continued presence of the virus will invoke the acquired immune response consisting of the humoral and cellular branches. The humoral response induces B lymphocytes to produce antibodies against viral antigens, typically proteins. In most cases, a portion of these antibodies will be neutralizing antibodies that bind virion surface proteins and block the virus from interacting with its receptors on host cells. The cellular immune response activates our T lymphocytes which protect us in several ways. Certain activated T cells directly kill viral-infected cells, some function to assist B cells with antibody production, and others secrete cytokines that have many antiviral effects, including stimulating inflammation. The acquired immune response helps us recover from the infection, but this process typically takes 1–2 weeks to develop which means we often get sick before there is sufficient immunity to rid our bodies of the new pathogen. Fortunately, exposure to the pathogen also creates immune memory by stimulating the formation of long-lived memory B and T cells. Once we recover, the immune memory cell population persists for many years to decades, though they can dwindle over time with aging. If that same pathogen is encountered again then the immune memory cells reproduce and differentiate into activated B and T cells very quickly. Now the immune protection occurs within hours to a few days rather than taking the 1–2 weeks as on initial infection. Because this memory response, also known as the anamnestic response, is so rapid we rarely get sick on subsequent exposures as the immune response stops the pathogen before it reproduces enough to cause disease. Vaccinations work by mimicking pathogen exposure to elicit the protective memory immune response but without causing disease.

All of the initial viral vaccines were either live attenuated viruses or killed viruses (Fig. 10.1). Live attenuated viruses are mutant derivatives of the wild-type pathogen that have lost the ability to cause severe disease. Attenuated

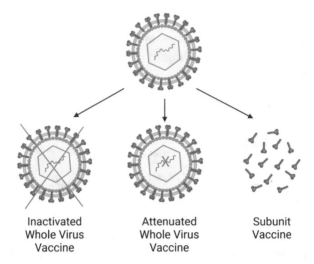

Inactivated   Attenuated   Subunit
Whole Virus   Whole Virus   Vaccine
Vaccine   Vaccine

Fig. 10.1 Types of vaccines. Viral vaccines are categorized as inactivated vaccines, attenuated vaccines, or subunit vaccines. Inactivated vaccines use whole virions and chemically treat the virions to render them noninfectious. The virions still retain their morphology and antigenicity which allows them to induce neutralizing antibodies. Since the inactivated vaccine virus is "killed" it cannot cause the original viral disease. Attenuated vaccines use a mutated version of the pathogen that has greatly reduced virulence. These weakened viruses still infect cells and replicate to a limited degree which generally induces a robust and long-lived B and T cell immune response. However, attenuated vaccines may still cause mild disease in normal recipients and should not be used in immunocompromised patients or pregnant women in whom they might cause serious disease. Subunit vaccines use only a portion of the virion, typically one or more surface proteins that will invoke neutralizing antibodies against the whole virion. The vaccine can consist of purified viral protein or can be the gene (DNA) or mRNA that encodes the target protein. The pathogen gene is often administered by cloning it into a harmless virus to make a chimeric delivery virus while mRNA is more commonly delivered via lipid nanoparticles. Subunit vaccines contain neither the whole virion nor the complete viral genome, therefore, they cannot cause the viral disease. (Created with BioRender.com)

viruses make excellent vaccines because they still invade the cells of the recipient and replicate, albeit weakly. This low-level infection evokes both a humoral and cellular immune response, producing B and T memory cells just as for an authentic infection. The combined immune response to attenuated virus vaccines tends to be strong, highly protective, and long-lived after a single dose of the vaccine. On the negative side, these vaccines contain live, replicating viruses that may cause mild disease symptoms in a subset of recipients. Furthermore, such vaccines can't be given to immunosuppressed individuals or pregnant women due to the possibility that even the weakened virus can cause serious disease in these people and/or harm the developing fetus. Rarely, even immunocompetent individuals can suffer disease from attenuated vaccines due to viral reversion events. While replicating, the attenuated virus can acquire mutations and become dangerously virulent again as is well-documented for the oral polio vaccine. Consequently, while quite effective, attenuated vaccines can have safety concerns and may not be the ideal approach for some viruses.

In contrast to the attenuated vaccines, inactivated virus vaccines chemically treat the virions which destroys their ability to replicate and cause disease. Typically this involves treating the virions with formaldehyde which leaves the individual virus particles intact, just "killed". Because the injected virus is already "dead" there is no risk of infection or reversion with this type of vaccine and it can be used on anyone. Fortunately, when injected into someone, even though it can't replicate, the killed viruses still present their surface antigens in a way that elicits neutralizing antibodies and immune memory. However, killed vaccines often don't produce immunity that is as robust or long-lasting as live vaccines. Due to this reduced immunogenicity, killed vaccines may require multiple injections and/or periodic booster shots to generate and maintain immunity sufficiently high enough for protection against the target organism. For example, the hepatitis A vaccine is an inactivated vaccine that requires two shots to gain full immunity.

To simplify and standardize vaccine composition and manufacturing, as well as reduce possible side effects caused by the complex antigenic mixes present in vaccines made with whole bacteria and viruses, the concept of subunit vaccines (Fig. 10.1) arose in the 1970s. A subunit vaccine uses only a portion of the pathogen in the vaccine rather than the whole pathogen as in attenuated and inactivated vaccines. Typically for viral vaccines, the portion used is one or more surface proteins from the virus that are important targets for neutralizing antibodies. Obviously, this requires knowledge of what proteins are located on the viral surface and which one(s) can be neutralized by antibody binding to prevent virion attachment to or uptake by host cells. However, once the correct target protein on the virus is identified, the gene for the protein can be cloned into an expression vector. At this point, the expression vector is transferred into bacteria or yeast cells where the viral protein is commercially prepared in large quantities. Proteins produced in this fashion are purified, quantified, and standardized to create highly uniform batches for clinical use. Defined quantities of the final purified viral protein become the vaccine, and when injected into people it induces antibodies and memory B cells. Upon any future exposure to the virus, the memory B cells will quickly produce the antibody which will neutralize the virus and protect from disease. Because the recipient is only receiving purified viral protein, there is no chance for the vaccine to cause the viral disease, making subunit vaccines quite safe. However, similar to inactivated vaccines, the subunit vaccines may require multiple rounds of inoculation to produce a strong and effective immune response. To date, there are subunit vaccines against two viruses that are licensed in the United States, the hepatitis B virus (HBV) and the human papillomaviruses (HPVs). In both cases, the vaccine is a viral surface protein, the S protein for HBV, and the L1 capsid protein for HPVs.

The Road Is Long with Many a Winding Turn[1]

The concept that there are only three basic types of vaccines, attenuated, inactivated, and subunit, may suggest that it is a simple process to create a new vaccine against any desired virus. This is far from the case as a typical vaccine may take 5–10 years and $500 million to go from initial identification of a new virus to a vaccine available to the public, assuming that an effective vaccine is even feasible. A priori, what vaccine type will prove most amenable for any given virus is not usually obvious, thus multiple vaccine types must be considered. Even before beginning vaccine development, there must be basic science studies of the virus's properties coupled with medical studies to understand how the virus infects and causes disease. This so-called "discovery" phase provides critical information on the virus-host interaction that informs subsequent decisions about what might or might not constitute a safe and effective vaccine. Much of the work is performed in academic or government labs as hundreds of independent scientists each pursue their own research agendas. This period can last several years and might explore dozens of potential vaccine candidates. For example, if an attenuated vaccine is desired then strategies to create, identify, and verify an attenuated strain must be developed and implemented. Similarly, if an inactivated vaccine is the goal then methods to inactivate the virus while retaining antigenicity must be developed and tested. Lastly, to produce a subunit vaccine the appropriate viral surface protein(s) must be identified, cloned, and expressed in a way suitable for future purification and large-scale production. Failure at any step along the way can eliminate that potential vaccine candidate and require restarting with an alternative approach.

Ultimately, the top candidates will go into the pre-clinical phase where they are tested in several animal models to ensure safety and immunogenicity, i.e. the ability to evoke an immune response. Many factors will be examined

---

[1] Lyrics from "He Aint Heavy He's My Brother" by the Hollies, 1969.

including dosage, number, and timing of the inoculations needed for a maximal immune response; type of immune response developed (B vs T cell); duration of immunity; the effect of different adjuvants (additives that boost the immune response); and the range of adverse effects. Once conditions are established for an effective immune response there may also be challenge studies where vaccinated animals are exposed to the virus to see if the immune response actually provides protection. Many promising vaccine candidates are lost due to failure in one or more of these categories, and usually only a small number of candidates make it through the pre-clinical stage into the human trial stage.

Human trials consist of four phases that test the safety first and then the efficacy of the vaccine. In phase I, a small number of healthy adults, usually less than 100 people, are administered the vaccine to test for safety and an immune response. The number of participants in phase I is small since even vaccines that were safe in animals may have unexpected adverse effects in humans. Minor adverse effects such as mild fever, soreness at the injection site, and even fatigue or vomiting are usually acceptable, but serious events like high fevers, seizures, or severe allergic reactions would be grounds for failing the vaccine candidate. Additionally, animal and human immune systems are not necessarily equivalent in their ability to respond to antigens, consequently, vaccines that produced strong immunity in test animals may show a meager response in human subjects. Only candidates that show effective immune stimulation without significant adverse events pass phase I and move on to phase II. In phase II, the candidate vaccine is tested on a larger group typically consisting of several hundred subjects. The gold standard for phase II is a randomized and controlled study that includes a placebo group that receives a mock vaccination. Phase II is similar to phase I in that it is again testing safety and immunogenicity, but on a larger scale where less frequent adverse effects may be detected and the immune response can be measured in a more heterogeneous and diverse group of individuals. The larger number of subjects also provides the opportunity to use subgroups of trial participants to test parameters that can affect the immune response such as various dosages of the vaccine material, use of multiple inoculations, and the most effective interval between inoculations. If safety issues arise or the immune response is not sufficient there may be attempts to reformulate and retry the vaccine candidate which can add extra time to this phase. The majority of vaccine candidates never make it past phases I and II.

With luck, skill, hard work, and good science a small number of vaccine candidates reach the phase III step where they are tested in a large cohort involving thousands to tens of thousands of people. The goals of this phase are to determine if there are rare adverse effects that didn't show up in the smaller trials, to assess the immune response in a much larger population sample, and to see if the vaccine can protect from the pathogen of interest. As in phase II, phase III studies use a randomized assignment of participants in either the vaccine or placebo group. Additionally, the best studies are double-blinded where the samples are coded such that neither the researchers nor the recipients know which is the vaccine group and which is the placebo group. Only after the study is completed will the code be opened and the data analyzed. This double-blind protocol ensures that there will be no bias, intentionally or unconsciously, that could falsely influence data collection about adverse effects and disease symptoms. In these phase III studies, the participants lead their normal lives where they may be exposed to the pathogen of interest in the community just like anyone else who is not part of the study. Subjects are followed to determine if the incidence and/or severity of infection by the pathogen of interest is reduced in the vaccine group compared to the placebo group. Depending on the type of pathogen and the prevalence of the disease in the subjects' communities, these studies can take from months to years to give statistically valid assessments of the vaccine's protective efficacy. Alternatively, some studies allow the direct challenge of the vaccine recipients with the pathogen, and their clinical responses are compared to the standard responses seen in individuals with community-acquired infections. While this approach can yield much faster information regarding vaccine efficacy, it carries a much higher risk for the participants and raises serious ethical concerns if the pathogen can cause severe or fatal outcomes. Regardless of the trial format, for a vaccine to go forward it must show significant ability to prevent disease and/ or reduce disease severity with no contraindicating adverse effects.

Successful vaccines are eventually licensed and marketed to the public, but that doesn't stop the testing. The widespread distribution of the vaccine to hundreds of thousands to millions of individuals is often considered phase IV. Information is collected about the efficacy and adverse effects to ensure that the vaccine is protective in all recipient population groups and to monitor for very rare adverse effects that were too infrequent to detect in the smaller numbers of phase II and phase III participants. For example, in 1998 a vaccine for rotavirus was marketed under the name Rotashield by the American pharmaceutical company Wyeth. Rotavirus causes diarrheal disease in infants and is a significant cause of infant mortality in many regions of the world, therefore a vaccine was urgently needed. The Rotashield vaccine successfully passed all three phases of clinical trials without significant problems, but within a year of release evidence accumulated that the vaccine may have caused a bowel obstruction known as intussusception in roughly 1 in 12,000 infants. Due to this health concern, the vaccine was withdrawn from the market in 1999, and it would take another 7 years before a new and safe version of the vaccine was approved for use in the United States. Similar monitoring occurs for all vaccines to ensure public safety and to detect any problems as rapidly as possible.

Anti-Vaxxers: The Force Awakens

From the time of their inception, vaccines have provided an enormous public health benefit by eliminating or reducing the incidence of many deadly diseases. Vaccines work in two ways, by directly protecting the vaccinated individual and by helping to create herd immunity. Herd immunity is the concept that pathogen spread through a population is reduced as more and more of the population develops immunity, either through natural infection or vaccination. When a new virus is introduced into an immunologically naive population then the virus can spread very effectively because everyone an infected person interacts with is susceptible. Epidemiologists talk of the viral reproduction number, R_0, which is the average number of individuals who become infected from a single infected person. R_0 reflects both the inherent transmissibility of the virus as well as the social, environmental, and host factors that contribute to or restrict spread. In the initial naïve population the R_0 will be its highest since everyone is susceptible to infection. However, as the virus spreads through the population individuals either succumb to the infection or recover and gain immunity. Over time there are fewer and fewer remaining vulnerable individuals who might come in contact with an infected person. This reduces the opportunities for virus spread, and eventually, there are so few susceptible individuals that viral spread is greatly curtailed and herd immunity is achieved. Mathematically, the percentage of the population that needs to be immune to achieve herd immunity (HI) is inversely related to the R_0 and is expressed by the formula HI = $100 \times (1-1/R_0)$. For example, if the R_0 is 2 then the HI is $100 \times (1-1/2)$ which equals 50%. An HI of 50% means that only 50% of the population needs immunity for viral spread to be greatly curtailed. In contrast, for a virus such as measles with a high R_0 of near 10, then the HI will be $100 \times (1-1/10)$ and 90% of the population needs to be immune to stop viral transmission. Historically, anti-viral immunity was only achieved through illness and recovery, resulting in many people dying before herd immunity was generated to stop pandemics. In the twentieth century, science and medicine changed that dynamic with vaccines that created widespread herd immunity without the illness, suffering, and deaths of all the preceding centuries of human existence. Maintaining herd immunity is critical to protect infants and children who haven't yet been vaccinated, the immunocompromised who are highly susceptible, and the elderly whose immunity may have waned with age. Yet the remarkable vaccine achievements of the past 100 years are challenged today by an increasingly vocal anti-vaccination sentiment that threatens to disrupt herd immunity.

It probably won't come as a surprise, but the objections to vaccines are as old as vaccines themselves. Following Edward Jenner's seminal work with cowpox as a vaccine for smallpox, vaccination spread quickly in England in the early 1800s and later in the United States. Almost immediately critics arose who condemned the practice because they distrusted the science, objected for religious reasons, or simply believed that any attempt to impose vaccination violated individual freedom and liberty. By the mid-1800s, England legislated mandatory smallpox vaccination for children with penalties for refusing to comply. This edict galvanized the resistance into formal opposition such as the Anti-Vaccination League and the Anti-Compulsory Vaccination League. The continued growth and strength of these movements eventually led to a modification of the law described in the Vaccination Act of 1898. This new law created the conscientious objector option which allowed parents to opt-out of the vaccination requirement, and it also removed the previous penalties for noncompliance. Similar anti-smallpox vaccine movements were also occurring in the United States as many states began to enact compulsory vaccination laws. Local pockets of opposition coalesced with the formation of the Anti-Vaccination Society of America in 1879, the New England Anti-Compulsory Vaccination League in 1882, and the Anti-vaccination League of New York in 1885. These large organizations waged legal battles against city and state laws that culminated with a 1902 challenge against the city of Cambridge, Massachusetts. Due to a large smallpox outbreak, the city mandated that all residents be vaccinated, a mandate rejected by one resident who went to court on the basis that the law violated his civil rights. The case ultimately went to the Supreme Court who made a landmark decision in 1905. The court ruled that the state's responsibility to ensure public health took precedence over the rights of the individual, a ruling that is foundational for many public health measures implemented since that initial case.

As the early twentieth century progressed, public acceptance grew for science in general and vaccination in particular. With the continued outbreaks of serious bacterial and viral diseases, such as diphtheria and polio, the public began to clamor for vaccines. The breakthroughs in cell culture and associated technologies enabled a steady stream of new vaccines during the "golden era" of vaccines (the 1950s to the early 1970s). Many once prevalent and/or deadly childhood diseases were virtually eliminated soon after their respective vaccines were introduced. As the incidence of these previously common diseases dramatically decreased, familiarity with these illnesses quickly faded from the public consciousness and new generations of parents and children arose who had never seen or experienced these horrible infections. Without any personal experience with these diseases, the necessity for the numerous childhood vaccinations became unclear to many parents, and the dangers of the mostly minimal side-effects became magnified. This lack of perspective concerning the risk/benefit ratio of vaccines, along with misinformation and misunderstanding about vaccine science, made many people increasingly wary of vaccinations in general. The first major backlash occurred in Great Britain in the mid-1970s. Two case reports suggested a connection between the vaccine for pertussis (whooping cough), a bacterial disease, and neurological complications in infants and young children. These reports led to a drastic decline in vaccination from around 80% of children inoculated in 1974 to only 31% vaccinated by 1978. Sadly, as vaccine levels declined and herd immunity diminished, Great Britain suffered a major whooping cough outbreak from 1977–1979 in which there were over 100,000 cases with 5000 hospitalizations and 36 deaths. A similar alarm was raised in the United States in the early 1980s, but by then there were carefully conducted studies that concluded that the risk for permanent neurological damage from the pertussis vaccine was 1 in 310,000 individuals. While not zero, the danger from the vaccine was much less than the danger from the disease seen in the British outbreak. With this new data and strong advocacy for the vaccine from the American Academy of Pediatrics, resistance to this vaccine never reached impactful levels in the United States.

Modern anti-vaccination sentiment really began with a 1998 report published by Dr. Andrew Wakefield in the journal *The Lancet*. The study purported to link the measles, mumps, and rubella (MMR) vaccine with the development of autism in 8 children. With the burgeoning rates of autism in the Western world, many parents in Europe and North America seized on the MMR vaccine as a root cause of autism. The growing criticism of the MMR vaccine following the Wakefield report led to its dwindling acceptance. As protection from measles began to wane, some American communities were left vulnerable to this highly contagious virus. While measles was declared eradicated in the U.S. in the year 2000, this virus is still prevalent in many other countries. Consequently, international travel has periodically reintroduced the disease to the U.S. leading to several measles mini-epidemics over the last 20 years. The ongoing tragedy of these outbreaks is that Wakefield's study is absolutely untrue. Numerous subsequent large and carefully conducted studies by major scientific organizations have completely negated any connection between the MMR vaccine and the development of autism. Furthermore, Wakefield's study was later found to not only be poorly done and incorrect but also fraudulent. Later investigations found evidence that at least some of the data in his report was simply made up, an act of scientific fraud that led *The Lancet* to retract his paper in 2010 (retraction by a journal indicates that the journal no longer believes that the data and/or conclusions of a study are valid). Wakefield's medical license in Great Britain was also revoked and he may no longer practice medicine in that country, though he remains an anti-MMR activist.

If Wakefield's contention that the MMR vaccine contributes to autism has been repudiated, then why does a connection between vaccines and autism persist in the minds of vaccine opponents. Many sociological factors have been proposed for why this myth is maintained, but two scientific issues are relevant for our virological focus. First, Wakefield's focus on MMR has been redirected to a more general vaccine concern. Vaccine opponents have raised concerns about multiple vaccine components, including adjuvants, preservatives, and stabilizers. This broader concern has taken much longer to investigate, allowing anti-vaccine objections to linger for years as each new proposed culprit had to be studied. Second, the public often accepts the fallacy that temporal correlation equals causation, i.e. the belief that if two events are linked in time then the first event must cause the second event. In the case of vaccines, many people believe that if a child receives a vaccine and shortly later develops symptoms of autism then the vaccine must have caused the autism. The problem with this assumption is that it ignores the fact that autism is typically diagnosed during the first five years of life, a time when children are having frequent vaccinations. Therefore, no matter when autism first manifests, there is likely to have been a vaccination in the recent past that could be blamed. So yes, vaccines and autism are linked in time because they both occur in early childhood, but such temporal linkage is never proof of causation. For example, you may go to the bus stop each morning and wait for the bus to come by and pick you up. These two events are clearly linked in time, but there is no causality. Your act of going to the bus stop doesn't cause the bus to come by as the bus is coming down the street regardless of whether or not you are waiting at the stop. It is often difficult for people to accept that temporal correlation is meaningless without rigorous scientific studies, thus the autism-vaccination myth continues to resonate with the public.

As damaging as the pertussis controversy and Wakefield's fraudulent study were to public trust in vaccines, they did have one positive effect, a renewed focus on vaccine safety. Starting with the original smallpox vaccine, every subsequent vaccine carries some risk of adverse effects. Most of the common adverse effects are minor annoyances such as fever or soreness at the injection site that only persist a few days, but rare adverse effects can be significant. Fortunately, we know the average incidence rate for all these adverse events, and the chance of significant adverse consequences from any vaccine is far less likely than the chance of illness or death from the disease the vaccine prevents. Consequently, the benefits of vaccines to individuals and society greatly outweigh the rare harm to some individuals. Still, even the remote chance for serious adverse effects from certain vaccines has spurred both regulatory and scientific efforts to ameliorate any vaccine-derived injuries. On the regulatory side, the National Childhood Vaccine Injury Act (NCVIA) of 1986 created the National Vaccine Injury Compensation Program (VICP) and the Advisory Commission on Childhood Vaccines (ACCV). Through a national trust fund generated by an excise tax on vaccines, the VICP provides direct compensation to individuals harmed by certain vaccines, thus seeking to avoid contentious lawsuits. This program not only reduces the time needed to receive financial assistance for adults or vaccine-injured children but also helps protect vaccine manufacturers from financial liability that could force them into bankruptcy and out of business. The government has a national interest to ensure that the critical supply of vaccines does not disappear due to the lack of manufacturers with the expertise and facilities to produce and distribute large numbers of vaccine units. The NCVIA helps protect this interest by stipulating that manufacturers cannot be sued for unavoidable side effects that occurred with properly prepared vaccines accompanied by appropriate directions and warnings. Manufacturers can still be sued, but only for defective vaccines or failure to provide adequate warnings of known risks. By limiting the manufacturer's liability and encouraging families to seek financial help through the VICP, the known burden of injury or death to rare vaccine recipients is carried at the national level rather than by the families or the vaccine producers. To help provide oversite and advice on the VICP, the ACCV consists of nine appointed members who make recommendations to the Secretary of Health and Human Services. Three members must be healthcare professionals with expertise in childhood diseases and vaccines, three members come from the general public though at least two must be parents of children who were injured or killed by vaccines, and three members must be attorneys with one having experience representing victims of vaccine injury and one who has represented vaccine manufacturers. This composition ensures that all sides are represented and helps protect the integrity and transparency of the VICP process.

A subsequent major advance in vaccine oversight was the Center for Disease Control's (CDC) creation of the Vaccine Adverse Event Reporting System (VAERS; https://vaers.hhs.gov/) and the Vaccine Safety Datalink (VSD) in 1990. Both systems function to identify vaccine-associated safety issues, though they work in different ways. VAERS is an open system where anyone (patient, family member, friend, healthcare provider, etc.) can report a suspected vaccine-related adverse effect or just examine the information within. In a typical year, there are roughly 30,000 events reported to VAERS. The goal of this system is to collect all of the possible adverse events associated with any vaccine so that this information is in one database, accessible to anyone, and searchable for trends or suspicious correlations. However, it is critical to note that just because something is posted as an adverse event that occurred following a vaccine it doesn't prove that the event was vaccine-induced; remember, temporal correlation does not prove causation! If you were skateboarding a week after receiving a vaccine and broke your leg you could post the broken leg as an adverse event, yet it's highly unlikely the vaccine caused this injury. As with the autism concern, there are many medical issues posted on VAERS that would have occurred even without a recent vaccination, hence any single claim must be viewed with appropriate skepticism. Where VAERS is really useful is looking for adverse effects that show up at a statistically significant level in recipients of a particular vaccine compared to the general population. Some adverse events are so rare that they are not identified in phase I-III trials and only become noticeable in phase IV when the general public is reporting them to VAERS. Having this system allows scientists and clinical researchers to scour huge data sets for the anomalies that reveal new vaccine side effects. Potential problems detected are then further studied and evaluated to assess whether or not they are truly vaccine-induced. Minor side effects may simply be added to the known list of adverse events, but more serious ones may require the development of new or improved versions of the vaccine.

Like VAERS, VSD collects information on patients and vaccines, but unlike VAERS the VSD is a closed system that is not open to the general public. For VSD the data come from the electronic medical records of large healthcare providers such as Kaiser-Permanente. Collaborating providers supply the data to the CDC who monitors the data for adverse event trends connected with current vaccines. Outside investigators can apply for access to the database for specific projects that adhere to strict rules for data protection and patient confidentiality. Since 2005 the CDC uses the VSD system for Rapid Cycle Analysis where vaccine data, particularly for new vaccines, is monitored for adverse events in real-time. This allows any significant adverse effects to be detected as quickly as possible and allows the vaccine to be quickly removed from the market while further studies are conducted. Between VAERS and VSD, the safety and efficacy of vaccines are highly scrutinized and any concerns should be detected with the most minimal impact on the public.

In conjunction with the regulatory support and attention to early detection of adverse events, scientific changes and advances have also worked to improve vaccine safety. Individual components found in many vaccines, most prominently mercury and aluminum, have caused concern. Mercury in vaccines is present in the form of an organic compound called thimerosal that is added for its antimicrobial activity against the growth of bacteria and fungi. Mercury toxicity has been well known for centuries, even being the basis for the phrase "mad as a hatter" which reflected the neurological damage incurred by hatters in the 1800s who routinely used mercury in hat making. By the 1960s and 1970s, there was growing alarm that industrial pollution was leading to mercury in the environment. In particular, ocean fish were found to have high levels of methylmercury that were causing neurological problems in people with a fish-rich diet. This concern about mercury ingestion caused anxiety regarding the possible consequences of mercury accumulation in children who were receiving multiple thimerosal-containing vaccines yearly. While thimerosal contains ethylmercury that is less toxic and more rapidly cleared from the body than methylmercury, the concerns warranted additional studies of the effect of thimerosal in vaccines. No adverse consequences from thimerosal were ever established, yet on the side of caution, the United States removed thimerosal from all childhood vaccines in 2001. Other vaccines, like the yearly influenza shot, still use thimerosal, but the amount present per vaccine is less than the amount of mercury in a three-ounce can of tuna fish. Furthermore, ethylmercury is rapidly excreted resulting in little or no build-up of this material in vaccine recipients and no identifiable risk associated with its presence in a few current vaccines.

Another vaccine additive that is often questioned as dangerous is aluminum because aluminum toxicity does occur at high doses. Aluminum is used as an adjuvant that stimulates and strengthens the immune response against the target antigens in a vaccine. Typically present as an organic compound such as aluminum hydroxide, aluminum phosphate, or aluminum potassium sulfate, aluminum-based adjuvants have been used safely in vaccines since the 1930s. It is well documented that aluminum and other adjuvants do contribute to the vaccine reactions seen at the injection site (redness, swelling, and pain) or systemically (fever, chills, and body aches), as these reactions reflect immune activation. However, these minor effects are an acceptable trade-off to ensure a robust immune response that provides effective protection from the target disease, and no serious consequence of aluminum used in vaccines has ever been demonstrated. The safety of aluminum stems largely from the minute quantities used in each dose of a vaccine. The average adult ingests 7–9 milligrams of aluminum daily which is between 10 and 50 times more aluminum than in a single dose of a vaccine. In a year a typical adult will ingest over 2500 milligrams of aluminum but would only receive a few milligrams from vaccines even if they received several vaccines that year. In small children, the numbers are much different but still are not cause for alarm. In the first six months of life, a breastfed baby will ingest roughly 7 milligrams of aluminum while a formula-fed infant will ingest nearly 37 milligrams of aluminum. In contrast, during this same period infants will receive about 4.4 milligrams of aluminum in total from all the vaccines they receive. This entire contribution of aluminum from vaccines is not considered a significant increase compared to the total received from food. As with thimerosal, there is no scientifically accepted data demonstrating any serious adverse effects from the minute quantities of aluminum in vaccines compared to the much larger quantities consumed in food and water. Unfortunately, all the science and regulatory protection in the world will not allay the fears of many anti-vaxxers. For now, it is important to acknowledge that vaccines are not without risk but to stress that the benefits from vaccines greatly outweigh their known risks. In the future perhaps there will be even safer and more effective vaccines that will be universally accepted.

Vaccine Technology – A Glimpse of the Future

The three basic classes of vaccines, attenuated, inactivated, and protein sub-unit, are the mainstays of vaccine development, but several new and innovative approaches are in development, particularly regarding new strategies for subunit vaccines. Viral subunit vaccines are typically the safest as they consist of only viral protein(s) and cannot cause the pathogen's disease. However, each protein is a unique entity varying widely in its stability, ease of production, and requirements for purification. Consequently, each new protein vaccine requires extensive time and effort to scale up for satisfactory commercial production. Additionally, protein vaccines stimulate a strong antibody response but often produce little or no T cell response, so they are not evoking the full range of immune response that would be induced by a natural infection. Newer strategies attempt to eliminate these current issues by creating more convenient platforms that can be rapidly adapted to any virus once an appropriate viral protein target is identified. The two main platforms under development are nucleic acid-based vaccines and chimeric virus vector vaccines. Instead of administering a viral protein, nucleic acid vaccines use a segment of either viral DNA or RNA as the material that is delivered to the person or animal being vaccinated. The concept for nucleic acid vaccines derived from an unforeseen observation made in the early 1990s. Researchers found that the injection of pure DNA into the muscles of lab animals yielded antibodies against the protein encoded by the DNA. This result was surprising because many steps had to occur before antibodies could be generated. First, the injected DNA had to enter cells and many scientists were skeptical about the ability of cells to take in large, free DNA molecules. Without uptake, the DNA would simply stay outside the cells where it would be inert until eventually degraded. Furthermore, for protein expression to occur any external DNA that is taken up by the cell needed to move from the cytoplasm into the nucleus where DNA could be transcribed into mRNA by RNA polymerase II. Fortunately, both the uptake of DNA and its transport to the nucleus do occur to a degree, and upon reaching the nucleus the remainder of the events occur naturally. DNA in the nucleus is easily recognized by transcription factors and RNA polymerase II which results in mRNA production. Once the mRNA is made in the nucleus, it is automatically transported back to the cytoplasm for translation into protein by the cell ribosomes. Foreign proteins expressed this way are presented to both B and T cells to evoke not only antibody production by B cells but also activation of the T cell network, including the important cytotoxic T cells. In animal models, DNA vaccines often result in a strong joint immune response against this foreign protein encoded by the DNA. If this foreign protein is an appropriate viral surface protein, then the resultant B and T cell immune responses will protect the vaccine recipient should they ever encounter the real virus.

Similar to protein subunit vaccines, a DNA vaccine requires knowledge of the proper viral protein to target. However, once a target is identified, DNA-based vaccines have several potential manufacturing advantages over traditional protein vaccines. First, vaccine proteins must be expressed in some organism, usually bacteria or yeast, where they can be produced in large quantities for mass use, and this production can be problematic. A protein may express poorly, it could be toxic to the host organism, it might not fold into the correct three-dimensional shape (tertiary structure), or it can lack certain post-synthesis modifications that normally occur in human or animal cells but not in bacteria or yeast cells. Additionally, a protein may be very difficult to purify due to issues with insolubility or instability. Since every protein is unique in its properties, any of these factors can make commercial-scale production problematic, and it can take months to years to work out an adequate solution to these problems. In contrast, DNA molecules are much more homogeneous in their physical properties, are readily expressed in high quantities in bacteria, and can be purified by simple and standardized procedures that are not affected much by sequence differences. Consequently, once the right target is identified, a DNA vaccine should theoretically be much faster to produce for distribution than a corresponding protein vaccine. For the DNA vaccine, the viral gene expressing the target protein is cloned into a suitable bacterial plasmid. Once cloned, the production and purification of the plasmid DNA are the same regardless of what viral gene the plasmid carries. Therefore, all the time and effort required to design the production and purification process for each new protein vaccine is eliminated, and new DNA vaccines could be developed much more quickly. A second potential advantage of a human DNA vaccine is that the target protein would be produced inside the human cell where it should fold and be modified in an authentic form that may be more highly antigenic than the version produced in bacteria or yeast. And finally, DNA is relatively stable compared to many proteins so the storage and shelf life of DNA vaccines may exceed those of a corresponding protein vaccine.

While the technical simplicity and other potential advantages of DNA vaccines make them an attractive platform for commercial development, other problems have prevented any such vaccine from being licensed for human use although there are already veterinary applications. One technical problem has been the efficiency of immune induction. While cells do take up the DNA and express the protein, this seems to occur only at very modest levels that often produce a less than robust immune response. Much effort has gone into exploring additives and adjuvants that can boost the immune response, but such materials also increase localized vaccine reactions such as redness and soreness at the injection site. This problem is not a complete deal-breaker for DNA vaccines as the search continues for an effective immune booster with minimal adverse effects. More importantly, two major safety concerns have limited the development of this vaccine platform. First, there is the risk of adverse consequences from the vaccine DNA integrating into the chromosomes of the host cells. Any foreign DNA introduced into a cell can, at a low frequency, randomly integrate into the cell DNA. Should a vaccine DNA integrate near a proto-oncogene the vaccine DNA might activate the proto-oncogene by promoter insertion just as can occur with retroviral integration. Alternatively, the integration event could inactivate an anti-oncogene by disrupting its sequence or might cause general chromosomal instability, either of which may promote tumor formation. None of these events has occurred in animal or human trials which suggests that the risk is small, yet this risk cannot be discounted as tumorigenic adverse effects still might arise when inoculating many more people than in trials.

A second theoretical safety issue is the possibility that vaccine recipients might develop an immune response to the DNA in the vaccine. Normally we don't produce antibodies against DNA as it is a natural constituent of our cells. However, in the autoimmune disease of systemic lupus erythematosus (SLE) about 30% of the afflicted individuals do produce anti-DNA antibodies that are believed to contribute to organ damage, particularly in the kidneys. It is potentially worrisome that exposure to significant quantities of pure DNA in a DNA vaccine, especially with repeated inoculations of the same vaccine or multiple different vaccines using this platform, could induce anti-DNA antibodies that would attack the individual's own DNA in a fashion resembling that seen in a subset of patients with lupus. The potential seriousness of either of these adverse effects, cancer or autoimmune disease, continues to dampen enthusiasm for the DNA vaccine platform, shifting the focus more to RNA vaccines.

Conceptually, RNA vaccines are analogous to DNA vaccines. In both cases, the idea is for the cells of the vaccine recipient to synthesize the viral protein and present it to the immune system. For antiviral RNA vaccines, the starting RNA is the mRNA that encodes the viral protein of interest. As for DNA, the mRNA is easily prepared in research or manufacturing facilities for incorporation into a vaccine. While less stable than DNA, mRNA can be packaged into nanoparticles to enhance stability and improve uptake by the cells. Once inside the cells, the mRNA is directly translated in the cytoplasm and it doesn't need to travel to the nucleus for transcription as does DNA. The elimination of these extra steps possibly accounts for some of the greater efficiency for immune induction seen with RNA compared to DNA. More importantly, mRNA has less risk for the adverse effects seen with DNA. RNA does not integrate into host DNA so initiation of cancer via alterations in the cell DNA is highly improbable with this type of vaccine. In addition, because mRNA degrades rapidly the exogenous RNA from the vaccine is much less likely to induce anti-RNA antibodies that could have autoimmune consequences. Since it has all the advantages of DNA and none of the potentially serious adverse effects, RNA has become the favored platform, though DNA vaccines are still being explored. Before the emergence of SARS-CoV-2, there had already been clinical trials for RNA vaccines directed against influenza, HIV-1, rabies, cytomegalovirus, and Zika virus that showed promising results. The RNA vaccine technology, although not yet used in any approved vaccine, was very mature, well studied, and primed for application to other viruses. The urgency of the COVID-19 pandemic spurred heroic efforts to exploit this RNA method for the development of SARS-CoV-2 vaccines. Previous studies with the original SARS virus and the MERS coronavirus had already identified the viral spike protein as the prime vaccine target, thus eliminating much of the time-consuming characterization usually needed for new viruses. Once the SARS-CoV-2 genomic sequence was available it was relatively easy to develop the appropriate RNA for a vaccine candidate. With remarkable speed, two RNA vaccines expressing the SARS-CoV-2 spike protein were independently developed and tested by Pfizer and Moderna in less than a year. By conducting overlapping rather than sequential clinical trial phases the testing process was condensed without sacrificing any of the requisite safety and efficacy standards. The phase III clinical trials demonstrated that both vaccines were safe and highly effective at preventing serious disease and death, allowing these vaccines to be rapidly put into worldwide distribution to help curtail the pandemic. The dramatic success of these inaugural mRNA vaccines presages a future where this technology will be widely employed to create vaccines against new diseases as well as potentially replacing older traditional vaccines.

Another promising vaccine advance is the use of chimeric viruses as delivery vehicles. Chimeric viruses are artificial constructs composed of nucleic acid from two viruses, the vector virus and the target pathogenic virus. The vector virus provides the bulk of the virus genome along with the capsid (with or without an envelope depending on the specific virus) and serves as the delivery vehicle. The vector virus will either be harmless for humans or will be a formerly pathogenic virus that has key genes deleted to remove its virulence. The gene from the pathogenic virus that encodes the target protein for the immune response, for example, the SARS-CoV-2 spike protein, is then cloned into the vector virus genome. Now, this chimeric virus is the vaccine and acts in much the same way as an attenuated virus. The chimeric virus will infect cells very efficiently and will express the cloned target protein in these cells. Just as in a natural infection or with the DNA and RNA vaccines, expression of the target protein in this fashion elicits a strong B and T cell immune response that provide long-lasting protection. The key advantage of chimeric viruses over the DNA and RNA vaccine platforms is that virus infection is vastly more efficient than uptake of DNA or RNA alone, taking much less of the source material to generate robust protection. Also, once conditions are established for the growth and purification of the basic vector virus then any target gene can be added without affecting the production process. So, like DNA and RNA vaccines, chimeric vector vaccines should be easy to produce and flexible enough to use with many different target genes without major changes in the overall manufacturing process. Animal and human trials with different vector virus platforms have been conducted against human viral pathogens including HIV-1, influenza, and papillomaviruses, while several chimeric vaccines are already licensed for veterinary use. As happened with the RNA vaccines, the COVID-19 pandemic thrust the chimeric virus technology to the forefront and it was quickly adapted to develop SARS-CoV-2 vaccines. The pharma companies AstraZeneca and Johnson & Johnson, as well as the Gamaleya Research Institute of Epidemiology and Microbiology (Russia), each produced chimeric vaccines based on adenovirus as the vector that delivers the spike protein gene. All three chimeric vaccines are safe and effective at generating protective immunity, confirming that this platform is suitable for future human vaccines.

In summary, the DNA, RNA, and chimeric virus vector platforms all have multiple properties that make them clinically and commercially attractive to pursue for future vaccine development. All three platforms are excellent candidates for potentially replacing current vaccines with safer and more effective versions that can help eliminate some of the concerns of individuals who doubt the safety of vaccines. Additionally, each of these platforms should provide faster and more flexible responses to new and emerging pathogens as was proven for the SARS-CoV-2 virus causing the pandemic of 2020. Hopefully, these new additions to our anti-viral armamentarium will help make the twenty-first century the safest and most disease-free century in human history.

Additional Reading

1. History of Vaccination. Stanley Plotkin. Proceedings of the National Academy of Sciences, 111(34):12283–12287. 2014.
2. Vaccine Rejection and Hesitancy: A Review and Call to Action. Tara C Smith. Open Forum Infectious Diseases, 4(3). 2017.
3. Novel vaccine technologies for the 21st century. John R. Mascola & Anthony S. Fauci. Nature Reviews Immunology 20:87–88. 2020.

Definitions

Adjuvant – A substance added to vaccines to increase the immune response.

Anamnestic response – The immune memory that provides a very rapid immune response on second or subsequent encounters with the same antigen or pathogen.

Anthrax – A disease of cattle and sheep caused by the bacterium *Bacillus anthracis*.

Antibiotic – A compound with the ability to kill or inhibit the growth of bacteria.

Antigenicity – The ability of a foreign substance, for example, an antigen or pathogen, to be recognized by antibodies.

Aseptic technique – The practices and procedures for maintaining sterility during cell culture.

Basal Eagle Medium (BME) – A chemically defined growth solution for cell culture.

Biological safety cabinet – A device that creates a clean and safe environment for working with biological samples.

Bivalent vaccine – A vaccine that elicits an immune response against two antigens or pathogens.

Chimeric virus vaccines – Vaccines that combine parts of two viruses with one virus serving as the vector and the other virus providing the gene(s) whose protein(s) will evoke the immune response.

Cholera – A severe gastrointestinal disease caused by the bacterium *Vibrio cholerae*.

CO_2 incubators – A device for incubating cell cultures where the temperature and CO_2 concentration can be regulated.

Cytomegalovirus – A double-stranded DNA virus of the herpesvirus family.

Diptheria – An inflammatory disease of the throat and upper respiratory tract caused by the bacterium *Corynebacterium diphtheriae*.

Double-blind – An experimental procedure where neither the experimenter nor the subjects know who is in the test group versus the control (placebo) group.

Dulbecco's Modified Eagle's Medium (DMEM) – A chemically defined growth solution for cell culture.

Embryologist – A scientist who studies the growth and development of embryos.

Epidemiologist – Someone who studies and is an expert in the science of disease transmission and control.

Ethylmercury – A compound of mercury that is present in some vaccines in the form of thimerosal.

Formaldehyde – A chemical compound used to inactivate some viruses for vaccine production.

HeLa cells – An immortalized cell line derived from the cervical cancer tissue of Henrietta Lacks.

Immunogenicity – The ability of a foreign substance, for example, an antigen or pathogen, to evoke an immune response.

Intussusception – An inversion of the intestines.

(continued)

(continued)

Jaundice – A liver disease where there is an excess of the pigment bilirubin which causes a yellowing of the skin and eyes.

L1 protein – The major capsid protein of papillomaviruses.

Lymph fluid – The clear liquid containing lymphocytes that flows in the lymphatic system.

Methylmercury – A toxic compound of mercury that accumulates in animals such as fish.

Morphology – The shape, appearance, and structural features of organisms.

Nanoparticles – Extremely small particles between 1 and 100 nanometers in diameter.

Nucleic acid vaccines – Any vaccine that uses DNA or RNA as the main ingredient.

Pertussis – The formal name for whooping cough, a respiratory infection caused by the bacterium *Bordetella pertussis.*

Placebo – A harmless, therapeutically inactive substance used as a control when testing drugs or vaccines.

Plague – A febrile infection caused by the bacterium *Yersinia pestis.*

Primary cultures – A cell culture that is established by taking cells directly from an animal. Cells in a primary culture will eventually senesce and die.

Rotashield – A vaccine against rotavirus infection that was removed from the market due to adverse effects.

Splitting cells – The process of subdividing a culture of cells into several cultures to stimulate new cell division and increase the total number of cells.

Subunit vaccine – A vaccine that only uses a portion (typically a protein) of a pathogen to evoke the protective immune response rather than the whole pathogen.

Systemic lupus erythematosus (SLE) – The formal name for the disease lupus.

Thimerosal – A compound containing ethylmercury that is used as a preservative in some vaccines.

Tissue culture – The growth of tissues or cells in media to propagate them outside the host organism.

Trypsin – An enzyme used to dissociate cells from tissues or cell culture containers (dishes, flasks, etc).

Trypsinization – The process of using the enzyme trypsin to detach cells from tissues or cell culture containers (dishes, flasks, etc).

Tumorigenic – Able to cause tumors in an animal.

Typhus – A febrile disease with a rash caused by rickettsial bacteria.

Viral reproduction number (R_0) – The number of new infections directly generated by one infected individual.

Yellow fever – A mosquito-transmitted disease characterized by fever and jaundice that is caused by a flavivirus.

Abbreviations

ACCV – Advisory Commission on Childhood Vaccines
BME – Basal Eagle Medium
CO_2 – carbon dioxide
CT – computed tomography
DMEM – Dulbecco's Modified Eagle's Medium
HeLa – **He**nrietta **La**cks
HI – herd immunity
MERS – Middle East respiratory syndrome
MMR – measles, mumps, and rubella
MRI – magnetic resonance imaging
NCVIA – National Childhood Vaccine Injury Act
PET – positron emission tomography
SARS – sudden acute respiratory syndrome
SLE – systemic lupus erythematosus
VAERS – Vaccine Adverse Event Reporting System
VICP – National Vaccine Injury Compensation Program
VSD – Vaccine Safety Datalink

11

New and Emerging Viruses

Keywords Zoonotic disease • Bats • Pandemic • Influenza • Coronavirus • SARS-Cov-2

> *There is nothing so patient, in this world or any other, as a virus searching for a host.*
> Mira Grant, Countdown

One Planet, Many Viruses, Countless Opportunities for Infections

As we've progressively tamed the common human viral pathogens with vaccines, more of our attention has turned towards new or emerging viruses. While SARS-CoV-2 paralyzed the world in 2020, it is neither the first nor last new virus to emerge as a human infectious agent. Some so-called new human viruses, such as the anelloviruses and recently found members of the polyomaviruses group (Chap. 4), aren't actually new and have likely been associated with humans for much of our evolutionary history. These viruses simply weren't discovered until recent decades due to their elusive properties and lack of obvious clinical impact. Truly new human viruses are zoonotic diseases where animal viruses jump from their original host species into humans. Several of our modern human diseases, such as smallpox and measles, are examples of zoonotic jumps that occurred several millennia ago. In both these cases, the original animal viruses became established in human populations and evolved to become endemic human viruses. Such zoonotic jumps continue to happen regularly, with estimates of as many as 1–2 such transmissions per year over the last century. Luckily, most of these zoonotic infections are transient in human populations, and the invading viruses are unable to establish themselves as permanent human diseases. Infections by these crossover animal viruses are generally limited to a small number of individuals because the animal virus is inefficient at replicating in humans and/or is poor at transmitting from person to person. Additionally, sometimes these zoonotic outbreaks are recognized early and viral spread is controlled by public health measures. Due to either limited capacity to spread or our active intervention, most zoonotic infections eventually die out of human populations. Nonetheless, these transient zoonotic outbreaks can cause significant human illness when they occur. Recent examples include the 1993 Hantavirus infections in the Four Corners region of the American Southwest, the 2003 SARS outbreak in Asia, the 2012 MERS coronavirus cases in the Middle East, and the periodic occurrences of the Lassa and Ebola viruses in Africa. In each case, these zoonotic outbreaks caused human disease and deaths. Fortuitously, these outbreaks were all limited and the human infections eventually disappeared, though the viruses remain circulating in their animal hosts and can always be reintroduced into humans.

Another type of emerging viral disease is when a virus establishes itself in a new geographical niche such as the incursion of West Nile Virus (WNV) into the United States. While mostly causing asymptomatic infections, 1 in 5 people develop a febrile illness and 1 in 150 suffer a life-threatening encephalitis. WNV is a zoonotic disease endemic in parts of Africa, the Middle East, and southern Europe. Birds and small mammals serve as the reservoir, and the virus spreads from animal to animal (or from animals to humans) via mosquitos. Sometime in the 1990s, the virus appeared in the United States, presumably carried by mosquitos or birds who were inadvertent passengers on cargo ships from endemic areas. The first human cases were detected as a clustered outbreak in New York City in 1999. For such a cluster to occur there must have been a high enough number of infected animals and mosquitos in the area to make spillover into humans a likely event. This suggests that the virus was already well-established in the local animal and mosquito populations by that time. Over the next several years the virus spread westward across America with migrating bird populations, resulting in waves of human infections that slowly advanced from the Atlantic coast to the Pacific coast. WNV is now firmly established throughout the United States, adding another virus to the list of known zoonotic pathogens that can adventitiously infect citizens of this continent. Thankfully, the number of cases per year is small and direct human-to-human transmission doesn't occur so these infections remain isolated and self-limited events.

In contrast to these limited zoonotic outbreaks, there are notable exceptions where animal viruses become excellent human pathogens that permanently circulate within the human population. For example, the human immunodeficiency virus (HIV-1) originated from the simian immunodeficiency virus (SIV) in chimpanzees (Chap. 6). In humans, SIV evolved into a new virus, HIV-1, which is quite adept at human infection and person-to-person spread. Because HIV-1 circulates silently among men and women and only causes significant disease years after initial infection, public health control measures weren't enacted quickly or stringently enough. By the time the disease and its viral cause were identified, HIV-1 had already dispersed globally. Like other zoonotic diseases that successfully established themselves as human pathogens, HIV-1 is now a permanent, endemic human disease. Similarly, SARS-CoV-2 originated in some yet-to-be-determined animal (probably a bat). As a respiratory virus causing acute disease, once introduced into humans it disseminated rapidly and effectively around the planet, unlike HIV-1 with its slow progression. Given the properties and global prevalence of SARS-CoV-2, it seems inevitable that this virus will become another endemic human disease.

What is most troubling about zoonotic diseases is that we don't know where or when the next such jump will come from or whether it will have a major impact on human health. Therefore, understanding the causes and conditions that favor zoonotic transmission is important so that we may someday do a better job of preventing and mitigating future outbreaks and pandemics. Our biosphere is teeming with viruses that infect every species of plants, animals, and unicellular organisms on the planet. It was once thought that viruses mostly diverged along with their host species in a process of co-evolution. As primordial plant and animal lineages split into new species, their viral pathogens would co-diverge as well and eventually become distinct viral species. This type of linked evolution would result in viral family trees closely matching the family trees of their hosts. Instead, modern interrogation of viral genome sequences reveals that cross host species transmission is far more frequent than once believed. All animal virus families show extensive evidence of jumping across host species lines, sometimes even across distantly related species such as from fish to birds or mammals. This active cross-species transmission implies that there are thousands of animal viruses that potentially could jump to humans to cause disease. Thus, the few identified zoonotic infections that occur each year represent just a small fraction of possible animal virus invasions for which humans are at risk. Among the viral families, DNA viruses usually are more host restrained while RNA viruses more readily cross species boundaries. Fidelity differences between the polymerase enzymes that DNA and RNA viruses use to replicate their genomes may account for the greater ease by which RNA viruses jump across species. DNA polymerases generally have high fidelity and make very few errors when replicating their viral genomes. In contrast, RNA polymerases typically have much lower fidelity and introduce numerous errors (i.e. mutations) into each new genome made. This greater number of mutations in each generation of RNA viruses makes it more probable that some individual virus will have an adaptive mutation that enables it to survive and thrive in a new host environment. All it may take is one unique mutant virus to gain a foothold in a new species and spread to establish a permanent jump into this new host. As there are many more RNA virus families than DNA virus families, RNA viruses constitute the biggest risk group for future novel human outbreaks, epidemics, and pandemics.

If viruses, especially RNA viruses, can move readily from their original host species to new hosts, what are some drivers that facilitate jumping across species? First, there must be some physical event that introduces the animal virus to a human being. Depending on the virus and its normal mode of transmission this can take several routes. Live animals can shed viruses that we inhale or introduce into our nasopharyngeal region from our hands just as in human-to-human spread of common respiratory viruses. Animals can also transmit viruses through bites or scratches. Similarly, butchering animals for food exposes the worker to the animal's blood and secretions which can harbor viruses able to infect the handler through cuts and abrasions or by ingestion of undercooked meat. Lastly, many viruses spread via an insect vector, hence direct contact with the animal reservoir isn't necessary, just exposure to an infected insect. For most city-dwelling Americans with infrequent insect exposure and whose animal contact is limited to domestic pets, animal virus risk is relatively low. The risk increases for people who work with or around animals, domestic or wild, such as farmers and hunters. In countries where hunting and trapping wild animals is a major food source exposure risk to novel animal viruses is greatly enhanced. One practice identified as particularly risky is wildlife farming where animals, often exotic, are captured for live sale in urban markets. Bringing together diverse animals in close contact and with poor hygienic conditions is a stressful environment ideal for facilitating cross-species jumping. As these viruses pass from their original host into new species there is a selection for viral mutants that are more fit for these new hosts. This expanding cadre of mutant viruses increases the risk that some strain will be an effective human pathogen that can jump to an unsuspecting worker or customer. Additionally, in so-called "wet markets" the animals are not only offered whole but are slaughtered, butchered, and sold in pieces that contain blood, secretions, and internal organs that can be a source of infection. Combine this with human crowds mingling in close proximity to the animals and zoonotic spread is almost inevitable. Fortunately, the vast majority of these incidental human infections are just background "chatter", i.e. part of our normal daily virus exposure that produces no disease, and these harmless viruses are called level 1 pathogens. Nonetheless, out of all these pathogen exposures, there will be some that do cause clinical symptoms in the infected person. If the infection does not spread beyond the initially infected individual then these are defined as level 2 pathogens. These level 2 viruses are poorly adapted to humans and while potentially serious for the infected individual they pose little risk to the community at large. The real concern is level 3 pathogens where the index case transfers the virus to other humans, indicating a virus that has become sufficiently human-adapted to infect and transmit under community-acquired conditions. If the virus is very successful in transmission then it might cause epidemics or pandemics and is designated a level 4 pathogen. Unfortunately, the ease and prevalence of international travel provide a mechanism for the rapid dissemination of level 4 pathogens such as SARS-CoV-2 that roared out of China and infected the entire world in a few months. Until the practice of wildlife markets is reduced or eliminated worldwide they will remain a major concern as breeding grounds for future zoonotic outbreaks. In response to SARS-CoV-2, in early 2020 the Chinese government proposed banning the consumption and trade of wildlife animals in hopes of preventing future pandemics from originating in their country. Optimistically, other nations will also eliminate this type of threat though this strategy is not always feasible in the short term. In many countries and regions, these markets are the primary source of food for the local populace and are significant economic engines that provide jobs and income. Alternative food supplies and new job creation are critical first steps that must precede market closures to prevent local misery while protecting global health.

A second major concern for enhancing the risk of zoonotic infections is the environmental impact of deforestation and urbanization. Given the enormous breadth and diversity of known and unknown viruses lurking in animal populations, some experts assert that any change to existing ecosystems favors the crossover of viruses into new host species, including humans. The consensus among public health authorities is that human incursion, especially in tropical forests, is a major potential source for emerging viruses. A study published in 2020 examined 6801 ecological communities on six continents and found that human disruption of ecosystems was associated with increased disease outbreaks. Removal of forests reduces the natural biodiversity of the ecosystem and replaces diversity with larger populations of limited species that thrive in human-dominated environments. These human-tolerant species include rodents, bats, and small mammals, all of which are prime carriers of viruses that can readily jump to humans. Converting forests to farmland further increases risks by introducing livestock and pet animals to the transitional environment. This exposes the domestic animals to the wild animal populations, where the domestic animals can acquire new viruses that can be passed on to their human handlers. Transmission among domestic animals also provides an opportunity for novel viruses to persist and accumulate mutations that can facilitate their jump into humans. There are numerous well-documented examples of viral outbreaks occurring at these interfaces between developed and undeveloped land yet the practice continues. Stopping human expansion into undeveloped tropical areas is difficult, but as with the wet market situation, more care and planning are necessary to prevent excessive deforestation and to mitigate the potential of viral transmission into expanding urban areas. Implementation of surveillance programs will also be important to detect and identify new pathogens at the early stages of outbreaks where containment measures can be more successful at preventing spread beyond the local community. While the cost of such programs will be high, the costs are modest in comparison to the economic devastation and human misery inflicted by a single pandemic of the magnitude of SARS-CoV-2.

A second environmental issue that is raising concerns among viral epidemiologists is global warming. Rising temperatures can change the geographic distribution of animals by affecting their food sources, favoring the survival of certain species over others, and creating situations where previously isolated species may now intermix. Any of these effects can have significant impacts on different ecosystems, and again, any such disruptions may encourage viral spillover from their original hosts into new, adventitious hosts, including humans. It is difficult to predict where or how global warming might lead to new viral outbreaks, but many experts suspect it is inevitable that such outbreaks will occur. As with deforestation, it is unclear if the political will and the resources are there to combat global warming, so the rest of this century may see an even greater number of new human viral pathogens than during the twentieth century.

The Unique Role of Bats

Bats. Often considered creepy, scary, and repugnant, bats have a dark history of association with evil and disease. From vampire tales to rabies carriers, these flying mammals have long been feared even though most are benign insect eaters. While much of our fear is irrational, there are real reasons for concern about bats. In recent years, their role as disease carriers has expanded with the recognition that bats host numerous zoonotic viruses that they shed via urine and feces. Some of these viruses are highly lethal such as Marburg, Ebola,

Hendra, and Nipah, while others are less deadly but more easily spread like influenza and coronaviruses. There are over 1400 species of bats populating every continent except Antarctica, including 40 species living in the United States. These animals are quite adaptable, living in habitats from tropical forests to urban communities. To date, 137 different viruses have been found in bats, and 68 of them are capable of infecting humans. This may be a vast underestimate as most bat species have not been thoroughly evaluated for their viral populations. Only rodents collectively carry more zoonotic viruses than bats, but bats have several significant features that make them prime reservoirs for viral transmission to other animals and humans. First, bats tend to live in huge and often tightly packed colonies that provide ideal conditions for viral spread and maintenance within the bat population. Second, while rodents have lifespans of only a few years, bats live between 20 and 40 years. This is an unusually long lifespan for a small animal as there is generally a correlation between an animal's size and its longevity. Since their lifespan is very long, chronically infected bats have a greater chance than most rodents to encounter other animals or humans and pass on their viruses. And finally, bats are the only flying mammal. Flying allows them to forage over large territories that can easily bring their urine and feces into contact with animals, wild or domestic, or with humans. These intrinsic bat characteristics are compounded by human activities that promote bat contact such as farming encroachment on previously wild habitats and capturing bats for sale and consumption in wet markets. Additionally, these same human activities are bat stressors, and recent research demonstrates that stress promotes viral shedding in bats. Consequently, the more we invade their space and alter their habitats, the more we increase our risk of exposure to their bat viruses. It is no accident that many of these novel viral zoonotic diseases show up in locations of the world where urban and undeveloped areas collide. By our very actions, we are helping to create the scenarios that are most dangerous for humans.

It is not only their ecology and behaviors that make bats a major source of zoonotic viruses, it is also their unique physiology and peculiar immune systems. Bats are incredibly disease tolerant and often persistently carry multiple viruses without any apparent consequences to the host animal. Numerous recent studies, including the complete sequencing of the genomes of several bat species, have begun to provide insight into this unusual biology. One important element is that bats have a very robust interferon response. Interferon is the major anti-viral signaling messenger of the innate immune system produced and released by cells in reaction to viral infection (Chap. 5). Surrounding cells detect interferon through their surface receptors, and interferon binding to the receptors triggers a signaling cascade within the cell that turns on many genes. These activated genes express numerous proteins that collectively create an internal anti-viral state that helps protect these cells from subsequent infection. Compared to other mammals, bats have a "super-charged" interferon response though interestingly different species of bats accomplish this by at least three different mechanisms. Certain species have evolved many more copies of their interferon genes than other animals, some have a much more efficient trigger for activating interferon production, and several maintain a stockpile of pre-made interferon rather than only synthesizing it after infection occurs. Regardless of the mechanism involved, bats appear to produce high levels of interferon more rapidly than other animals, and this intense response quickly limits a virus's ability to spread within the bat. Normally, interferon also triggers an inflammatory reaction which has anti-viral benefits but can also cause significant collateral damage to the host. Bats avoid the negative inflammatory effects of their strong interferon response because they also have multiple genetic defects that impair their inflammatory mechanisms. The enhanced interferon response coupled with reduced inflammation enables bats to control viruses rapidly after infection without suffering from massive inflammatory damage. This renders bats very disease tolerant compared to other mammals but often results in their failure to completely clear infections. Instead, bat viruses can persist in the host animal and be shed for years. Coupled with their biological and social characteristics, these unique immunological features enable these flying mammals to host numerous viruses that can be a grave danger to humans and animals.

The Virus-Host Interface

Scientists estimate that there are roughly 1.6 million unknown viruses worldwide that infect mammals and birds with up to half of those viruses potentially able to infect humans. So what protects humans from this vast array of animal viruses? The ecology of birds, bats, and other wild animals that limits their contact with humans is an important element protecting us from zoonotic infections, as without contact no animal virus can jump into humans. But equally or more important are the molecular issues that animal viruses face in becoming a human pathogen. The first major barrier is the presence of a suitable human cell surface molecule that can function as a receptor for the animal virus since attachment is the obligate first step for all viral infections. This attachment is mediated by a virion surface protein, generically referred to as the viral attachment protein, that binds to a molecule on the surface of the host cell. The virion-receptor interaction is highly specific and depends on the virion attachment protein and its receptor having complementary structures that fit together with great precision. Without an appropriately shaped receptor, the virus cannot attach to and enter a cell. So unless humans have a receptor molecule that is identical or highly similar to the normal receptor for the animal virus there will be no possibility of infection. Lacking an appropriate cellular receptor, any human exposure to an animal virus is a dead end with no clinical impact, and this excludes many animal viruses from ever being human pathogens. Consequently, the farther away the original host species is from humans on the evolutionary tree the less likelihood there is of virus jumping, for example, no plant viruses are known that can infect humans. Conversely, primates and other mammals are the most common source of human zoonotic infections as these species are more closely related to us, and as close relatives, we share many of the cell surface molecules that are utilized by viruses as receptors. However, even subtle structural differences in related molecules between humans and animals can prevent effective virus binding and render the human molecule nonfunctional as a receptor for the animal virus. Using a lock (the receptor) and key (the virion) analogy, even a very small positional change in one internal pin of a lock will prevent the key from functioning. Similarly, small variations between the animal and human versions of a receptor molecule often provide a barrier to prevent the cross-species jump. Sadly for us, this barrier is not necessarily permanent due to viral mutation. While the original viral attachment protein might be unable to form a proper fit with the human receptor, mutations in the viral protein can generate altered structures that now mesh nicely with the human receptor and restore infectivity. This can occur in a single jump or may develop progressively as the virus hops through several animal species before finally entering humans.

Influenza viruses provide a well-understood example of how mutations in viral attachment proteins can influence cross-species transmission. The attachment protein for influenza viruses is the hemagglutinin protein (HA protein) which is found on the surface of the virion. The receptor for the influenza virus is a host molecule called sialic acid (SA) which comes in two forms: SA-alpha-2,3 and SA-alpha-2,6. Avian influenza viruses use SA-alpha-2,3 while human seasonal influenzas use the SA-alpha-2,6 form that is prevalent on cells in our upper respiratory system (Fig. 11.1). Normally exposure to an avian influenza virus won't cause human disease because the virus can't infect our upper respiratory tract cells which lack the SA-alpha-2,3 needed for viral attachment. However, we do have SA-alpha-2,3 on cells in our lower respiratory tract, occasionally allowing a human to contract an avian influenza infection if enough avian virus is deeply inhaled and reaches the cells with

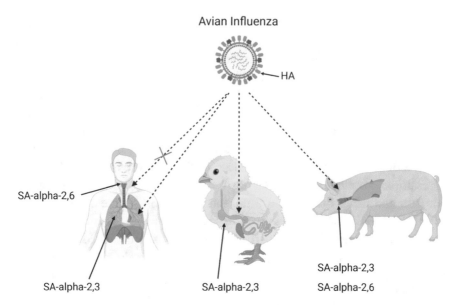

Fig 11.1 Receptors for avian influenza virus. Influenza viruses use their attachment protein, the viral hemagglutinin (HA) protein, to bind to receptors on host cells. Sialic acid (SA) in the alpha-2,3 form (SA-alpha-2,3) is the receptor that interacts with the avian influenza HA protein, while human influenza viruses use SA-alpha-2,6 as their receptor. The alpha-2,3 form of SA is found on the cells in the digestive tract of avian species, the lower respiratory tract of humans, and the upper respiratory tract of swine. Cells in the upper respiratory tract of humans have the SA-alpha-2,6 form of sialic acid and this receptor is not suitable for avian influenza viruses. However, cells with the SA-alpha-2,3 form that are present in the lower human respiratory tract can be infected by avian influenza if the virus reaches this region. Swine have both the SA-alpha-2,3 and the SA-alpha-2,6 forms on cells in their upper respiratory tract so they are susceptible to both avian and human influenza viruses. (Created with BioRender.com)

SA-alpha-2,3. Now the avian virus can infect and spread in the lower respiratory tract, and such infections can be dangerous to the infected person. However, the avian virus is poorly released from these lower respiratory tract infections so the infected person does not shed significant amounts of virus in coughing or sneezing. Additionally, any small amount of released avian virus still can't infect the upper respiratory tract of nearby humans since this region lacks the correct avian influenza receptor. As a result, there is typically little or no person-to-person transmission of isolated avian influenza cases in humans. The cross-species problem arises during viral replication in the initially infected person. Replication errors can result in mutations to the viral HA protein that allow it to recognize and bind the SA-alpha-2,6 receptor prevalent in the human upper airway. Now, this mutant avian virus can spread from the lower respiratory tract into the upper respiratory tract of the infected person where it causes typical influenza. The released mutant avian virus is easily transmitted to other humans since it uses the human upper respiratory tract SA-alpha-2,6 receptor, resulting in an avian influenza strain that spreads effectively between humans. Such mutation-induced changes in attachment proteins for other viruses can also lead to receptor switching that facilitates cross-species transmission and disease.

Once an appropriate human receptor is found and the animal virion attaches to a human cell the animal virus still faces severe hurdles to productive infection and spread. The virion must be taken into the cell and uncoated, viral genes must be expressed into new proteins, the genome must be replicated, and new virions must be assembled and released. At each of these steps of the viral life cycle there can be host immune factors seeking to limit or destroy the virus. As with the viral attachment protein-host receptor interaction, attacks by these defense factors usually require precise binding between viral and host proteins or between viral nucleic acid and host proteins. Failure to evade any of these interactions in a human cell may thwart the animal virus's ability to reproduce and spread within the human host. For example, there are numerous host restriction factors (see Chap. 5) that protect our cells by attacking invading animal viruses. One such restriction factor is the APOBEC3 family (A3 family) which has broad antiviral activity against both RNA and DNA viruses. Humans have seven A3 genes and each member of this family functions in part by attacking viral RNA or DNA and converting cytosine (C) nucleotides to uracil (U). This activity creates multiple mutations in the viral genomes that can render the virus so defective that it fails to reproduce. In particular, the A3s preferentially attack cytosines preceded either by thymine (T) or another C, called TC and CC pairs, respectively. To avoid the action of A3s, human viruses such as papillomaviruses, polyomaviruses, and some herpesviruses have evolved such that their genomes have reduced numbers of these vulnerable pairs, thus giving the A3s fewer targets to attack. An animal virus newly infecting human cells wouldn't have been under selective pressure to counteract A3s and would likely have numerous TC and CC pairs in its genome, making it highly susceptible to destruction by this defense mechanism. Similarly, all established human viruses have countermeasures to defeat other restriction factors as well as mechanisms for avoiding or delaying the human innate immune response which is very effective at suppressing viral infections. Because animal viruses haven't evolved to thwart the human immune system they typically are more susceptible to the various human restriction factors and less able to escape our robust innate immunity, thus rendering them less able to reproduce in our cells. Still, random mutations that are present in the virus population or which occurred during the initial transient replication in an infected human can produce a mutant animal virus able to bypass our host defenses.

Mutations can also help an animal virus adapt to using human proteins to facilitate viral replication and reproduction. While animal viruses struggle to avoid our host defenses, they are often dependent on certain host proteins to perform specific life cycle steps. A typical virus must interact with dozens to hundreds of host cell proteins that are requisite for performing the various steps in the viral life cycle. The lack of even one of these critical interactions can completely curtail viral replication and/or progeny virion production. If an animal virus needs a particular host protein that is lacking in human cells or if the human version of the requisite protein is structurally too divergent from its animal counterpart then again viral reproduction will be blocked and the species barrier can't be crossed. The polyomavirus family provides an illustrative example of this concept. Mouse and human polyomaviruses are highly related and are quite similar in their genomes and proteins, yet mouse polyomavirus won't replicate in human cells and human polyomaviruses won't replicate in mouse cells. To replicate their genomes each of these viruses uses a viral protein known as T antigen (see Chap. 7). T antigen binds to the viral DNA and then recruits the host DNA polymerase alpha (DNAPα) enzyme that the virus uses to start the synthesis of new copies of its genome. The interaction between T antigen and its host species DNAPα is highly specific. Mouse polyomavirus T antigen binds mouse DNAPα but won't bind human DNAPα. Conversely, the human polyomavirus T antigen will bind human DNAPα but won't bind mouse DNAPα. The species specificity of the T antigen-DNAPα interaction provides a barrier that prevents easy cross-species jumping of these viruses. Again, this barrier could be overcome if mutations of the mouse polyomavirus T antigen enable it to engage the human DNAPα protein. There are many such critical viral protein-host protein interactions in every virus infection, thus the ability of any foreign virus to reproduce in human cells has a low likelihood of success. Given the multitude of things that can prevent cross-species transmission, it is no wonder that successful zoonotic diseases are relatively rare events that represent only a fraction of the ongoing human exposures to animal viruses. Yet because viruses mutate so rapidly the threat of a successful spillover is never absent. And when animal viruses do successfully cross into humans they can have an enormous impact as seen for HIV-1 (Chap. 6) and the new SARS-CoV-2.

Once and Future Threats: Influenza and Coronaviruses

Seasonal flu, caused primarily by the influenza A virus, sweeps across the globe every year. As a respiratory virus, coughing and sneezing spread the virions out into the environment in aerosols that can be inhaled directly by those nearby or be deposited on surfaces. Virions on surfaces remain viable for hours up to a few days during which they can be picked up on the hands of unsuspecting passersby who then transmit the virus to their nose, eyes, or mouth. Combined with travel by cars, buses, trains, boats, and aircraft, the virus spread silently and invisibly around the world. While more prevalent in the colder months, possibly just because people are inside more which facilitates spread, flu is present throughout the year and never fully goes away. Year in and year out influenza kills 250,000–500,000 people globally, including 20,000–40,000 people in the U.S., yet we hardly pay any attention to these deaths in most years. Most of us just get our flu vaccination and go about our business without much concern or distress as the flu reappears each fall and winter. Primarily a killer of the very young and the very old, the rest of us tend to downplay this disease even when we occasionally suffer through a miserable case of flu. It's only when the scientists and the media warn of an impending newly emerging flu strain that most of the world pays any attention to this virus. Over the last 200 years, mankind experienced a new influenza pandemic roughly once every 10–25 years. Each new pandemic strain is very distinct from the currently circulating seasonal strain and their viral surface proteins present as new antigens. Because these new strains are antigenically novel, most people will have no pre-existing immunity, either from previous infections or from vaccines, that can protect against the pandemic strain. This lack of any pre-existing immunity means that the global population will be very susceptible to infection and the new strain is likely to cause many more infections than in a typical year. If the new pandemic strain transmits well then it usually replaces the current seasonal strain, becoming the new seasonal strain. The last new pandemic strain appeared in 2009 as the H1N1 swine flu that replaced the previous H3N2 strain, and H1N1 is now our circulating seasonal flu. H stands for the viral hemagglutinin protein and N for the viral neuraminidase protein, both of which are major surface proteins on the influenza virion (Chap. 4). Thankfully, the current H1N1 strain is not highly pathogenic and the annual fatality rate has been averaging less than 31,000 deaths per year in the United States over the last decade. However, if a new strain is also highly virulent then devasting outbreaks such as the 1918 influenza pandemic could occur.

We know from history that new influenza strains continue to arise, so where do these new strains come from and what is it about influenza viruses that make them potentially dangerous? The short answer to the first question is birds. Sequence comparisons of influenza viruses from different species suggest that the influenza virus originated in avian species and eventually spread into mammals. There are 18 known types of the H protein and 11 types of the N protein, and all type combinations are represented in influenza viruses found in avian species. As wild birds migrate within and between continents there is ample opportunity for the spread and mixing of these different influenza strains among the bird populations. One of the important and relatively unique properties of influenza viruses that helps generate new strain types is genetic reassortment. Genetic reassortment is a mixing of genetic information between two viruses that can only occur for viruses that have segmented genomes. Since the majority of viruses that infect humans do not have segmented genomes the process is mostly observed with influenza viruses. We humans have genomes with 23 DNA segments (our chromosomes), while influenza A virus genomes have 8 RNA segments. Each influenza virus RNA segment encodes 1 or 2 proteins with the H and N proteins being coded from different segments. When two different influenza strains infect the same animal there will be individual cells that have both viruses inside and reproducing. As new virions assemble within the doubly infected cell every new virion will get one copy of each of the 8 genome segments, but segments can come from either parent virus (see Fig. 4.1). For example, if an H1N1 virus and an H3N2 virus infect a single cell then some new virions will receive both their H and N segments from the same parental virus to regenerate the original parental H1N1 and H3N2 types. Alternatively, some virions will be hybrids that get their H and N segments from different parent viruses to generate the new H1N2 or H3N1 types that are called reassortants. And it isn't just the H and N genome segments that get reassorted. All 8 genomic segments can mix giving rise to 256 possible reassortants for every dual infection. Just as human sexual reproduction scrambles the chromosomes from the parents to generate unique offspring, these genetic reassortments by influenza viruses can produce highly novel and distinct new viruses with vastly different properties than either parental virus. When this reassortment process generates a virus with a unique combination of surface antigens this is called an antigenic shift, a potentially dangerous occurrence since most humans would likely have no existing immunity against this new virus. Both the H2N2 virus and the H3N2 viruses that caused the 1958 and

1968 pandemics, respectively, were reassortants between avian and human viruses. Fortunately, most such reassortants remain poorly infectious in humans and aren't significant dangers.

Another important factor in the evolution and spread of avian influenza viruses is their method of transmission. Unlike the respiratory disease in humans, influenza in birds is primarily a gastrointestinal disease. This is because the SA-alpha-2,3 molecule that the avian viruses use as their receptor is primarily found on cells in the gastrointestinal tract in birds (see Fig. 11.1). Upon exposure to the virus, birds develop a gastrointestinal infection that often produces little disease but results in large numbers of virions being shed in their feces. For waterfowl, these shed virions can contaminate ponds and lakes where healthy birds then contract the virus through ingesting the virus-infested water. This leads to widespread transmission across bird species over large geographic areas as birds travel and migrate. Importantly for humans, domestic poultry and farm animals, such as pigs, who also drink from these waters can become infected. The pig is particularly critical as an intermediate vector for influenza since pigs have both SA-alpha-2,3 and SA-alpha-2,6 (the human influenza virus receptor) molecules on cells in their respiratory tract (see Fig. 11.1). This makes the pig susceptible to both avian and human strains of influenza as well as swine influenza. The ability to be doubly or triply infected makes pigs a perfect breeding ground to generate novel reassortants that could have human pandemic potential. If a person contracts one of these new reassortant viruses from an infected pig, then the virus can spread as a typical respiratory infection in humans as did the 2009 H1N1 swine flu.

In addition to the dramatic genetic shifts caused by reassortment, influenza viruses have a highly error-prone RNA polymerase which causes mutations to accumulate rapidly during their genome replication. By some estimates, the error rate is so high that each newly replicated genome likely has 2–3 mutations. This constant mutational barrage can slowly change the antigenic presentation of the H and N virion surface proteins, a process known as antigenic drift. Antigenic drift from one flu season to the next can alter the H and N proteins sufficiently to prevent our neutralizing antibodies (elicited by previous exposures or vaccinations) from effectively recognizing these altered proteins. Reduction in the efficacy of our antibody protection necessitates getting a new vaccination every year against the current prevalent antigenic type of influenza in circulation. The vaccine manufacturers work nonstop to predict the coming major antigenic types and develop a new vaccine each year.

Creating, manufacturing, distributing, and administering millions of new influenza vaccines annually is a massive project that places a burden on the health care system and makes it hard for many people to stay compliant with their vaccinations. Until some clever researcher(s) finally solves the riddle of creating a long-lasting influenza vaccine we'll be "stuck" with yearly shots to maintain our protection.

It is this unusual combination of reassortment ability and high mutational rate that make influenza viruses so malleable and medically problematic. While we can generally handle genetic drift through our vaccination programs, it is genetic shifts that keep influenza researchers on edge. This problem was recognized decades ago and led to the creation in 1952 of the World Health Organization's (WHO) Global Influenza Surveillance Network to try to detect and interdict new strains before they become pandemic. This network was renamed the Global Influenza Surveillance and Response System (GISRS) in 2011 and continues to operate today. From a modest beginning of only 26 participating laboratories, there are now over 150 research facilities in 113 countries that comprise the network and who work collectively to monitor influenza viruses around the world. This constant tracking helps identify genetic drifts and is used to inform vaccine development for each coming year. In conjunction with animal surveillance agencies, such as the United States Department of Agriculture (USDA), a second function is the early detection of new and novel animal influenza strains to prevent or limit their spread to humans. In many cases, this involves the culling of tens of thousands (sometimes even millions) of infected domestic birds or swine in the outbreak sites to stop viral spread among the animal population and to prevent cross-over into humans. While the economic impact of these animal kill-offs can be huge, it is nothing compared to the devastation that could arise from a serious human influenza pandemic; we have only to look at the 2020 coronavirus outbreak to see how the entire world is suffering from one small new virus.

GISRS has been and continues to be a successful global model for how international collaboration and scientific excellence can help combat our viral enemies yet nothing as extensive exists for coronaviruses. Unlike influenza, coronaviruses were of marginal human concern until the SARS outbreak of 2003. At that time the two known human coronaviruses were considered merely nuisance viruses that caused common colds similar to the illnesses produced by many other relatively innocuous human viruses such as rhinoviruses. When SARS erupted it had a devastating and shocking mortality rate that was previously unknown for a human coronavirus. As expected, such a dangerous new virus grabbed headlines worldwide and immediately raised grave concerns among scientists, clinicians, and public health officials who feared a deadly global pandemic. However, since most infected individuals exhibited severe symptoms it was relatively easy to track the viral spread as victims inevitably presented to medical authorities. This allowed effective contact tracing and quarantine procedures that rapidly confined the virus, prevented a global pandemic, and squelched the local outbreak. Regrettably, the containment efforts were so successful that public concern about future novel coronaviruses faded quickly. After an initial burst of new funding and research on SARS and other coronaviruses, interest and attention waned as SARS never reappeared. Even the 2012 MERS coronavirus outbreak in the Middle East was a small blip that aroused only minimal concern as it too was quickly controlled. While coronavirus experts understood the potential danger of other unknown coronaviruses and continued their work, the rest of the world relaxed. Even the initial reports from China in late 2019 of a troubling new respiratory infection outbreak failed to alarm most of the world as it seemed small, distant, and unlikely to affect our lives. How wrong we all were.

From just a smattering of cases in Wuhan, China in late 2019 and early 2020, by March of 2020 the novel COVID-19 coronavirus was a worldwide threat as nation after nation began lockdowns along with other infection control measures. This new coronavirus was quickly identified as a close relative of the original SARS virus based on the sequence similarity of their RNA genomes, leading to the new virus being formally designated as SARS coronavirus type 2 (SARS-CoV-2). So now there were three novel coronavirus outbreaks in less than 20 years, all of them causing potentially deadly respiratory infections in contrast to the mild colds caused by the previously known human coronaviruses. The immediate question was where did these new viruses come from? Part of the answer to this question is believed to be bats although the chain of transmission to humans is still murky. Of the over 1400 bat species, the small fraction of species tested generally carry multiple coronaviruses. The observed prevalence of coronaviruses in tested bat species suggests that there are between 10,000 and 15,000 different coronaviruses collectively lurking in bats worldwide. This is a huge reservoir of inherent biological and genetic variation that can expedite viral adaptation to other hosts besides bats.

In addition to the large number of different coronaviruses in bat populations, this viral family has three other features that promote even greater genetic diversity. First, coronavirus genomes are among the largest of the RNA viruses with a linear genome of between 26,000 and 32,000 bases, a large size that facilitates genomic plasticity. Genome plasticity refers to a genome's ability to acquire and survive significant changes such as duplication of genes or insertion of new genes. Viruses with small genomes have very limited capacity for such major genomic changes since every portion of their genomes contains essential genes and cannot tolerate much disruption. In contrast, coronaviruses have a region at one end of their genomes (by convention the right-hand end as the genome is typically depicted) that seems very malleable to large changes. The length of this region expands or contracts in different coronaviruses and accounts for the 6,000 base difference in size between the coronaviruses with the smallest and largest overall genomes. Importantly, this region contains an unusual diversity of genes with different coronaviruses having novel genes that are not consistently present across all coronaviruses. These novel genes are still poorly characterized but likely impart new and important biological properties on the coronaviruses that possess them, thereby broadening the pathogenic potential of this virus family. Coupled with this genomic plasticity, coronaviruses have one of the highest rates of replication error leading to numerous mutations accumulating in each replication cycle. Coronaviruses also undergo extensive recombination between genomic copies during replication that further scrambles their genetic repertoire. This triad of gene variation, high mutation rate, and recombinational capacity makes these viruses particularly adept at spreading from animal to animal and adapting their genomes quickly to adjust to replication in new hosts. These genetic characteristics combined with vast numbers of different coronaviruses existing in bat populations suggest that there will inevitably be future spillovers into human populations. What is still less clear is the pathway these coronaviruses will take into humans. The initial thinking was that bats spread these viruses to some other animal and that humans contracted the infection from the secondary animal, not directly from a bat. This is the route for SARS and MERS, viruses that passed through civets and camels, respectively, before infecting humans. The evidence for an intermediate animal for SARS-CoV-2 is still lacking and there may never be confirmation of the original source animal(s). The closest relative to SARS-CoV-2 to date is a bat coronavirus found in a bat species from Yunnan province in China (96.2% identical at the genome level), and initial reports of potential intermediate animal hosts have not been confirmed. Interestingly, there is accumulating evidence for the possible direct bat to human transmission of coronaviruses. This suggests that both direct and indirect routes may be feasible which would further expand the threat of these viruses.

A second important question concerning these novel coronaviruses is why are they much more lethal than endemic human coronaviruses? The answer to this question is still an ongoing mystery. Both SARS and MERS had extraordinary fatality rates of 9% and 36%, respectively. Although the untreated fatality rate for SARS-CoV-2 appears much lower and is in the 1–2% range, this is still extremely high compared to the endemic human coronaviruses that are virtually never fatal in healthy people. So what are the factors that make these bat coronaviruses such fearsome human pathogens? While much about the pathogenic mechanisms of these viruses remains unknown, some important clues are emerging. One essential element that facilitated human infection by the SARS and SARS-CoV-2 viruses was the ability of these viruses to use the human ACE2 protein as the viral receptor. Coronavirus virions have surface projections composed of the spike protein which function to interact with host receptors. Specifically, a region of nearly 200 amino acids in the spike protein, called the receptor binding domain (RBD), directly interacts with the ACE2 protein. Both SARS and SARS-CoV-2 have acquired mutations in the RBD that enhance spike binding to human ACE2. Importantly, the SARS-CoV-2 spike protein binds ACE2 10–20 times stronger than does the spike protein from the SARS virus. Furthermore, mutations in the SARS-CoV-2 spike protein also introduced a new cleavage site for a cellular protease enzyme called furin. After coronavirus virions bind to ACE2 through the spike protein RBD, the spike protein must be cleaved by cellular enzymes. This cleavage triggers the fusion of the viral membrane with the host membrane which releases the virion nucleocapsid into the cell. The strong ACE2 binding and the new cleavage site in the SARS-CoV-2 spike protein seem to contribute to more efficient infection of human cells. Enhanced infection is important both for viral spread within an infected person's body and for person-to-person spread; more effective transmission may account in part for the massive global spread of SARS-CoV-2 compared to SARS. However, another issue in the global spread of SARS-CoV-2 is the preponderance of asymptomatic cases. For both SARS and MERS, most infected individuals became significantly ill and viral shedding peaked after the onset of clinical symptoms. With this type of presentation, it was easy for people to self-identify as being ill and to quarantine themselves before they infected many others. In contrast, the majority of SARS-CoV-2 cases are mild or asymptomatic, and even in seriously ill patients, the peak of viral shedding comes before the onset of severe clinical symptoms. Because of this early shedding and the frequent lack of symptoms, many SARS-CoV-2 infected individuals can unknowingly spread the virus widely. These silent spreaders unwittingly thwarted containment measures and facilitated a worldwide outbreak within a few months after this new human pathogen emerged. Now that the virus is established in humans this train of transmission will continue unabated and the virus will spread freely until herd immunity is reached through community infections and/or vaccination. At that time SARS-CoV-2 will likely persist as an endemic virus that never completely goes away.

Beyond the role of the spike protein in facilitating infection of human cells, other features of SARS-CoV-2 that contribute to pathogenesis are still being defined. Several studies have found mutations in other viral genes besides the spike protein gene, but if and how such mutations affect viral virulence is uncertain. While we lack a clear mechanistic understanding of how this virus evokes many aspects of its disease, we at least have a general outline of the clinical presentation. Starting as a respiratory infection, SARS-CoV-2 can also damage a wide variety of other organs, including kidneys, hearts, and even brains. Damage may accrue directly from viral infection of these organs as the ACE2 receptor is common on many tissue types. Alternatively, much of the local damage in the lungs, as well as the peripheral damage, appears to result from an overly aggressive immune response to the infection. One hallmark of serious SARS-CoV-2 infections is the so-called "cytokine storm" characterized by a massive and uncontrolled release of the immune response proteins known as cytokines (see Chap. 5). An overload of cytokines can lead to massive inflammation, disruption of blood clotting, and tissue destruction, potentially destroying organ function wherever these effects occur. Such cytokine-induced damage in the lungs results in loss of oxygen exchange capacity which causes an inability to breathe without a ventilator; lung failure is ultimately the cause of death for many severe SARS-CoV-2 infections. Exactly why severe disease occurs more frequently with advancing age is still speculative, but the trend is definite and makes populations over 60 years of age the most vulnerable to fatality. Individuals with pre-existing conditions such as hypertension, diabetes, obesity, immunodeficiency, and heart disease are also particularly at risk for severe disease although the explanation for their increased risk is also not yet established. It will undoubtedly take years of further study to fully understand the pathogenic mechanisms of this virus. In the meantime, this new virus continues its relentless march across the globe with currently over 190,000,000 million infections and more than 4,000,000 deaths by mid-2021, with cases rising again due to the high contagious delta variant. On a more hopeful note, vaccines were quickly produced and therapeutic drug development continues to move forward. With effective prevention and treatment, someday soon this horrible pandemic will be reduced to a manageable disease akin to seasonal flu.

Will SARS-CoV-2 be the last or the worst of the zoonotic viruses to wreak havoc on human populations? Undoubtedly not. Bats and the coronaviruses they carry will remain an important source of viral spillover into humans but this is not our only concern. Let's not forget that there are hundreds of thousands of unknown viruses from other viral families that could potentially cross into humans. This extraordinary number alone presages that the future inevitably holds more pandemics and potentially ones much more lethal than COVID-19. While future zoonotic spillover and possible pandemics may be inevitable, there are mitigation strategies that desperately need implementation. Funding for basic virus research is critical so that we gain sufficient biological and molecular understanding of each viral family to inform the effective and rapid development of antivirals and vaccines as new viral outbreaks occur. The more we know about the structures and pathogenic mechanisms of each viral family, the quicker we can apply this knowledge towards therapeutics for new viral diseases. Coupled with basic research, there is a great need for a unified global surveillance program to detect and interdict new viral pathogens before they can become established epidemics and pandemics. Some attempts at global surveillance have existed such as the PREDICT program sponsored by the United States Agency for International Development (USAID). Created in 2009 in response to the avian flu outbreak of 2005, PREDICT was an international consortium that sought to identify and characterize novel viruses present in animals from designated "hotspot" locations around the world. Unfortunately, U.S. funding ended in 2019 and was not renewed by the Trump administration causing this program to close in March of 2020 just as the COVID-19 pandemic was exploding. Furthermore, the program was way underfunded with a budget of only around $20 million per year. Consider that the U.S. authorized over $2 trillion for COVID-19 relief in 2020 which is 100,000 times the annual budget for PREDICT. The failure to invest in science and preventative measures, not only by the U.S. but the entire world, resulted in a massive global economic disaster that will take years to recover from. We can only hope that the lessons learned during this current pandemic will help prevent future viral outbreaks and the resultant economic hardship, suffering, and deaths.

Additional Reading

1. Evolution and Emergence of Pathogenic Viruses: Past, Present, and Future. 2017. Intervirology 60: 1–7.
2. Characteristics of SARS-CoV-2 and COVID-19. 2020. Nature Reviews Microbiology.

Definitions

A3 family – A group of related cytosine deaminase enzymes that have antiviral activity.

ACE2 – The human protein used as a receptor by the SARS and SARS-CoV-2 viruses.

Antigenic drift – A slow and gradual change in the surface antigens of an influenza virus due to mutations caused by viral replication errors.

Antigenic shift – An abrupt and major change in the surface antigens of an influenza virus due to reassortment.

Cytokine storm – A massive and uncontrolled release of cytokines produced in response to some infections, either viral or bacterial.

Delta variant – A mutant form of the SARS-CoV-2 that is more easily transmitted.

DNA polymerase alpha (DNAPα) – The enzyme used to initiate human DNA synthesis.

Furin – A human protease enzyme that can cleave the SARS-CoV-2 spike protein.

Genomic plasticity – The ability of a genome to acquire and survive significant changes such as duplication of genes or insertion of new genes.

Hantavirus – The family of single-stranded, negative-sense RNA viruses that are mostly found in rodents.

Hendra virus – A single-stranded, negative-sense RNA virus of the paramyxovirus family that is carried by bats and causes a highly fatal disease in horses and humans.

Hypertension – High blood pressure.

Lassa virus – A single-stranded, negative-sense, segmented RNA virus of the arenavirus family that is carried by West African rodents and which causes a highly fatal disease in humans.

Middle East respiratory syndrome (MERS) – A human respiratory illness caused by a coronavirus endemic in camels.

Nasopharyngeal – Relating to the region that includes the throat and nose.

Nipah virus – A single-stranded, negative-sense RNA virus of the paramyxovirus family that is carried by bats and which causes a highly fatal disease in humans.

Receptor binding domain (RBD) – The region of a viral surface protein that is important for interaction with the host cell receptor.

SA-alpha-2,3 – The form of sialic acid that serves as the receptor for avian influenza.

SA-alpha-2,6 – The form of sialic acid that serves as the receptor for human influenza.

Sialic acid (SA) – The cell-surface biomolecule that serves as the receptor for influenza viruses.

Spike protein – The virion surface protein of SARS-CoV-2 that interacts with the host receptor (the ACE2 protein).

Sudden acute respiratory syndrome (SARS) – A human respiratory disease caused by a coronavirus endemic in bats.

Viral attachment protein – A protein on a virion surface that interacts with the host cell receptor to mediate attachment.

Abbreviations

A3 – APOBEC3 family
ACE 2 – angiotensin-converting enzyme 2
APOBEC – apolipoprotein B mRNA editing catalytic polypeptide-like
DNAPα – DNA polymerase alpha
GISRS – Global Influenza Surveillance and Response System
HA – hemagglutinin protein
RBD – receptor binding domain
SA – sialic acid
USAID – United States Agency for International Development
USDA – United States Department of Agriculture
WNV – West Nile virus

12

Beyond Antibiotics – Are Phages Our Allies?

Keywords Bacteriophage • Phage therapy • Antibiotic resistance

The field of bacterial viruses is a fine playground for serious children who ask ambitious questions.
Max Delbrück, recipient of the 1969 Nobel Prize for Physiology or Medicine

From Fact to Fiction

In the 1920s, the island of St. Hubert in the West Indies had an outbreak of the bubonic plague caused by the bacterium *Yersinia pestis*. In the pre-antibiotic era diseases such as the plague spread rapidly with a high fatality rate due to the lack of any effective treatment. To assist with the crisis, Dr. Martin Arrowsmith arrived from the United States with a bacteriophage preparation to try and save the island population. Bacteriophages, often referred to as phages, are viruses specific to bacteria. These viruses can infect and kill bacterial cells just as human viruses infect and kill our cells. However, phages don't infect plant or animal cells, so they can't directly harm human cells. It was speculated that administering *Yersinia*-killing phages to patients would destroy the bacteria in their bodies and cure the infection. To test this prediction, Arrowsmith wanted to treat half the patients with phages and leave the other half untreated as the control group. If more patients in the phage-treated group survived he could establish the efficacy of his therapy. However, the deaths and suffering were so extreme that he eventually treated everyone and lost the opportunity to verify the therapeutic value of his experimental

V. G. Wilson, *Viruses: Intimate Invaders*, https://doi.org/10.1007/978-3-030-85487-4_12

approach. While this is a fictional account from Sinclair Lewis' 1925 novel (the eponymous *Arrowsmith*), the study of bacteriophages is a legitimate scientific field that arose in the early decades of the twentieth century and informed Lewis' writing. The knowledge that phages could lyse and kill bacteria quickly led to the hypothesis that viruses might have a clinical role in treating bacterial diseases. Importantly, nearly 100 years ago the Arrowsmith novel outlined one of the central and confounding issues of phage therapy: how do you prove that this therapy is safe and effective? As promising as this approach was for curing bacterial infections, due to the scientific and technical limitations of the era its therapeutic benefits were not easily established. Consequently, phage therapy mostly languished for decades after the novel's publication, and we are only beginning to revisit this approach in the twenty-first century as a viable clinical therapy for certain bacterial diseases.

A Brief History of Phage Therapy

The word bacteriophage literally means bacteria eater and stems from the Greek word "phagein" meaning "to eat". The credit for the discovery of bacteriophages goes to the British physician and bacteriologist Frederick Twort. Twort was the first to culture the leprosy bacillus (*Mycobacterium leprae*) and in 1913 he made the key discovery that this bacterium required vitamin K as an essential nutrient for growth. In subsequent work with *Staphylococcus*, he noticed that a few colonies grown on agar plates had small clear sections where the bacterial cells were either dead or failed to grow. When he carefully picked samples from these clear sections and added them to intact colonies the recipient colonies began to develop clear sections. Remarkably, this process was repeatable indefinitely by sequential transfer of the clear samples from colony to colony, suggesting that the "killing factor" was either extremely active or was self-renewing. By performing filtration experiments, Twort showed that the killing factor was extremely small and thus unlikely to be another bacterium. Based on these findings he postulated that the killing factor was an enzyme, although we now know that it was likely a bacteriophage that could infect and lyse *Staphylococcus* cells. Twort published his findings in 1915 in the prominent British medical journal *The Lancet*, but he did not pursue these results further. The actual study of phages began two years later with a peripatetic and self-taught microbiologist, Félix d'Hérelle. The young d'Hérelle spent time in both Canada and France where he did his early schooling and was exposed to scientific thinking on both continents. Although he abandoned formal education after high school, he had an innately curious and scientific mind that led him to pursue independent study in many areas. He spent his 20s studying bacterial diseases of plants and animals from Canada to South America with periodic returns to France to work at the Pasteur Institute.

In 1917, while studying dysentery, d'Hérelle observed results similar to those of Twort. d'Hérelle was growing the bacterial dysentery agent (*Shigella*) in liquid media where the multiplying bacteria filled the solution and turned the media turbid. During handling of these cultures, he inadvertently introduced an unknown agent that killed the *Shigella*, turning the turbid solution clear again as the bacteria died and lysed. As with Twort, d'Hérelle quickly showed that the killing agent passed through bacteria-retaining filters so could not be another type of bacteria. Because even minute amounts of his filtered agent could spread throughout a culture of dysentery and be passed indefinitely to consecutive cultures, he reasoned that this must be a new type of self-reproducing microbe that he termed bacteriophage in a 1918 report. Without any real physical or biological understanding of the nature of his new phage, d'Hérelle intuited that these agents could be potential therapies for bacterial diseases. By the next year, 1919, he had isolated another phage from chicken feces and showed that he could cure typhus-infected chickens by administering his phage preparation. With confidence from these animal studies, he performed a human trial later that year and successfully cured a patient's dysentery using his *Shigella*-killing phage preparation. Many scientists were skeptical and concerned about using phages in humans, while others quickly embraced this approach. During the 1920s there were many attempts to treat a variety of bacterial diseases with phages. Early successes led to great acclaim for d'Hérelle who received an honorary doctorate from the University of Leiden, was awarded the Leeuwenhoek medal (given only once every 10 years), and was nominated multiple times for the Nobel Prize, although he was never chosen as a recipient.

d'Hérelle continued his phage work for the next decade with studies and trials in several countries, including the Soviet Union where he worked with Professor George Eliava, founder of the Tbilisi Institute. Because of their early success, phage therapy became a mainstay in parts of the Soviet Union. In the rest of the world, however, phage therapy diminished and ultimately vanished from medical usage. Even with d'Hérelle's support, phage therapy in Europe and the Americas never became an accepted practice. An important factor that contributed to the decline of phage therapy was a lack of reproducibility in clinical usage. The limited understanding of phage biology and the paucity of techniques for analyzing and characterizing phages confounded many early studies. Without methods for knowing exactly what was in phage preparations, it was impossible to ensure purity and consistency from batch to batch.

While it was quickly recognized that different phages existed and that each only attacked a single type of bacteria, early researchers had no conception of the immense diversity of phages and their high degree of specificity for particular bacterial strains. For example, a phage preparation that effectively cured dysentery in one person might completely fail in another patient because the individuals were infected with slightly different strains of the *Shigella* bacteria. Furthermore, the effect of mutations on both the bacterial targets and the phages was poorly appreciated. Frequently during treatment, some bacteria would mutate and acquire phage resistance. The resistant organisms were then unaffected by the phages and multiplied readily to maintain the disease. Likewise, the passage of phages through bacterial cultures to produce working stocks for clinical use provided opportunities for the phages to mutate. Therefore, different batches of phages could have very different efficacies against their target bacteria. All of these factors contributed to inconsistent clinical results with certain trials showing great success while others failed dismally. This unpredictable variation in results undermined the medical profession's confidence in phages as a reliable therapeutic approach. Vocal critics arose and their adamant denunciations of phage therapy further eroded public confidence.

Even with the failures and lack of reproducibility, work on phage therapy might have persisted had it not been for the development in the 1930s of sulfonamides and penicillin as potent antibacterial agents. By the time World War II ended in 1945, there was widespread acceptance by physicians and the public that antibiotics were the frontline therapy for bacterial infections. The remarkable efficacy and the broad spectrum of penicillin against several life-threatening bacterial diseases quickly established its reputation as the "miracle drug". Unlike phages that were poorly characterized and difficult to standardize due to their genetic malleability, penicillin was a chemical compound that was stable, relatively easy to manufacture, and could be prepared in pure and consistent batches for clinical use. At last, here was the answer to the litany of bacterial diseases that had plagued medical science for centuries: a drug that killed bacteria consistently and efficiently. Researchers in academia and industry enthusiastically joined the pursuit for additional antibacterial drugs, and the next 50 years was a golden era with the discovery and commercialization of different classes of antibiotics that each attacked a unique aspect of bacterial physiology. Within each class of antibiotics, there are multiple related compounds and many chemical variations giving rise to second, third, fourth, and even fifth-generation versions of some antibiotics. This rapid proliferation of antibiotics obviated the need to explore the inconsistent and still controversial phage therapy, and phages as a clinical tool nearly disappeared. Only in parts of the Soviet empire did phage therapy survive and prosper where it is still used routinely for the treatment of specific bacterial diseases. Sequestered behind the Iron Curtain and suffering through the isolationist Stalin Era, phage therapy work begun by d'Hérelle and George Eliava at the Tbilisi Institute continued while the rest of the world adopted antibiotics.

A Phage Primer

Although studies on phages as a therapeutic mostly ceased after antibiotics were introduced, scientific interest in phages remained high as researchers attempted to understand the nature of phages and how they function. Notable in the 1940s and 50s was the work of Max Delbrück, Salvador Luria, Alfred Hershey, and their colleagues, the so-called Phage Group. This group used phages to probe the basic elements of genetics and molecular biology, helping to set the foundations for modern biological science. Seminal findings from their collective studies included proof that mutations are spontaneous and random, that phage mutation can lead to new characteristics such as the ability to infect previously resistant bacteria, that DNA is the hereditary material (not proteins as some believed), that phages can transfer DNA between bacteria (called transduction), and that phages can recombine to exchange genetic information when infecting the same cell. These and many other findings became the critical underpinnings for future work in bacteriology, virology, and the biochemistry of nucleic acids. In 1969 the Delbrück, Luria, and Hershey trio received the Nobel Prize in Physiology or Medicine for their insightful and groundbreaking work with phages.

From the purely observational beginnings of phages as an undefined bacteria-killing agent, the last 100 plus years of research have elucidated a great deal about their fundamental physical, biochemical, and biological properties. One important early advance was the development of electron microscopy in the 1930s which allowed phages to be visualized for the first time in 1940. These early images revealed a complex morphology and confirmed that phages were biological entities and not simply biomolecules with toxicity to bacteria. We now know that phages are viruses specific to prokaryotes (bacteria and archaea) and that phages share both the life cycle stages and many molecular features with the eukaryotic viruses that infect plants and animals. The most commonly characterized type of phages has the general structural characteristics shown in Fig. 12.1. Like viruses, bacteriophages have a capsid or head composed of protein that surrounds and protects the nucleic acid genome. Genome sizes of different phages vary widely from a few

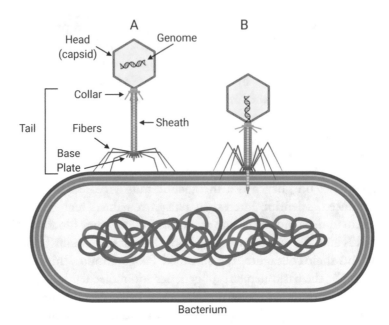

Fig. 12.1 Phage structure. The figure shows two phages, one phage just after attachment to the surface of a bacterium (**A**) and a second phage injecting its genome into the bacterium (**B**). Depicted is a tailed phage that consists of two parts, a head structure (the virus capsid equivalent) containing the phage genome and a complex tail structure composed of a collar, a sheath, a base plate, and fibers as shown. The tail fibers are the appendages that interact with receptor molecules on the surface of a bacterium to initiate the binding of the phage. After the phage-receptor interaction forms, the tail contracts which drives the sheath through the bacterial membrane and cell wall. Penetration creates a pathway through the sheath for the phage nucleic acid to leave the head and enter the cytoplasm of the bacterium where it can begin to replicate. (Created with BioRender.com)

thousand bases encoding only 3–4 genes to a recently found massive phage genome of over 700,000 bases likely containing hundreds of genes. Linear, double-stranded DNA is the most common genome type as it is found in around 95% of known phages. However, it is not known if the preponderance of linear, double-stranded DNA genomes among known phages reflects a true dominance of this genome type in nature. Possibly this observation is just a sampling bias based on the limited number of phages characterized to date, mostly those in the human environment. Among the other 5% of known phages are ones with linear genomes of single-stranded DNA, single-stranded RNA, or double-stranded RNA genomes, as well as circular genomes of single-stranded or double-stranded DNA. Whether or not these other genome types are more prevalent in yet undiscovered phages remains to be determined. Nonetheless, the same genome types observed with eukaryotic viruses also exist in the realm of phages. This is not unexpected as phages and viruses share an evolutionary history and certain viruses evolved from phage ancestors. Like viruses, phages have spread widely among their host species, and it is likely that every type of bacteria on Earth has one or more phages that can infect it. This makes phages the most numerous and ubiquitous biological entities on Earth.

In addition to their breadth of genome types, phage genomes seem to have a somewhat modular organization with widespread evidence of module swapping between different phage types. If two different phages infect the same bacterium they can easily exchange genetic material yielding hybrid genomes with mixed modules. Along with genetic exchange, mutations can occur throughout phage genomes making their genomes incredibly rich in genetic diversity. Several large-scale genome sequencing studies indicate that the realm of phages contains a huge number of novel genes not found in other organisms. So not only are phages the most populous organism on Earth inhabiting every conceivable niche in the biosphere, but they are also the greatest source of novel genetic information. Estimates suggest that we have only characterized a small fraction of the total phage world, so mining new phage genomes for unique genes with useful properties for clinical applications or as biotechnology tools will likely be a very fruitful endeavor in the coming years.

While all phages have a nucleic acid genome and a capsid structure to contain the genome, many, but not all phages, have a multipart tail structure composed of a collar, a sheath, tail fibers, and a base plate all made from proteins (Fig. 12.1). For these "tailed" phages the tail is critical for phage infection as it functions both in attachment to and penetration of the bacterial cell. The tail fibers are the component that interacts with receptor molecules on the surface of a bacterial cell. As for eukaryotic viruses, the interactions between phage tail fibers and their receptors are highly specific and determine what bacteria are infected. The specificity is so great that often phage infection is limited to only specific substrains of a bacterial species and not the entire species. Contributing to this exquisite specificity is the fact that bacterial surface receptor molecules can mutate so that they no longer are bound by the phage tail fibers, thus making the bacterium resistant to infection by that phage. However, the tail fibers of phages can subsequently mutate to restore their interaction with the mutant receptors so that the mutant phages now regain infectivity for the formerly resistant bacteria. Consequently, there is a continual evolutionary battle between bacteria and phages leading to an enormous population variation in phage fibers and their cognate bacterial receptor molecules.

Once phage tail fibers encounter an appropriate receptor, binding of a phage to the bacterial cell surface triggers a flexing of the tail fibers that brings the base plate close to the surface of the bacterium (Fig. 12.1). Phage degradative enzymes help dissolve the rigid bacterial cell wall followed by contraction of the sheath. Acting much like a hypodermic syringe, sheath contraction drives the shaft through the bacterial membrane and delivers the phage genome into the cytoplasm. Once accessible to the cellular enzymes and raw materials, phage nucleic acids express their genes by producing mRNAs and translating these into new phage proteins. As the requisite phage proteins accrue then replication of the phage genome begins followed by assembly into new phage particles. Tailed phages eventually produce an enzyme called endolysin that degrades the bacterial cell wall leading to cell lysis with the eventual release of hundreds to thousands of progeny phages. Conceptually these life cycle steps for bacteriophages are identical to those of eukaryotic viruses although phage reproduction is much faster and typically takes only minutes as opposed to hours or days for viruses.

In addition to lytic phages with the life cycle described above, certain types of temperate phages can undergo an alternative life cycle known as lysogeny. In lysogeny, the infecting phage has a choice and may not initially reproduce to kill the host bacterium. Instead, under specific conditions, the phage genome inserts itself into the bacterial genome and becomes a resident DNA sequence in the host DNA. The integrated phage DNA is called a prophage, analogous to the provirus form of a retrovirus inserted into an animal genome. Like a provirus, the prophage persists in the host DNA and is replicated along with the host genome when the bacteria grow and divide. In this fashion, the prophage can spread widely into all the descendants of the originally infected bacterium. Since phage integration is generally an innocuous event for the bacterium, it is a mechanism for the prophage to survive and spread within a bacterial population even when conditions aren't favorable for lytic reproduction. Interestingly, if the prophage carries genes that the bacterium can use it may change the properties of the bacterium. For example, some phages carry genes that encode proteins that act as toxins to eukaryotic cells. When the toxin is expressed from a prophage and excreted by the bacterium it can convert a harmless bacteria into a human pathogen.

In contrast to retroviruses, prophages do not produce progeny in the integrated state. While proviral integration for retroviruses is a permanent event, lysogenic prophages must excise themselves from the bacterial genome and return to their unintegrated state to replicate. The excision process is usually triggered by conditions that now favor the production of new progeny phage. Once released from the bacterial genome, the freed phage genome begins the typical lytic cycle where progeny phages are assembled and the bacterium is lysed to release the progeny. Each progeny phage will infect another bacterium and in each new bacterium, the phage will enter the lytic phase or the lysogenic phase depending on the cellular and environmental conditions. Collectively, the lysogenic phages and the purely lytic phages wage a constant assault on Earth's bacteria. Importantly, without these ubiquitous and voracious bacterial predators, our planet would rapidly be overrun with bacteria, thus phages play an essential ecological role in maintaining microbial homeostasis to the benefit of humans and all other life forms.

The Antibiotic Problem

Who among us hasn't used antibiotics? Most of us have used oral or topical antibiotics multiple times in our lives to treat ongoing infections or as a prophylactic to prevent infection from wounds, surgical procedures, and even certain dental procedures. Antibacterial topical creams, soaps, and cleansers are ubiquitous in the marketplace and are commonly found in American households. Furthermore, some studies suggest that 80% of Americans take prescription antibiotics at least once per year for a variety of ailments. Antibiotic access and usage are similar in most other Western countries, so the collective human use of antibiotics worldwide is enormous and the benefits are equally large. The number of lives saved worldwide since antibiotics were introduced is difficult to estimate, but approximately 200,000–300,000 lives are saved by antibiotics each year in the United States alone. This life-saving ability of our modern armamentarium of antibiotics is a major contributor to our longevity and likely adds 5–10 years to our life expectancy at birth. Indisputably these drugs have had a momentous impact on human health and have mostly freed us from the terrible morbidity and mortality associated with bacterial infections.

As great as the human usage of antibiotics is, agribusiness across the world uses more antibiotics than do humans. Historically, prophylactic administration of antibiotics in animal feed was routine for many years to prevent disease and promote growth. Animals ingesting daily antibiotics are generally healthier and typically grow larger than their antibiotic-free counterparts resulting in more food production and significant profit enhancement for the producer. Because of these benefits, antibiotics have been widely used in cattle, swine, and poultry production. In recent years the United States and many European nations have reduced or eliminated antibiotic usage in poultry and livestock except to treat specific diseases, and where possible are employing antibiotics not used in medical applications. However, the overall usage of these drugs is still prevalent in many countries and continues to contribute heavily to total antibiotic consumption. Similarly, though to a lesser extent, we use antibiotics in agriculture to protect crops from bacterial diseases. For example, fruit trees in some regions are commonly sprayed with tetracycline and streptomycin to prevent bacterial infections, while streptomycin is also used on specific vegetable crops such as beans, celery, peppers, tomatoes, and potatoes. As with livestock, agricultural use of antibiotics improves crop yield resulting in more effective food production and greater financial return for the farmer, so antibiotics remain a staple for agribusiness.

Between human and agricultural use, an enormous amount of antibiotics are applied or consumed each year with great benefit to human health, food production, and the economy. On the downside, the billions of humans, animals, and plants exposed to antibiotics every year are each host to extensive bacterial communities. Through antibiotic use, each of these entities potentially becomes an incubator for antibiotic-resistant organisms to develop. Furthermore, residual antibiotics in the foods we eat can expose us to chronic, low levels of antibiotics even when we aren't actively taking an antibiotic. Within any bacterial population, there will be millions of individual bacterium with a diverse array of random mutations in their genome (acquired by replication errors or environmentally-induced DNA damage). As an antibiotic kills off susceptible members of the bacterial population, any rare mutant that just happens to be resistant to the antibiotic in use will survive, thrive, and propagate. These resistant strains can become established permanent residents of the bacterial community. Especially when antibiotics are present continuously at low levels in livestock feed, are prescribed unnecessarily, or when human patients do not take them as directed this enhances the opportunity for selection of resistant strains. Additionally, this widespread antibiotic usage has caused massive leakage of these drugs into the environment. Antibiotics sprayed on crops inevitably contaminate the soil and water. Likewise, animals and humans ingesting antibiotics can excrete the drugs in their urine and feces where they also leach into the soil and water supply. Add to this the fact that unused human antibiotics are often flushed down the toilet or thrown away in the garbage, again with the potential to pollute soil and water. The net result is that random sampling of soil and water around the world often detects antibiotic contamination. Since soil and water have rich, complex, and extensive microbial communities, this inadvertent antibiotic accumulation in the environment creates selective pressure for bacterial strains that are antibiotic-resistant just as happens in humans and animals.

To compound the resistance problem, bacteria are notoriously good at sharing their genes within and across species through mobile genetic elements that facilitate horizontal gene transfer (HGT). There are three major HGT mechanisms: conjugation, transformation, and transduction (Fig. 12.2). In conjugation, bacteria physically come together in a mating process that allows them to transfer or exchange mobile genetic elements such as plasmids or transposons, even transferring parts of their genome. For example, bacteria can carry an antibiotic resistance gene on plasmids which are extrachromo-somal DNAs that are generally present in multiple copies per cell. During conjugation, a resistant bacterium can pass copies of the plasmid to a non-resistant bacterium so that now the donor and the recipient bacteria are both antibiotic-resistant. In this fashion, resistance can spread widely and quickly through the population. Transformation is an even more general gene transfer

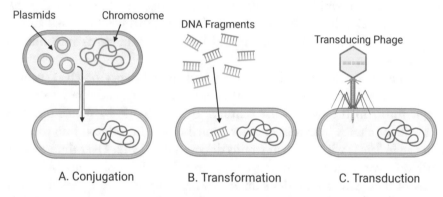

Fig. 12.2 Mechanisms of resistance transfer. (A) Conjugation is a mating process between two bacteria where a physical connection forms between the two cells. The connection allows nucleic acids to pass from one bacterium (the donor) to a recipient bacterium. In the example shown the upper bacterium is the donor. This donor bacte-rium contains its DNA chromosome as well as small, circular DNA plasmids carrying an antibiotic resistance gene (orange segment). Transfer of one or more plasmid copies to the recipient cell will endow the recipient bacterium with antibiotic resistance. **(B)** As bacteria die and lyse, their DNA fragments (both the plasmids and the chromosome) are released into the environment. Transformation is the process where bacteria ran-domly take up these DNA fragments in the environment and can incorporate those fragments into their genomes. If a bacterium incorporates a DNA fragment containing an antibiotic resistance gene (orange fragment) then that bacterium transforms into an antibiotic-resistant strain. **(C)** Transduction is the process of bacteria acquiring new genes when they are infected by a lysogenic phage. During infection, the phage genome will enter the bacterium and be integrated into the bacterial genome. If that phage carries an antibiotic resistance gene (orange segment) then that gene will become part of the bacterial genome and will confer resistance on the bacterium. Collectively these three gene transfer mechanisms enable bacteria to rapidly acquire and swap antibiotic resistance. (Created with BioRender.com)

mechanism where bacteria can randomly take up DNA fragments released from dying bacteria (such as bacteria killed by phages). This process occurs readily across species so an antibiotic resistance gene present on a DNA fragment released by a dying soil or water bacterium could be picked up by a human pathogen to create a dangerous antibiotic-resistant strain. Lastly, transduction is an HGT mechanism directly mediated by lysogenic phages. Sometimes when the prophage excises it accidentally captures a segment of the adjacent bacterial DNA, and that bacterial DNA then becomes part of the phage genome. If that captured bacterial DNA contains a functional antibiotic resistance gene then that gene is now part of the phage genome. When that phage infects a new bacterium and becomes a prophage the antibiotic resistance gene will be expressed in the recipient bacterium and its progeny making them all resistant to the antibiotic. Collectively, transduction, conjugation, and transformation enable the rapid and widespread transfer of genetic information both within and across diverse bacterial species. Because of these mechanisms, even harmless environmental bacteria that have acquired antibiotic resistance can inadvertently spread their resistance genes into human pathogens.

The pervasiveness of antibiotic usage and the efficiency of HGT caused antibiotic resistance to develop in the twentieth century, and it has become a major threat in the twenty-first century. This antibiotic resistance problem was evident almost immediately after penicillin was introduced into medical practice in 1941, with resistant strains discovered in 1942. Over time, as new antibiotics were introduced resistant bacterial strains eventually arose for many of these new antibiotics similar to what happened with penicillin. For the 60 years following the introduction of penicillin the strategy was simply to stay ahead of resistant strains by finding new antibiotics or chemically tweaking existing antibiotics to overcome bacterial resistance. As long as science produced new generations of antibiotics the bacterial resistance problem was circumvented and clinicians had effective therapeutic options for most infections. Unfortunately, the antibiotic discovery pipeline slowed considerably in recent decades. From 1980–1999 there were 50 new antibiotics brought to market, but from 2000–2019 there were only 21 new antibiotics approved by the FDA, and 5 of those aren't readily available on the market. This dearth of new antibacterial drugs means that as strains become resistant to current antibiotics there will be fewer and fewer alternative drugs to treat serious infections. This is already evident from the fact that roughly 700,000 people worldwide die each year from antibiotic-resistant bacteria, including 35,000 people yearly in the United States. Even these current fatalities reflect only the proverbial tip of the iceberg as around three million patients each year in the United States are infected with bacterial strains resistant to at least one antibiotic. Fortunately, most of these resistant strains can still be treated successfully with alternative antibiotics although there is a steady rise in multidrug-resistant (MDR) strains of some bacteria, notably *Staphylococcus aureus*. However, even without fatalities, this proliferation of mono- and multidrug-resistant strains means increased hospitalizations, more patient sequela, and greater healthcare costs, making the personal and economic impact of antibiotic resistance enormous. Horrifyingly, the United Nations predicts that the number of deaths due to antibiotic-resistant bacteria could rise to 10 million a year by as early as 2050.

There are multiple reasons for this decline in new antibiotic discovery, but a primary one is simply economics; antibiotics just aren't that profitable. Various estimates suggest that bringing a new drug to market takes well over 10 years and costs more than a billion dollars. This process includes multiple steps including the initial discovery research to identify a lead compound, *in vitro* toxicity assessment, animal testing, and the three phases of human trials. Most potential new drugs fail along this pathway and never even make it to market, so the aggregate investment in time, resources, and capital to achieve even one approved new drug is staggeringly costly. This burden is acceptable and profitable for drugs with a large market, a long market lifespan, and pro-longed usage such as drugs for chronic diseases like diabetes, high cholesterol, or arthritis. For these chronic diseases, millions of patients use the medications daily for months to years, and a successful drug may dominate the market until its patent expires, thus ensuring robust and steady sales that exceed the development costs. In contrast, the number of patients using any one antibiotic is much smaller than for chronic disease drugs, and antibiotics are typically given for short durations of days to weeks rather than months to years. To compound the problem, the widespread emergence of bacterial resistance can render an antibiotic useless, so that sales may disappear long before the patent expires. Consequently, the overall market for an antibiotic is significantly smaller and riskier than for chronic disease drugs. Because of these issues, pharmaceutical companies in recent decades have shifted more of their research focus towards highly profitable drugs with less investment in new antibiotics whose sales may not recoup development costs. With only a trickle coming out of the antibiotic development pipeline, this leaves us all in the precarious position of facing an increasing threat from antibiotic-resistant bacteria with a decreasing array of effective drugs. For example, the CDC currently lists 4 resistant bacteria as urgent threats (*Acinetobacter*, *Clostridioides difficile, Enterobacteriaceae,* and *Neisseria gonorrhoeae*) and 10 as serious threats, including *Salmonella, Staphylococcus aureus* (particularly MRSA – methicillin-resistant *S. aureus, Streptococcus pneumonia,* and *Mycobacterium tuberculosis.* If antibiotics fail, where will we turn?

Phage Therapy Redux

The shortage of new antibiotics coupled with the rise of resistant and multidrug-resistant bacteria has inspired scientists and clinicians to reexamine the mostly discarded concept of phage therapy for bacterial infections. Because of their transducing ability, phages help spread bacterial resistance but maybe they can also be part of the solution to the problem. As posited nearly 100 years ago, why not use phages to kill off pathogenic bacteria and rid the body of harmful infections? Once you know what bacteria are causing the infection, the patient can be given lytic phages that can attack and destroy that particular bacteria. The administered phages should infect the target bacteria, replicate, and lyse the offending bacteria to eliminate the disease. Since phage infection of bacteria is mechanistically unrelated to antibiotic resistance, phage therapy is equally effective on antibiotic-sensitive and resistant strains making it a viable therapy when antibiotics fail. Importantly, because phages don't infect eukaryotic cells, they pose no direct harm to the person being treated. Also, unlike antibiotic drugs, phages are highly selective killers of bacteria. Antibiotics often have a broad bacteriocidal activity that not only eliminates the target pathogen but also indiscriminately kills harmless commensal organisms, especially in the gastrointestinal tract. The collateral damage to the normal microbiome not only leads to intestinal distress in a subset of patients but can also predispose to an inappropriate expansion of certain indigenous bacterial populations when the normal homeostasis of the microbial flora is perturbed. This can result in the overgrowth of potentially pathogenic organisms like *C. difficile* which are normally restrained by other commensal bacteria in the gastrointestinal tract. Such disruptive effects on the microbiome are not seen with phage therapy as only the target bacteria are killed.

In addition to high selectivity, phages possess several other properties that make them attractive therapeutics. First, phage particles are physically very stable and can be freeze-dried without loss of viability; this makes them easy to store and transport which is always a valuable feature for a biopharmaceutical. Second, since phages replicate they can be administered at a very low dose to reduce possible systemic effects of introducing a foreign substance. Once in the patient, the phage quantities will increase to therapeutic levels as the phages infect the pathogen and reproduce. Because of this reproductive ability, even a one-time administration of the phage solution may be sufficient for treatment, though in practice phage solution administration is usually repeated several times to ensure continued effective levels in the body. In contrast, antibiotics must be taken at frequent and regular intervals throughout the therapeutic regimen to maintain their effective killing concentration in the patient, a dosing requirement that often leads to poor patient compliance. In addition to their favorable dosing schedule, phages have good bioavailability and rapidly circulate throughout the body to reach the target pathogen wherever it is growing. Importantly, phages show better effectiveness than antibiotics against bacterial biofilms that form in some parts of the body. Biofilms are thick layers of bacterial growth that can form on surfaces during any infection and are very difficult to eradicate. As the bacteria grow and pile up they excrete compounds that "glue" the cells together creating a solid and slimy film. Biofilms are particularly resistant to antibiotics as the antibiotics cannot penetrate the outer layers of the film to reach and kill the interior bacteria. However, phages can attack biofilms one layer at a time. The phages will infect the outer cells and lyse these cells to reveal the next layer. By repeating this process over and over, phages kill off layer after layer until they demolish the entire film. This capacity to reduce biofilms is an exciting finding and may become another application for phage therapy beyond treating antibiotic-resistant infections. Lastly, whether treating resistant infections or biofilms, once the phages have eliminated their target the phage numbers will no longer increase as they cannot reproduce without susceptible bacteria to infect. Without new bacteria to infect the remaining phages stay extracellular where they are susceptible to our bodily mechanisms for removing foreign bodies. These residual phage particles are either degraded into constituent amino acids and nucleotides or are excreted in our bodily wastes. Either way, the patients have recovered and the phages are gone from their bodies. And unlike antibiotics, excreted phages have no known environmental impact.

While using phages to treat intractable bacterial infections seems like a simple and elegant approach with great promise, there remain challenging biological, clinical, and regulatory issues with phage therapy that have confounded widespread adoption of this methodology. Biologically, the major issue is the genetic instability of both phages and their target bacteria. Given the rapid reproductive cycle of bacteria and their phages, on the order of minutes, there is ample opportunity for extensive mutations to occur during an infection and phage treatment. If a bacterial mutant arises with an altered receptor molecule that is no longer recognized by the phages then that mutant pathogen becomes resistant to phage infection, a scenario that has already been observed in some clinical trials. The phages will kill the sensitive bacteria while the resistant bacterium and its progeny multiply and continue the infection unabated causing the phage therapy to fail. To avoid this pitfall, many investigators suggest using a phage cocktail containing 2 or more different phages that attack the target pathogen. In this approach, even if a mutant bacterium arises that is resistant to one of the phages, it will still be susceptible to the other(s) and will continue to be destroyed. The greater the number of different phages in the cocktail, the less likely are the odds that the pathogenic bacteria could simultaneously develop resistance to all of the phages. This strategy works effectively in the lab on test cases, but can only be employed if there are multiple lytic phages available that target the patient's pathogen, an option that isn't always available for certain bacteria.

Genetic instability is also an issue for the phages themselves as their genomes will mutate during replication. To prepare a phage stock for patient treatment, the phages must be grown within a bacterial host where they can reproduce. Mutant phages will undoubtedly arise in this population, and some may acquire properties that detract from their ability to infect and kill the target pathogen. Alternatively, such mutations could even occur as phages amplify within the patient during the treatment of the infection. Mutant phages that arise would continue to infect bacteria and reproduce, and as the mutants accumulate they might compete with the therapeutic phages and reduce the effectiveness of the treatment. This will always remain a potential problem since phages are biological entities that can't be produced and maintained in a constant and unvarying form, unlike antibiotics that are invariant chemical compounds. This particular concern makes regulatory approval challenging and could be a critical limitation to the widespread usage of phage therapy.

In addition to the biological constraints on phage therapy, numerous practical considerations have limited clinical applications of this treatment methodology. As mentioned above, there must be at least one and preferably multiple known lytic phages specific for the pathogen in question, a condition that is not fulfilled for all pathogens. For example, *C. difficile* is an opportunistic pathogen that is normally present at low levels in the gastrointestinal tract of many people. Antibiotic treatment can disrupt normal microbial homeostasis due to the widespread killing of beneficial commensal organisms which allows *C. difficile* to proliferate to harmful levels. Once established at high levels, *C. difficile* causes chronic intestinal infection with diarrhea, inflammation, and sepsis that can be debilitating and even fatal. There is no universally effective treatment yet for *C. difficile* infections, and this bacterium would be a prime candidate for phage therapy except that as of 2020 no lytic phages for *C. difficile* have been discovered. This doesn't mean that lytic phages for this organism don't exist, but until one is isolated phage therapy remains unavailable for this significant pathogen.

Another clinical concern is potential phage therapy side effects. Phages don't infect human cells and should be unable to harm cells via direct destruction, yet there are other problems associated with their administration. Phages are large, complex assemblies of proteins and nucleic acids that can evoke negative responses from host immune defenses that see phages as foreign invaders. Phages have been shown to induce both innate and acquired immunity with concomitant inflammation and cytokine release. Alternatively, phage administration into certain organ systems may simply upset physiological conditions and produce negative effects. In phase I trials when phages were introduced into patients in large quantities there have been a variety of side effects detected depending on the route of administration. For oral delivery, abdominal pain and indigestion occurred, but with no serious side effects in the test patient group. In contrast, actual clinical cases using intravenous delivery have produced severe side effects ranging from disruption of liver function to hypotension. Although the effects were manageable in most cases, in others the therapy was discontinued. As these were mostly single patient trials it is difficult to know if the adverse effects were directly due to the phages, the ongoing infection, or some combination. Additional routes of administration such as inhalation, topical application to the skin, or direct application to surgical sites are also being explored. Each route of administration may have its own potential for negative side effects that might prevent therapeutic usage, so much further study of this issue is sorely needed. In addition to immune responses creating potential adverse side effects in patients undergoing phage therapy, the immune response may also render the therapy ineffective. Patients may already have antibodies against particular phages from prior natural exposures or can develop anti-phage antibodies during phage therapy. Just as for our viruses, if there are neutralizing antibodies against the phage particles then these antibodies would prevent the phages from infecting the target bacteria. This would preclude the destruction of the pathogen and lead to therapeutic failure. Even if neutralizing antibodies don't impair the initial therapy, the eventual development of such antibodies would likely make the reuse of the same phages in a subsequent repeat infection problematic. This neutralizing antibody issue may limit phages to use as a "boutique therapy" that is a one-time, personalized treatment for a particular infection.

A third unresolved clinical issue is the debate over pathogen-specific phage therapy versus generalized phage cocktails. For the former, the patient's specific pathogen is isolated and tested against a phage collection to identify one or more phages that specifically kill the pathogen. These killing phages are then produced in therapeutic quantities and used to treat the patient. This approach is time-consuming and expensive since it involves maintaining a large phage collection, having the laboratory setup and skills to isolate the pathogen and screen the phage library, and possessing the capacity to generate sufficient quantities of the selected phages in a purified form for therapeutic use, all for one individual. Crafting a personalized therapy unique for each patient makes this approach slow and very costly with estimates of as high as $50,000 per person. The time and expense required for this approach make it too costly for common infections, and it would be much more amenable for specialized cases with hard-to-treat infections. Still, some researchers argue that the narrow specificity of phages for a given bacterium necessitates this approach to ensure that the treatment phages can actually kill the pathogen in a timely and effective way. In contrast, other researchers argue that bespoke phage treatment is unnecessary and that generalized phage cocktails are more practical, more economically feasible, and more suitable for broad-scale clinical use while still having good therapeutic efficiency. The generalized approach uses premade cocktails containing many different lytic phages that all attack that same bacterial species. Once the patient's pathogen is generally identified, say for example it is *S. aureus*, the physician will simply go to the freezer and remove an anti-*S. aureus* phage cocktail. The concept is that one or more of the phages in the cocktail will hopefully be able to attack and kill the patient's specific type of *S. aureus* regardless of whatever substrain it might be. These cocktails can be commercially manufactured, distributed, and stored in advance, eliminating the need to isolate the patient's pathogen, screen the library to find the perfect phage, and then prepare the phage for administration, thus saving time and cost. Of course, it is always possible that the patient has a strain of *S. aureus* that is not susceptible to any of the phages in the cocktail, so there is uncertainty about success with this generalized methodology compared to the personalized, pathogen-specific approach. Nonetheless, in parts of Eastern Europe where phage therapy is routinely practiced, it is the generalized scheme that is mostly in use with reasonable success. However, no phage therapy has yet been licensed in the United States or the European Union (EU) because none has been proven effective in randomized clinical trials (RCTs), the gold standard for evidence-based medicine. Still, based on the limited available data, both the personalized and generalized phage therapy approaches are technically feasible and show clinical potential in many circumstances.

A Glimpse of the Future

Even with the many issues and potential problems facing phage therapy, early promising results coupled with the growing emergence of antibiotic-resistant bacteria continue to spur efforts to test and evaluate the clinical efficacy of this method. However, phages are a biologic and must obtain regulatory approval before administration to patients, just like other biologics such as monoclonal antibodies or vaccines. In many countries, phages are considered a drug and must follow the same regulatory and approval requirements as for any new drug. In the United States, this requires approval by the Federal Drug Administration (FDA) and production under Good Manufacturing Procedures (GMP), while in Europe phages are designated a medicinal product subject to the regulations of the EU. These regulatory requirements that are standard for a typical manufactured drug become more problematic for genetically malleable biologics like phages whose genomes can change with mutations in ways that are difficult to control and standardize. Much of the struggle in this field involves creating and characterizing phage batches that meet regulatory statutes for reproducible administration into patients. Still, clinicaltrials.gov lists over 40 clinical trials involving bacteriophages that span the last two decades, roughly half of them involve phase I and phase II trials to evaluate the safety of administering phage therapy via different routes. The lure of potentially lucrative phage therapy has attracted several biotech companies in recent years, and they are investing heavily in developing this technology. As more and more data are collected about safety, dosing, and patient responses to therapeutic quantities of phages, we should soon see the proliferation of the phase III trials, the last hurdle to regulatory approval.

As we wait on medical applications of phages for the general public, there have been several high-profile compassionate use cases in recent years that highlight the amazing possibilities of this novel therapy. Compassionate use involves trying an experimental therapy that is not yet approved for public usage. To gain approval for compassionate use of phage therapy, or any other unapproved therapy, the patient must have a life-threatening condition for which there is no other alternative treatment. In this circumstance, the potential risk of the experimental therapy is acceptable as the patient is unlikely to survive and the experimental therapy is the last hope. The risk of any damage or harm caused by the therapy becomes insignificant compared to the chance for survival. Two life-saving compassionate use cases, one in the United States and one in Great Britain, illustrate the potent clinical utility of phage therapy. The British case began in 2017 and involved a young woman, Isabelle Carnell, who was born with cystic fibrosis (CF). Patients with CF produce excess amounts of sticky mucus that accumulate in the lungs, impairing breathing and promoting bacterial growth in the lungs often leading to fatal infections. Isabelle's lungs were eventually colonized by *Mycobacterium abscessus*, a relative of the bacteria that cause tuberculosis. Over time, the infection badly damaged her lungs, and by age 15 Isabelle needed a double lung transplant to replace her nearly non-functioning lungs. As for any transplant procedure, Isabelle needed immunosuppressive drugs before and after the surgery to prevent rejection of the donated lungs. While the surgery was successful, her chest incision became infected with *M. abscessus* that was already in her body. Now because of the immunosuppression required for her transplant, the organism rapidly spread throughout her body creating a life-threatening infection that did not respond to antibiotics. Desperately seeking alternatives to the failed antibiotics, Isabelle's mother and physician reached out to phage researchers to explore the possibility of phage therapy. They ultimately contacted Dr. Graham Hatfull at the University of Pittsburgh who curates a massive phage collection with over 15,000 different phages collected from all over the world. Hatfull and his team tested his collection against Isabelle's *M. abscessus* strain and eventually found three phages, Muddy, BPs, and ZoeJ, that all infected the target bacteria. (Note the phages often have peculiar and comical names because the discoverers can name them anything they choose.) Muddy was an appropriate lytic phage, but BPs and ZoeJ were lysogenic phages that needed creative genetic engineering to eliminate their lysogenic capacity and convert them to exclusively lytic phages. With three lytic phages available for a treatment cocktail there was less chance for Isabelle's pathogen to develop phage resistance, so intravenous administration of the cocktail began. The skin lesions showed signs of improvement within three days and after a month of twice-daily doses, there was a marked reduction of infection throughout her body. Treatment continued for six months to fully eradicate the infection without any significant side effects. As in the Arrowsmith novel, this is an uncontrolled experiment and not a definitive randomized clinical trial. Nonetheless, Isabelle's remarkable turnaround and recovery from this near-fatal infection is a strong testament to the power of phage therapy.

The second example is similar in that it also is an uncontrolled experiment on a single person with a dramatic rescue of that patient, Dr. Tom Patterson. Dr. Patterson is a professor of psychiatry at the University of California San Diego (UCSD). While visiting Egypt he contracted an infection with a multi-drug resistant strain of *Acinetobacer baumannii*, an organism that is now common in Asia and is slowly spreading globally. By the time he returned to San Diego, Dr. Patterson was comatose and in critical condition with multisystem organ failure. As with Isabelle Carnell, the lack of effective antibiotics forced Dr. Patterson's wife and physicians to consider phage therapy as a last resort. Patterson's pathogen was sent to the laboratory of Dr. Ryland Young, a phage expert at Texas A&M University, in hope that Dr. Young's research group could find one or more lytic phages that would attack and destroy the *A. baumannii* strain. Scientists at the Naval Medical Research Center in Maryland also joined the hunt for *A. baumannii* killing phages, and between the two groups, multiple lytic phages were isolated to make the therapeutic cocktail. Treatment started on a Tuesday, and by Saturday Dr. Patterson awoke from his coma and began to recover. Dr. Patterson eventually fully recovered without ill effects from the phage treatment and returned to his work at UCSD. While not formally provable, given his dire and unchanging condition for weeks before the phage therapy began, it seems highly likely that it was the killing activity of the phages that saved Dr. Patterson's life rather than some miraculous spontaneous recovery.

These two anecdotal cases powerfully illustrate the therapeutic potential of phage treatment, particularly for infections that are unresponsive to conventional antibiotics. Much more study is needed to evaluate the personalized and generalized approaches for phage therapy, to optimize the parameters for production and administration of these agents, to obtain regulatory approval for this form of therapeutic, and to gain widespread acceptance for phage applications in the medical community and the general public. In the meantime, optimism remains high that phage therapies will eventually become a standard component of clinical care both to treat routine infections and to address more complex infections that do not respond to other treatments. As antibiotic-resistant bacteria continue to emerge and spread in the world, these minute and ferocious predators of bacteria may become our best hope.

Additional Reading

1. The Perfect Predator: An Epidemiologist's Journey to Save Her Husband from a Deadly Superbug. S. A. Strathdee and T. Patterson. 2019. Hachette Books.
2. The Antibiotic Resistance Crisis, with a Focus on the United States. E. Martens and A.L. Demain. The Journal of Antibiotics 70:520–526. 2017.
3. Phage Therapy in the Year 2035. J.-P. Pirnay. Frontiers in Microbiology 11:1171. 2020.(continued)(continued)

Definitions

Acinetobacer baumannii – A species of rod-shaped bacteria normally found in soil and water. These bacteria are opportunistic pathogens for humans and are increasingly found as hospital-acquired infections.

Acinetobacter – The genus of short, rod-shaped bacteria found in the environment.

Agar plates – Petri dishes filled with agar (a gelatinous solid made from seaweed) and used to grow bacterial cultures.

Bacteriocidal – Capable of killing bacteria.

Base plate – A component of the tail structure of bacteriophages.

Bioavailability – The extent to which a drug can reach its intended target.

Biofilms – Thick layers of bacterial growth that form on surfaces and are difficult to eradicate.

Biologic – Biomolecules that are used at therapeutics, preventatives, or diagnostic reagents.

Biopharmaceutical – A biomolecule that is used as a pharmaceutical product.

Bubonic plague – A form of plague caused by the bacterium Yersinia pestis and is spread by fleas. The disease is characterized by swollen lymph nodes (buboes) in the groin, armpit, and neck.

Clostridioides difficile – A rod-shaped bacterium (also known as Clostridium difficile) that inhabits the human intestinal tract. Overgrowth of the organism following antibiotic therapy can cause a life-threatening infection.

Collar – A component of the tail structure of bacteriophages.

Commensal organism – An organism that lives in or on another organism without hurting or helping the host organism.

Conjugation – A bacterial mating process where two bacteria form a physical bridge that allows the transfer and/or exchange of nucleic acids.

Cystic fibrosis (CF) – A genetic disease characterized by excessive production of abnormally thick mucus.

Dysentery – A severe intestinal infection caused by *Shigella* bacteria.

Endolysin – A phage enzyme that degrades the bacterial cell wall to lyse the cell and release newly made phages.

Enterobacteriaceae – A family of rod-shaped bacteria found in the environment, on human skin, and in the intestinal tract of humans and animals.

Extrachromosomal – A piece of nucleic acid that is not part of the chromosome, e.g. a plasmid.

Horizontal gene transfer (HGT) – The spread of genes among bacterial populations through conjugation, transformation, or transduction.

Hypotension – Low blood pressure.

Leprosy – A disfiguring skin and nerve disease caused by the bacterium Mycobacterium leprae.

Lysogeny – The bacteriophage life cycle that involves the integration of the phage genome into the bacterial chromosome to form a prophage.

Lytic phage – A phage that kills its host bacterium during phage reproduction.

Microbiome – The microorganisms that inhabit a particular environment or part of the body.

(continued)

(continued)

Monoclonal antibody – A single type of immunoglobulin molecule that is produced in vitro. Monoclonal antibodies are used as therapeutics, preventatives, or diagnostic reagents.

Mycobacterium abscessus – A species of rod-shaped bacteria normally found in soil and water.

Mycobacterium leprae – The bacterium that causes leprosy.

Mycobacterium tuberculosis – The species of rod-shaped bacteria that cause tuberculosis.

Neisseria gonorrhoeae – The species of spherical bacteria that cause gonorrhea.

Penicillin – A broad-spectrum antibiotic initially isolated from mold.

Prophage – A phage genome that is integrated into the chromosome of the host bacterium.

Salmonella – The genus of rod-shaped bacteria that are often associated with food poisoning.

Sepsis – A life-threatening condition where harmful microorganisms are present at high levels in the blood or other tissues.

Sheath – The structure on tailed bacteriophages that penetrates the bacterial cell wall to deliver the phage nucleic acid into the bacterial cytoplasm.

Shigella – The genus of bacteria containing rod-shaped organisms that infect the intestinal tract.

Staphylococcus – A spherical bacterium that grows in grape-like clusters and can cause skin infections.

Staphylococcus aureus – A species of Staphylococcus that is commonly found on the human body.

Streptococcus pneumonia – The species of spherical bacteria that cause pneumococcal pneumonia.

Streptomycin – A type of antibiotic with broad activity against many types of bacteria.

Sulfonamides – A class of synthetic compounds with antibacterial activity.

Tail fibers – The protein structures that some phages use to attach to receptors on their host bacteria.

Temperate phage – Any phage that can undergo lysogeny.

Tetracycline – A group of antibiotic compounds characterized by having four ring structures.

Transduction – The process of bacteria acquiring new genes when they are infected by a lysogenic phage.

Transformation – The process where bacteria internalize DNA fragments released by other bacteria.

Yersinia pestis – The bacterium that causes the plague.

Abbreviations

CF – cystic fibrosis
EU – European Union
FDA – Federal Drug Administration
GMP – good manufacturing procedures
HGT – horizontal gene transfer
MDR – multidrug-resistant
MRSA – methicillin-resistant Staphylococcus aureus
RTC – randomized clinical trial
UCSD –University of California San Diego

13

Viruses in Gene Therapy and Cancer Therapeutics

Keywords Gene replacement • Gene editing • Viral vectors • Genetic diseases • Oncolytic viruses • Retrovirus • Adenovirus • Adeno-associated virus

> *In our view, gene therapy may ameliorate some human genetic diseases in the future.*
> T. Friedmann and R. Roblin. Science 175:949–955. 1972.

Gene Therapy Emerges

That cautious yet optimistic quote from Friedmann and Roblin forecasted a nascent technology that is now very much a reality. With over 25,000 genes, the human genome has many critical targets that can be damaged by DNA mutations to yield genetic diseases. Gene defects constitute more than 80% of all known diseases with an estimated 300 million people affected worldwide. Many of the serious genetic maladies were already recognized by the 1970s although very few were treatable in that era and most still lack effective therapies today. In 1972, ameliorating or correcting genetic diseases, including cancer, was just emerging as an ambitious possibility. The knowledge and the expertise were still insufficient, but it was clear that science was slowly and carefully advancing towards this goal. Among the first strategies conceived for gene therapy was replacing a defective gene with a wild-type copy. Humans have what is known as a diploid genome where every gene has two copies, one contributed by our father and one by our mother. For certain genetic diseases, the illness is recessive and only manifests when a person has two defective copies of a critical gene. Sickle cell anemia is a classic example involving the gene for hemoglobin, the protein that carries oxygen in our red blood cells (RBCs). Individuals with one wild-type copy of the hemoglobin gene and one defective copy are healthy because the single wild-type gene makes enough functional hemoglobin to supply the body's needs. Only when both copies of the hemoglobin gene are defective does sickle cell anemia result. These patients can suffer from anemia, severe pain, organ damage, and infections that all reduce the quality of life and can be fatal. If we could somehow permanently introduce a single functional copy of the hemoglobin gene into all of the patient's RBCs this would restore functional hemoglobin and cure the disease.

Another example of an even more lethal recessive disorder is a group of genetic defects known collectively as severe combined immunodeficiency (SCID) diseases. These disorders affect the white blood cells (either B or T lymphocytes) and cause a catastrophic deficiency in the immune response leaving the victim unable to effectively fight off even minor infections. Individuals with these disorders must remain in sterile environments ("bubble boy" disease) or they quickly succumb to even relatively benign microorganisms. One well-characterized SCID is caused by a deficiency of the enzyme adenosine deaminase encoded by the ADA gene. This enzyme is prevalent in lymphocytes and is responsible for converting the molecule deoxyadenosine into deoxyinosine. Deoxyadenosine is toxic to lymphocytes while deoxyinosine is harmless. An individual with two defective ADA genes (homozygous for the mutant gene) cannot make functional adenosine deaminase causing his or her lymphocytes to accumulate the toxic deoxyadenosine and die. Without adequate supplies of lymphocytes to fight infection these patients cannot survive. As with sickle cell anemia, heterozygous individuals with one normal copy and one defective copy of ADA are fine as the normal copy supplies sufficient adenosine deaminase to remove deoxyadenosine and keep the lymphocytes healthy. The first human gene therapy trial in 1990 attempted to cure this type of SCID by introducing a wild-type ADA gene into two patients with this genetic defect (see next section).

The alternative to gene replacement is to use gene-editing technologies to correct endogenous defective genes rather than introducing an entirely new exogenous copy of the target gene. This approach is particularly important for dominant-negative genetic diseases where the afflicted individuals have one mutant gene and one wild-type gene. The single defective gene is sufficient to cause the malady because the mutant protein produced by the defective gene is dominant and can subvert or disrupt the function of the wild-type protein. One example of a dominant-negative disease is progeria (also known as Hutchinson-Gilford syndrome). Progeria is a devastating illness that manifests as premature and rapid aging with patients rarely living past their early

20s. This is a multiorgan disease that presents in infancy with a multitude of symptoms including failure to grow, distinctive facial features (large head and eyes, small jaw, and thin nose), scleroderma (hardening of the skin), hair loss, kidney failure, loss of eyesight, atherosclerosis, and other cardiovascular problems. All of this results from a defect in a single gene for a protein called lamin A. Lamin A is a structural protein in the cell nucleus and is critical for the proper control of RNA and DNA synthesis. Inside the nucleus, lamin A creates a lattice that forms the shape of the nucleus and helps organize the chromosomes, a function that is essential for DNA replication and DNA repair. Even though each cell will make both wild-type and mutant lamin A, incorporation of the mutant lamin A into the lattice distorts the structure and renders it less functional. This is analogous to building a brick wall where every other brick is broken. The result is a malformed and weakened structure that doesn't provide adequate support. Likewise, all of the cells in a patient with progeria have defective nuclei and exhibit DNA damage whose accumulation over time produces the clinical symptoms of this disease. Adding a new wild-type copy of the lamin A gene wouldn't cure this illness as the defective lamin A is still present and exerting its disruptive effects. Therapy for this and other dominant-negative genes requires blocking defective protein production either by shutting down the expression of the mutant gene or by correcting the genetic mutation. Both of these options are amenable to gene editing rather than gene replacement.

The most widely used gene editing approach is the CRISPR/Cas9 system, usually just referred to as CRISPR. CRISPR is actually a defense mechanism that bacteria use to protect themselves from bacteriophages. In its natural form, CRISPR is a three-component system involving a DNA-cleaving protein (the Cas9 protein) and two types of RNAs, the crRNA and the tracrRNA. Bacteria encode many crRNAs, and each crRNA interacts with the tracrRNA plus the Cas9 protein to form an active complex. The sequence of crRNA guides the complex to a DNA sequence complementary to the crRNA. Once directed to the DNA by the crRNA, the Cas9 protein in the complex cleaves the target DNA. For gene therapy applications the crRNA and tracrRNA sequences are synthesized as one RNA known as the single guide RNA (sgRNA). As any sequence can be synthesized for the crRNA portion of the sgRNA, the sgRNA-Cas9 complex can be targeted to any desired DNA sequence, for example to a mutant gene sequence. Depending on how CRISPR is applied, it will cleave the DNA target sequence to destroy the defective gene or it can remove specific mutant bases to allow repair of the DNA sequence. CRISPR applications have successfully cured dominant-negative diseases in cell culture and in animal models of human diseases, demonstrating the utility of this system. An exciting report in 2021 showed great promise for progeria by correcting the lamin A mutation and significantly improving health in a mouse model of this illness. We aren't quite there yet for gene editing in humans, but studies are progressing rapidly and human trials are likely in the near future.

From Adversary to Deliveryman – Harnessing Retroviruses for Gene Therapy

Central to all gene therapies, whether gene replacement or gene editing, is the delivery of exogenous nucleic acid sequences into our cells. Both replacement DNAs for defective genes and DNAs encoding the gene-editing CRISPR components (Cas9 and sgRNA) are produced in the lab prior to being introduced into the patient's cells. For certain therapies, the patient's cells are harvested and the exogenous nucleic acids are added to the isolated cells in the lab before returning the cells to the patient's body (*ex vivo* therapy). Alternatively, the therapeutic nucleic acids are delivered directly to the patient, either into a specific target tissue (*in situ* therapy) or systemically (*in vivo* therapy). With either the *in situ* and *in vivo* approaches this is a complex problem that involves protecting the exogenous nucleic acids from damage and degradation in the body, targeting these materials to the correct cells, and gently breaching the cell membranes to introduce these nucleic acids into the cell nucleus where they can act on our genome, all without harming our cells. Different physical and chemical delivery systems have been developed and tested, but none has emerged as the superior method as they all suffer from technical limitations and less than optimal efficiency for delivering the nucleic acid cargoes into cells. The most efficient and widely used delivery system remains Mother Nature's creation: viruses. Viruses are particularly suitable and useful vehicles for gene therapy as they are highly evolved for carrying nucleic acids, for penetrating our cells, and for delivering genetic information into our cells. These viral properties are exactly what is needed for therapeutic applications, leading to the early adoption of viral systems by researchers and clinicians in this field. Such virotherapy harnesses our natural adversaries and seeks to use their unique properties to our advantage to defeat diseases. Although viruses are fraught with certain dangers and limitations, several different viral types have been adapted for gene therapy and are now widely used to deliver corrective nucleic acids into suitable patients. There are still problems to overcome, but viral vectors are constantly being tweaked and modified to improve delivery and safety. Until a more effective delivery system is invented or discovered, viruses are and will remain important components in gene therapy protocols.

Some might consider it ironic, but the birth of gene therapy depended on two oncogenic viruses, Moloney murine leukemia virus (MoMLV) and SV40 (see Chaps. 7 and 8). In 1990, Dr. William French Anderson and colleagues performed the first human therapeutic gene replacement experiment on a 4-year-old girl named Ashanthi DeSilva who suffered from ADA-SCID. Based on a decade of previous research, Anderson's group decided to use a modified retrovirus, MoMLV, as the vector to transport the wild-type ADA gene into their patient. A cancer virus may seem an odd choice for a human therapeutic application, but retroviruses have features that make them useful for gene delivery applications. In the early 1980s, Dr. Richard Mulligan at Harvard University was looking for a method to introduce exogenous DNA into cultured cells and he realized that retroviruses could be useful tools for this purpose. First, they readily infect many cell types, a property that is beneficial when trying to get new DNA in many different kinds of tissues and organs. Second, they integrate their genome into the host DNA such that any foreign gene carried by the virus will also be integrated and thus be permanently retained by the recipient cells, a key requirement for gene replacement therapy. Additionally, the highly active SV40 gene promoter was cloned into the MoMLV vector to direct the transcription of the inserted gene and ensure that this "transgene" would express sufficient levels of its protein product. After successful use of such vectors both in culture and in preliminary human safety trials, the Anderson group received approval to use the MoMLV-ADA construct in the first patient. Their approach was an *ex vivo* therapy where the patient's lymphocytes were removed and infected with MoMLV-ADA in culture to deliver the transgene. The resultant ADA-positive cells were transfused back into the young girl in 1990, partially restoring ADA activity and immune function. However, the levels were insufficient to cure the disease and Ms. DeSilva still required regular replacement therapy with purified ADA protein. Nonetheless, thirty years later roughly 30% of her lymphocytes are still the descendants of these MoMLV-ADA positive lymphocytes, confirming that long-term maintenance of a transgene can be achieved with retroviral vectors.

The integrative capacity of retroviruses that confers transgene persistence, coupled with their ability to infect a broad range of cell types, makes them attractive vectors for gene replacement therapies. Sadly, these same properties can create adverse events that have limited the widespread adoption of retroviral vectors for clinical applications. Since they readily infect numerous cell types, it is challenging to target them to a specific cell type *in vivo*. Systemic application of these vectors in patients could lead to infection and transgene expression in unwanted tissues; these so-called off-target effects might result in unanticipated adverse consequences for the patient. To avoid such effects the retrovirus vectors are more commonly used for *ex vivo* applications rather than *in vivo* administration. However, even the *ex vivo* approach doesn't eliminate the more serious problems that can arise through proviral integration in the target cells. During the *ex vivo* procedure, hundreds of thousands of cells are infected with the viral vector, and the retrovirus will integrate randomly into the chromosomes of each of these cells. Eventually, this cell population with the heterogenous integrants is returned to the patient where the cells multiply and express the therapeutic transgene protein. The problem that may arise is proto-oncogene activation by the integrated vector. As detailed in Chap. 8, leukemia viruses transform cells through the promoter insertion mechanism where they inadvertently integrate adjacent to a proto-oncogene and inappropriately activate its expression. Alternatively, vector insertion into the host DNA can disrupt a tumor suppressor gene which makes that cell vulnerable to acquiring cancer-causing mutations. Either of these types of insertion events can occur unknowingly in an *ex vivo* infected cell which is then reintroduced to the patient. A single aberrant cell returned to the patient may ultimately develop into cancer, a reality that has occurred in several clinical trials for another type of SCID, SCID X1.

SCID X1 is another deadly immunodeficiency, in this case, caused by a defect in the IL2RG gene. The IL2RG gene is located on the X chromosome, so males have one copy of this gene and females have two copies. If a male suffers a genetic defect in his single IL2RG gene then he develops SCID X1 while a female would need mutations in both copies of the gene to manifest the disease. The IL2RG gene encodes a protein called gamma c that is a subunit of several receptors for different interleukins. T lymphocytes without a functional gamma c subunit have poorly working receptors which disrupts interleukin binding and prevents normal T cell development. As with ADA-SCID, infants with this gene defect are severely immunocompromised and rarely survive a year due to severe infections. Hematopoietic stem cell transplant is a potential treatment, but these transplants require a matched donor, and even with a match have high morbidity and death. Because it is a life-threatening and difficult to treat disease caused by a single defective gene, SCID X1 was another attractive candidate for early gene therapy attempts. By the late 1990s, the IL2RG had been cloned and inserted into a modified MoMLV in preparation for administration to children with SCID X1. In 1999, 2 infants with SCID X1 who lacked suitable donors for a stem cell transplant were chosen for *ex vivo* IL2RG gene treatment. Hematopoietic stem cells were harvested from each child, infected with the IL2RG-MoMLV construct, and reintroduced to the patient. Both children responded well with increased T cell counts and improved immunity. Over the next several years 18 more patients underwent this IL2RG therapy delivered by MoMLV with similar results. Of the 20 total children treated, 18 of them acquired adequate immunity and were able to lead normal lives without the opportunistic infections typically seen in immunocompromised individuals. These results confirmed that gene replacement was an effective approach for repairing a recessive genetic defect and could provide dramatic clinical benefits for most patients with SCID X1. Tragically, however, 6 of the 18 cured children subsequently developed acute T-cell lymphoblastic leukemia several years after administration of the retroviral-infected stem cells. All 6 individuals were treated with chemotherapy which was successful in 5 patients while one succumbed to this cancer. Examination of the MoMLV integration sites in the leukemia cells from the 6 patients showed that the proviral DNA from each person was located near a cellular gene with oncogenic potential. This colocalization of the viral DNA with a cellular oncogene strongly supported a promoter insertion mechanism for the leukemias. In this scenario, the promoter and enhancer elements in the left end proviral LTR directed IL2RG transcription while the right end LTR inadvertently drove the expression of the adjacent oncogene (see Fig. 8.3). Although this oncogenic potential of the MoMLV vector was a known risk, the expectation was that it would be a rare occurrence, not an outcome for a third of the patients!

The unexpectedly high rate of leukemia in the patients with SCID X1 forced a reevaluation of the MoMLV vector system. As the promoter and enhancer located in the proviral LTRs made MoMLV too dangerous for clinical application, a modified MoMLV vector lacking these elements was created. When the proviral DNA formed it would still have LTRs on both ends, but neither LTR would possess a functional promoter and enhancer. These transcriptionally defective retroviruses were called self-inactivating (SIN) vectors. Without a promoter and enhancer, the right end LTR of a SIN vector would no longer activate the transcription of an adjacent cellular oncogene, thus eliminating this type of oncogenic risk. Of course the left end LTR would also lack the promoter and enhancer making it unable to express the critical transgene. This situation was easily remedied by simply attaching an exogenous promoter directly to the transgene so that it was transcribed independently of the LTRs.

Another approach for avoiding promoter insertion oncogenesis involved changing vectors. MoMLV is classified as a gammaretrovirus, and this type of retrovirus has a propensity for integrating its proviral DNA close to the transcriptional start site of cellular genes. This site preference ideally positions the proviral right end LRT sequence for activation of the adjacent cellular gene, including a possible oncogene. In contrast, the lentivirus group of retroviruses tends to integrate within genes, an event that typically damages the gene rather than activating its transcription. Consequently, lentiviruses such as HIV have become popular vectors for human gene therapy applications. It may seem counter-intuitive that an incurable human pathogen is used for clinical treatments, but it is easy to render HIV harmless while retaining its useful vector properties. Additionally, SIN versions of HIV have been generated to even further reduce the risk of oncogenesis from promoter insertion. Between 2010 and 2020, the safety-improved MoMLV and HIV vectors were used in gene therapy applications for dozens of patients. During this period there was only one report of a possible vector-related cancer. This cancer, acute myeloid leukemia (AML), occurred in one individual during a clinical trial using a SIN-HIV vector to treat sickle cell anemia. The leukemic cells had the HIV vector integrated, but the proviral DNA was not juxtaposed with a gene that has oncogenic potential, thus the leukemia may have been coincidental and unrelated to the gene therapy. Additionally, a 2019 study reported that a hematopoietic malignancy developed in 1 of 10 Rhesus monkeys treated with a standard lentivirus vector (not a SIN vector). Both the human and primate cases are troubling because they suggest that lentivirus vectors, even SIN versions, may still have an oncogenic potential. A further caveat is that retroviral-associated cancers often appear in patients years after the gene therapy. Because of the long lag period between treatment and the appearance of malignancy, patients treated in the last 10 years may still be in jeopardy for cancer. While the true risk of current retroviral vectors remains unquantified, research continues towards creating more innocuous derivatives that will one day fulfill all the hopes of gene replacement therapy without the fear of unintended disease.

Into The Fray – What Other Viruses Bring to Gene Therapy

The ability of retroviruses to integrate their DNA into our genomes makes them proficient vectors for permanently delivering replacement genes. Yet this same integrative property makes them potentially dangerous due to promoter insertion oncogenesis. This danger has limited their use primarily to *ex vivo* applications rather than direct administration into patients where viral infection would be more widespread and less controlled. Since the cells for these *ex vivo* procedures need to be harvested, infected, and returned to the patient, this approach is typically confined to easily obtained blood cells. This makes retroviral therapies particularly suitable for genetic diseases involving blood cells, but less so for other types of genetic diseases involving solid tissues. The need to address genetic diseases affecting cell types other than blood cells spurred the exploration of additional viral delivery systems. Most current research and clinical applications focus on two DNA viruses, adenovirus (AdV) and adeno-associated virus (AAV), although several other viruses have also been studied as possible gene therapy vectors.

Adenoviruses are medium-sized viruses with a linear, double-stranded DNA genome averaging around 35,000 base pairs. Isolated in 1953 from human adenoids (hence the name), human adenoviruses are a diverse group with more than sixty different serotypes. Most of the serotypes cause only mild respiratory disease and these are considered very benign viruses. In the 1980s when gene therapy was in its infancy, adenoviruses had four attractive properties as a vector: (1) a substantial fraction of this genome can be deleted without impairing the basic replication functions which allows ample space for inserting a therapeutic transgene, (2) the virus can infect a wide variety of cell types so it can be used to deliver transgenes to most organs, (3) the inherently minimal pathogenesis of adenoviruses precludes the risk of dangerous viral-induced disease in the vector recipients, and (4) adenoviruses lack integrative capacity and should not have the risk of promoter insertion-induced oncogenesis seen with retroviruses. Unfortunately, although the lack of integration reduced the oncogenic risk, it also meant that there would be no permanent persistence of the transgene such as occurred with retroviral vectors. However, the transience of the therapeutic gene wasn't necessarily viewed as a complete impediment to clinical use. There was considerable optimism that adenovirus DNA would persist in the host cell nucleus and express the transgene for weeks to months, and the virus could simply be readministered whenever the transgene levels dwindled below the therapeutic threshold. While not a "cure" for genetic diseases, this repeated administration approach could maintain patients in a healthy state similar to insulin injections for people with diabetes.

One of the early pioneers in adenovirus gene therapy was Dr. Ronald Crystal at the National Institutes of Health (NIH) in Bethesda, Maryland. After lengthy safety and efficacy studies in cell culture and rodents, Dr. Crystal and collaborators were ready to test first-generation adenovirus vectors to deliver a therapeutic gene. These first-generation vectors lacked a viral regulatory gene called E1 and some versions had a second deletion of the E3 gene (Fig. 13.1). Deletion of E1 greatly reduced both viral replication and

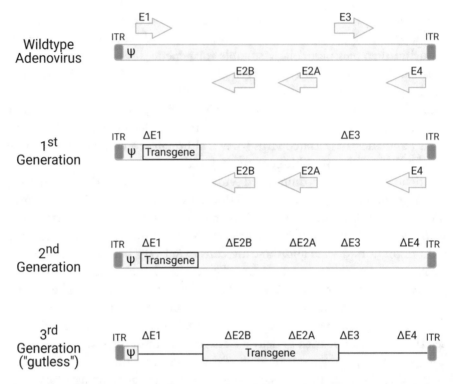

Fig. 13.1 Adenovirus vectors. Shown are the wild-type adenovirus genome and three generations of vectors derived from the wild-type virus. The arrows indicate the positions of the early genes E1–E4. The Δ symbol indicates a deleted gene and the Ψ symbol shows the location of the sequence needed for DNA packaging into adenovirus virions. The ITRs are the inverted terminal repeat sequences present at each end of the adenovirus genome. These ITRs are the only sequences needed for adenovirus-directed replication of DNA. Any exogenous therapeutic gene being delivered by the vector is the transgene. (Created with BioRender.com)

expression of the other viral gene which created a vector that could deliver the transgene but not reproduce and spread effectively. In 1993, they used an adenovirus vector to introduce a wild-type cystic fibrosis gene (the CFTR gene) directly into the lungs of a 23-year-old patient. Therapeutic levels of the CFTR protein were achieved, but the effect waned more quickly than anticipated. Re-administrations of the CFTR-adenovirus resulted in even more rapid diminishment of the CFTR expression, making the transgene effects too short-lived for any clinical benefit to the patient. Subsequent studies revealed that the problem was the host immune response. Even with the reduced viral gene expression, the vector virus continued to produce low levels of other viral proteins that efficiently evoked a killing immune response. Repeated administration of the CFTR-adenovirus simply exacerbated the immune response which quickly curtailed CFTR expression by destroying the cells infected with the CFTR-adenovirus. This disappointing result forestalled clinical usage of the first-generation adenovirus vectors, and most of the remainder of the decade was spent developing improved versions of the adenovirus vectors that displayed reduced immunogenicity.

By the late 1990s, there were second and third-generation adenovirus vectors available. Second-generation vectors had additional viral genes deleted to further reduce viral protein production in hopes of avoiding immune activation. While considerably less immunogenic than the first-generation vectors, second-generation vectors were still not ideal for clinical usage leading to the third generation of "gutless" adenoviruses (Fig. 13.1). As bizarre as it might sound, gutless adenovirus vectors simply had no adenoviral genes. All that the adenovirus genome needs to replicate and assemble into virions are short regulatory sequences (ITR elements) on each end of the linear viral DNA and the packaging sequence (Ψ). Each of the protein-encoding genes in between these elements can be removed and replaced with foreign DNA to construct a "nonviral" genome containing the desired transgene. When this recombinant DNA is delivered into so-called "packaging cells" in the lab, the DNA replicates and is incorporated into adenovirus virions that are indistinguishable from the wild-type virions. Without any viral DNA to express viral proteins, these transgene-containing virions can be administered to patients without inducing the anti-viral immune responses seen with first and second-generation vectors.

The future in gene therapy seemed bright for these gutless vectors until the tragic death of Jessie Gelsinger. Mr. Gelsinger was an 18-year-old who suffered from a mutation in his gene for the liver enzyme ornithine transcarboxylase (OTC). A deficiency in OTC causes an excessive and potentially fatal accumulation of ammonia in the bloodstream. While somewhat manageable with drugs and exercise, the long-term prognosis is not favorable for individuals with this genetic error. With a clear need for more effective treatments, this disease was considered a suitable target for intervention with novel approaches. In 1999, Mr. Gelsinger enrolled in an OTC clinical trial at the University of Pennsylvania, an institution that was pioneering gene therapy. A research group at the university's Institute for Human Gene Therapy was doing a phase I test to assess the safety of OTC delivered by a gutless adenovirus vector. Preliminary dosage studies in rhesus monkeys and macaques established that the adenovirus virions themselves had toxicity at high concentrations. This toxicity necessitated that the virus quantities chosen for the human studies be set far below those that induced severe reactions in small primates. As a further safety step, the dosages set for the human trials were first tested in larger baboons with no significant adverse reactions, clearing the way for human administration of the OTC-adenovirus. Trial participants were divided into six groups with each group receiving a different dose of the OTC-adenovirus infused directly into the right hepatic artery for delivery to the liver. Jesse Gelsinger was in the highest dose group and was the second person in the group to receive the infusion. Within 18 hours after the infusion, he began to demonstrate serious symptoms not seen in any of the other trial participants. Despite intensive medical intervention, he progressed to systemic inflammation and multi-organ failure that proved fatal on the fourth day after treatment. Ultimately Mr. Gelsinger's death was attributed to hyperactivation of his innate immune response, although why this excessive response developed solely in this patient remains unclear.

The profoundly unexpected and shocking outcome of Mr. Gelsinger's response to the OTC-adenovirus halted gene therapy trials across the country. Researchers and regulatory agencies suddenly realized that our understanding of the nuances and dangers of viral vectors was far from complete. Faced with this inherent safety issue, adenovirus vectors lost their status as a favored alternative to retroviruses for gene replacement delivery. Nonetheless, over the last decade adenoviruses have reasserted themselves as ideal for certain types of therapies and they are now used in roughly 50% of all the ongoing clinical trials involving viral vectors. In particular, two uses have garnered much attention: in anticancer therapy (discussed in a section below) and as a vaccine carrier. Adenovirus is appropriate for vaccine delivery because vaccination is typically by intramuscular injection rather than systemic administration and involves lower quantities of virions than for gene replacement. This combination of localized delivery and low dosage avoids the risk of a serious systemic immune reaction that was seen in the Gelsinger case. Importantly, the innate immune stimulation provoked by adenoviruses contributes to a robust localized response against the vaccine target protein carried by the virus. Several vaccines using adenoviral delivery are in development, and Johnson & Johnson brought the first one to market with its SARS-CoV-2 vaccine released in 2021. As established by Johnson & Johnson, delivering the SARS-CoV-2 spike protein by cloning it into adenovirus produced an effective protective immunity against the pathogen. Given the success of the spike protein-adenovirus vaccine, other vaccines based on this platform are likely forthcoming, thus adenoviruses may emerge as a major player in the vaccine arena.

A third major viral contributor to the field of gene therapy is the adeno-associated virus (AAV). This is a nonenveloped virus with a small (~4700 bases), linear, single-stranded, DNA genome. AAV was discovered as a contaminant of an adenovirus culture and is an unusual virus because it lacks reproductive autonomy. While AAV virions can infect many cell types, new virions are not produced unless the cells are co-infected with a helper virus such as adenovirus or herpes simplex virus. This dependency on a helper virus is an attractive safety feature for a gene delivery virus. In the absence of the helper virus, AAV virions carrying a therapeutic transgene can infect the target cells but can't produce new virions to spread within the patient or be shed to infect other people. Additionally, even in the presence of a helper virus, AAV is not known to cause any disease or have any clinical impact on infected individuals. The benign nature of AAV eliminates the potential danger that the vector itself could cause a serious viral disease in the recipient patients.

In addition to lacking disease propensity, AAV has two other properties that are useful for a gene therapy vector. First, there are 11 AAV serotypes and more than 1000 variants. While many of the serotypes have broad host cell ranges, there are differences in the host cell receptor usage by the serotypes and their variants. These differences have been exploited for specifically targeting gene delivery to certain types of cells. For example, AAV2 is the only serotype that infects the kidneys while AAV8 is the only serotype that infects the pancreas. By creating hybrids between serotypes it is possible to generate virions with enhanced targeting so that the transgene is mostly delivered to the specific cell type that needs the therapy. A second important feature of AAV is that "gutless" versions of the genome have been developed similar to the third generation adenovirus vectors. As for adenovirus, all the sequences that the AAV genome needs for replication and packaging into virions are located in two short ITR segments, one on each end of the genome. The entire coding region of AAV can be deleted and replaced with a transgene, and as long as the transgene is flanked by ITRs then the recombinant genome will replicate and assemble into virions in an appropriate packaging cell line. Unfortunately, since AAV has a small genome, the total transgene capacity is much less than for adenovirus, limiting what genes can be carried by this vector.

Another similarity between the gutless AAV and the gutless adenovirus vectors is that neither has a specific integration mechanism. Unlike retroviral vectors that directly integrate the transgene into the host genome, the gutless vectors remain extrachromosomal. While AAV vectors can still randomly integrate at a low frequency as can any exogenously added DNA, this should be a very low occurrence event compared to retroviruses where integration occurs in every infected cell. Consequently, the promoter insertion oncogenesis observed with retroviral vectors should be absent or very rare with AAV vectors. The downside to this lack of integration is that transgene maintenance is not permanent with AAV vectors. Initially, the concern was that the transgene DNA would only persist for days to weeks in the infected cells before being degraded. This short life span would necessitate frequent re-administration of the vector to maintain therapeutic levels of the transgene. Repeated virion delivery would be problematic as neutralizing antibodies against the virions would develop in the patient. The neutralizing antibodies would coat the AAV virions and prevent their entry into the target cells, thus preventing any subsequent therapeutic usage of the vector for that patient. This can even be an issue on the initial therapy as roughly 50% of adults have anti-AAV antibodies due to community-acquired exposures to this virus. Fortunately, in practice, the transgene DNA often persists for months to years. It appears that the transgene DNA usually circularizes and can remain in the cell nucleus stably for long periods, especially in cells that divide rarely or not at all. This combination of an adequate duration of persistence along with excellent safety and manipulatable targeting has spurred considerable clinical testing of AAV vectors for gene therapy applications.

From the early 2000s, there have been over 150 clinical trials utilizing AAV vectors. Potential therapies address genetic defects affecting many different organs and include diseases such as hemophilia, Parkinson's disease, muscular dystrophy, and macular degeneration. These studies use *in vivo* administration of the AAV vector and generally involve direct injection into the target tissue, although intravenous delivery is also widely used and there has even been aerosol delivery for a cystic fibrosis trial. Since most studies are early phase (phase I or phase II) or are still in progress, the actual clinical efficacy of the AAV gene therapies to correct the target disease is mostly in doubt. What is clear is that the safety of AAV vectors is quite high and the transgene expression persistence has been excellent, both of which are outcomes that will justify further development and testing.

In addition to the ongoing trials for many diseases, three AAV-based therapies have already reached the market. The first was a product, Glybera, that used an AAV1 vector to deliver the gene encoding lipoprotein lipase (LPL) to patients with a defective LPL protein. It was approved in Europe in 2012, but eventually taken off the market due to low demand and a million-dollar price tag! In 2017, the U.S. Food and Drug Administration (FDA) approved Luxturna, an AAV2-based therapy for the treatment of progressive blindness caused by Leber's congenital amaurosis. The following year Luxturna was also approved for use by the European Union (EU). At $450,000 per eye, this is quite pricy and the results have been variable, but no other treatment is available so gene therapy will continue to be refined and improved for this condition. The third product, Zolgensma, was approved by both the U.S. and Japan in 2019 and the EU in 2020. Zolgensma uses an AAV9 vector to deliver a transgene called survival motor neuron one (SMN1) and is used to treat spinal muscular atrophy (SMA) which is caused by an SMN1 genetic defect. SMA typically is fatal in early childhood due to respiratory failure and Zolgensma significantly extends both the quality of life and lifespan in these patients. Costing over two million dollars for the treatment, this is again an extraordinarily expensive therapy. While the cost for these three gene therapies may seem exorbitant, they at least demonstrate the power and potential of viral vector-based gene therapies. The near future will undoubtedly see more of these life-saving AAV-based products that can ameliorate or even cure illnesses that once seemed impossible to treat.

For all its current successes and future promise, AAV vectors are not without their inherent dangers. A 2020 report found that 6 of 9 dogs treated with an AAV vector in a hemophilia study showed integration of the AAV vector DNA into chromosomal DNA of their liver cells. Although no adverse effects were observed in these animals, it does raise the issue that the actual efficiency of DNA integration by AAV vectors is largely unknown. These canine results are especially troubling given the development in late 2020 of liver cancer in one of 54 human patients with hemophilia in an AAV clinical trial conducted by the company uniQure. The FDA halted the trial until this case can be thoroughly investigated. Early data on this patient's cancer suggest that it was not due to the AAV treatment, but more follow-up is needed. Even if the risk is small compared to retroviral vectors, if AAV integration can still occasionally generate oncogenic events then patients must be monitored long-term for such occurrences. Of more immediate concern, now there is evidence for toxicity from high-dose administration of AAV virions carrying transgenes. In both human and animal trials, high doses of the virus, especially when given intravenously, caused adverse toxicity. In some cases, the toxicity was due to an excessive immune response evoked by virions, reminiscent of the Gelsinger case with adenovirus. In other studies, there were direct toxic effects on the liver. Tragically, in 2020 there were three deaths due to liver issues in patients being treated with AAV8 for a neuromuscular disease called X-linked myotubular myopathy (XLMTM). As with the uniQure trial, the XLMTM trial was put on hold by the FDA until the role of AAV in these deaths is clarified. These recent safety concerns highlight the need for more research into unanticipated interactions of AAV with animal and human recipients of these vectors. Ultimately, for gene therapy to reach its full potential there must be additional scientific advancements to improve the safety and efficacy of the viral vectors. Additionally, there must be serious discussions regarding product cost and the ethical issues of how society will deal with healthcare expenses for rare diseases and their potential treatment complications.

Oncolytic Viruses – Pitting One Killer Against Another

Using our enemies as transgene delivery vectors to correct genetic errors is not the only medical use of human viruses. The ability of many viruses to lyse infected cells makes them potential tumor-destroying agents. If viruses can be targeted just to cancer cells and not healthy cells, then the viral killing activity might destroy tumor tissue with little or no harm to the patient. This concept of oncolytic virotherapy is actually much older than viral gene therapy as animal and human anticancer trials using viruses were first attempted in the 1940s and 1950s. As might be expected though, results from these early studies were often unsuccessful or inconsistent as little was known concerning the molecular properties of the viruses used or the tumors tested. Given the diversity of cancers and the large number of different viruses, finding the right combinations for effective clinical results is a complex problem. It wasn't until the 1990s and beyond when our understanding of viruses and tumor biology became mature enough to slowly make headway with oncolytic virotherapy. The last decade has seen enormous and exciting progress, and a much clearer picture is emerging about how oncolytic viruses work against cancer and what needs to be improved to optimize viruses for clinical usage. A variety of viruses are being explored for oncolytic therapy, both RNA and DNA viruses, with the current emphasis heavily on DNA viruses. The leading candidates being tested in clinical trials are adenoviruses and herpes simplex virus which together account for roughly 50% of finished and ongoing trials.

A key issue in the effective use of viruses to treat cancers is how to target the virus exclusively or predominantly to cancer cells rather than normal cells. Specific targeting prevents collateral damage to the patient's healthy tissues and eliminates unwanted viral disease. This remains a challenging aspect of oncolytic virotherapy although many ideas have been tested in recent years and definite advances have been made. One of the earliest strategies was conditional viral replication exemplified by a mutant form of adenovirus called ONYX-015 (Fig. 13.2). The ONYX-015 virus cannot produce the viral protein known as E1B-55K. E1B-55K degrades cellular p53 and prevents p53-induced apoptosis which destroys cells before adenovirus can replicate. Being deficient in E1B-55K makes the ONYX-015 virus unable to reproduce in normal cells that are all p53 positive. In contrast to normal cells, over 50% of all tumors lack functional p53 due to mutations in this gene. (The reason that tumors often have defective p53 is that p53 has anti-tumor activity and loss of this activity favors tumor development.) In tumor cells without active p53,

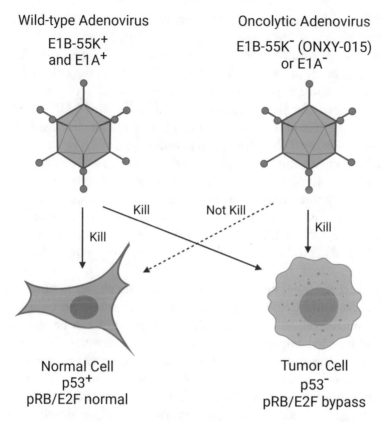

Fig. 13.2 Oncolytic adenoviruses. Wild-type adenoviruses express both the E1B-55K and E1A proteins (E1B-55K$^+$ and E1A$^+$). These viral proteins allow the wild-type adenovirus to overcome the cell-protective effects of the p53 and pRB proteins in normal cells, resulting in cell death (solid arrow). Wild-type adenoviruses also effectively kill tumor cells (solid arrow) as these cells frequently lack p53 (p53$^-$) or have bypassed the pRB/E2F checkpoint. Because wild-type adenoviruses not only kill tumor cells but also normal cells in the patient, they are not suitable for tumor therapy. In contrast, the oncolytic adenoviruses do not produce a functional version of either the E1B-55K protein (the ONYX-015 virus) or the E1A protein (E1A$^-$ viruses). Viruses lacking E1B-55K or E1A are defective for replication in normal cells because of the virus-inhibitory activities of cellular p53 and pRB. Consequently, the oncolytic adenoviruses do not kill normal cells effectively (dashed arrow). However, tumor cells often lack p53 (p53$^-$) or have bypassed the pRB/E2F checkpoint. Without these protective mechanisms, the tumor cells become susceptible to infection and killing by the oncolytic adenoviruses (solid arrow). This selective killing of tumor cells and not normal cells makes oncolytic adenoviruses potentially effective for therapeutic application against appropriate cancers. (Created with BioRender.com)

the ONYX-015 adenovirus replicates very effectively and kills the cancer cells with high efficiency while sparing the normal cells. Another conditional replication strategy with adenovirus uses a deletion of the E1A gene whose protein product binds the host pRB protein (Fig. 13.2). In the cell, pRB forms a complex with another host protein, the transcriptional activator E2F. Active E2F is necessary for the S phase of the cell cycle where DNA replication occurs, but the pRB-E2F complex is inactive until pRB is released. Adenovirus needs the cell in the S phase for viral replication, so its E1A protein attacks the pRB-E2F complex and causes pRB to release E2F which promotes S phase entry. Without E1A, adenovirus cannot force normal cells into the S phase which limits viral replication. However, tumor cells are abnormally proliferative and many have already bypassed the pRB-E2F restriction. Consequently, similar to the ONYX-015 virus, E1A deficient adenoviruses can replicate effectively and kill tumor cells while having little effect on the patient's normal tissues.

Two other strategies being explored for targeting viruses to tumor cells are receptor usage and transcriptional control. Tumor cells often express unique proteins on the cell surface, and these tumor proteins can be used as novel receptors for viral targeting. This requires that a virion surface protein be genetically changed such that it now binds to the tumor cell protein rather than its normal receptor. Such a change not only retargets the virus to the tumor cells but further reduces viral infection of normal cells since the virus will no longer interact with its authentic receptor. While simple in concept this is difficult in practice because the altered viral protein still has to function to form virions and elicit penetration of the virions into the tumor cells. Various research groups are experimenting with adenovirus receptor retargeting, and it is probable that someday there will be a collection of retargeted oncolytic adenoviruses for use with the receptors on different tumor types. Independently or in conjunction with receptor retargeting, modification of viral transcriptional control also seeks to allow viral replication only to tumor cells. A critical early step in viral reproduction is early gene expression as the viral early gene proteins regulate and perform viral genomic replication. Expression of these early gene products is under the control of viral promoters that have evolved to be functional in the virus's primary host cells. If you remove the viral promoters then viruses are greatly reduced in their ability to reproduce in the normal host cells. To redirect viral replication to the tumor, the viral promoter is replaced with a promoter that is only active in the tumor cell. Now the viral early genes are preferentially expressed in the tumor cells leading to viral reproduction and lysis of these cells. One example of this approach uses the human telomerase reverse transcriptase (hTERT) promoter.

This promoter has minimal activity in normal cells but is highly active in tumor cells. Replacing the normal adenovirus early promoter with the hTERT promoter allows active viral reproduction and killing in tumor cells while preventing replication in normal cells. Collectively, the various targeting strategies are greatly expanding the repertoire of tumor-specific adenoviruses and other oncolytic viruses.

Fortuitously, retargeting isn't always required for the use of oncolytic viruses. Even without targeting mechanisms, certain wild-type viruses just inherently replicate better in tumor cells than they do in their normal host cells. This is due to intrinsic tumor cell properties such as highly active proliferation, enhanced metabolic rate, dysregulation of multiple signaling pathways, and a general immunosuppressive environment often with defects in innate immunity. All of these properties that facilitate tumor growth likewise favor viral reproduction. Additionally, some tumor cells have higher than normal amounts of viral receptors and this promotes preferential viral attachment to the tumor cells. Because of these combined factors, even certain unmodified viruses often reproduce more effectively in tumor tissue than they do in normal tissue. With this enhanced reproduction in the tumor cells, even relatively benign viruses such as Reovirus (an RNA virus) may cause extensive damage to tumors while causing only minimal disease in the host. Whether or not such wild-type viruses can be harnessed to treat certain cancers under carefully defined conditions remains to be proven.

In addition to advances in targeting, 20 years of clinical trials provide critical insight into the anti-tumor activity of viruses. The initial concept was simply that the infecting viruses directly killed the tumor cells. After viral administration, the first round of infected tumor cells was lysed and killed to release new virions. These new virions infected and killed additional tumor cells and the infection/killing cycle continued until all the tumor cells were gone. If the virions made it into the bloodstream they might even circulate and encounter metastatic tumor cells and kill those distal cells as well as the cells in the primary tumor. While this direct viral destruction of infected tumor cells remains an important contributor to the overall anti-tumor activity of oncolytic viruses, we now appreciate that immunological activation is an equal if not more important contributor.

A critical feature of tumors is their ability to evade immune surveillance and hide from the host defenses that normally would kill transformed cells and protect us from cancer. Solid tumors accomplish this by creating an immunosuppressed local microenvironment. For example, tumor cells often have defects in innate immunity and associated interferon response that shield these cells from the anti-tumor activity of these first-stage host defenses. Some tumor cells also have impaired cell surface antigen presentation. With a reduced display of tumor-specific antigens on their surface, these tumor cells look more like normal cells and are difficult for circulating T cells and macrophages to recognize and attack. Furthermore, tumor cells typically secrete soluble factors that directly impair the functionality of various immune system cells, including T and B cells, and can prevent these cells from being recruited to destroy the tumor. This combination of mechanisms can make tumors invisible to the immune system and allows the tumor to survive and grow unimpeded by host defenses. You can imagine a tumor as an invading enemy army shrouded in fog. Reconnaissance pilots may find it hard to see the enemy army and what little glimpses do occur may be insufficient for the pilots to distinguish the foe from their own army. Similarly, the immune system fails to recognize tumors as foreign and declines to attack them.

As our knowledge and understanding increased, it became clear that direct viral killing of cancer cells was not the only mechanism at work. Along with viral-induced cell lysis, an important facet of the tumor-killing activity of oncolytic viruses is their ability to induce a host immune response in the tumor environment and "reexpose" the tumors to immune attack. Our immune systems are quite effective at recognizing viruses, so viral infection of and replication in tumor cells can trigger a localized immune response even when the tumor cell itself does not elicit such a response. The local immune response can produce inflammation and tumor cell killing even in bystander tumor cells that haven't been infected, thus spreading the killing activity beyond the infected cells. This effect can be enhanced by genetically modifying oncolytic viruses to remove their anti-immune defense genes, resulting in viruses that cannot suppress the immune response. Furthermore, when tumor cells are lysed they release all the internal viral and tumor-specific proteins. This collection of foreign proteins and cell debris is now visible and available to antigen-presenting cells (APCs). The APCs process these antigens and use them to activate and recruit T cells that enter the cancer and begin destroying tumor cells. The foreign antigens also evoke a B cell response leading to the production of antibodies that can bind tumor cells and trigger antibody-dependent killing mechanisms. Together, these local viral-induced immune responses greatly expand the anti-tumor activity of oncolytic viruses as these effects overcome the immunosuppressive microenvironment to reveal the tumor. As the immunosuppressive "fog" lifts, the immune defenses can readily marshall to locate and destroy the oncogenic cells. This oncolytic virus-induced immune stimulation can even generate systemic effects that help eliminate metastatic cells that have spread far from the source tumor. Once activated against tumor antigens, B and T cells circulate throughout the body and will attack tumor cells wherever they are found. Therefore, even if the oncolytic virus doesn't spread effectively from the initial tumor, the immune response it generates should provide systemic surveillance against rogue cancer cells.

Ideally, the direct viral killing of tumor cells coupled with virus-induced immune killing would completely eradicate tumors and their metastatic progeny. In practice, most oncolytic viral trials to date have only demonstrated partial success with most patients showing modest or no effect, while significant benefits are only seen in a minority of individuals. These initial results clearly indicate that significant improvements must be made in this technology before it becomes a routine component of our cancer treatment arsenal. One promising approach for improving the anti-cancer activity of oncolytic viruses is to use them in conjunction with immune-stimulating agents. In some cases, the purified agent is co-administered along with the virus and in other cases, the gene for the agent is cloned into the viral genome. Just as deleting immunosuppressive genes from viral genomes improves their anti-cancer activity, adding genes that encode immune-activating proteins should make these viruses even more effective at destroying tumors. Using either virally delivered or purified immune agents, the goal of the combinatorial approach is to boost the therapeutic effect of oncolytic viruses so that substantial tumor reduction or complete remission occurs. Numerous immune stimulators are being tested but most are still in the early stages of clinical evaluation and the results are still preliminary. Since conclusions are limited we'll only consider one illustrative example of an oncolytic virus coupled with an immune enhancer, Talimogene laherparepvec (T-vec). T-vec is a herpes simplex virus carrying a transgene that encodes a cytokine called granulocyte-macrophage colony-stimulating factor (GM-CSF). Licensed in the U.S. in 2015, T-vec is a treatment for metastatic melanoma. As its name implies, GM-CSF is a white blood cell growth factor that promotes stem cells to produce granulocytes and macrophages, thus expanding the available number of these immune cells. GM-CSF alone was a standard treatment for metastatic melanoma, but T-vec showed superior results with a significantly longer survival time than just the drug by itself. This was the first oncolytic virotherapy approved in the U.S. and just the third approved worldwide.

T-vec is an important advance, and hopefully is just the start of a new era for cancer treatment. It is already clear that oncolytic viruses have an excellent safety record in clinical trials with far fewer side effects than chemotherapy. If the specificity and effectiveness of oncolytic viruses can be maximized then their therapeutic benefits will be enormous. Although we are still a long way from the broad implementation of oncolytic virotherapy, we are likely on the threshold of many exciting new clinical applications using viruses to treat common cancers. Coupling the innate killing properties of viruses with clever application of immune modulation promises to be a powerful anti-cancer strategy. Many innovative treatment regimens utilizing oncolytic viruses are in trials and their entrance into clinical practice is just on the horizon. Viruses are never going away, but human curiosity and scientific ingenuity are limiting their harmful effects and repurposing them to our advantage. It would be hugely satisfying if our great adversaries, after having caused millions of deaths, could be transformed into an important ally in the war on cancer.

Additional Reading

1. Gene Therapy Leaves a Vicious Cycle. Reena Goswami, Gayatri Subramanian, Liliya Silayeva, Isabelle Newkirk, Deborah Doctor, Karan Chawla, Saurabh Chattopadhyay, Dhyan Chandra, Nageswararao Chilukuri, and Venkaiah Betapudi. Frontiers in Oncology 9:297. 2019.
2. History of Gene Therapy. Thomas Wirth, Nigel Parker, and Seppo Ylä-Herttuala. Gene 525:162–169. 2013.

Definitions

Acute myeloid leukemia (AML) – A type of leukemia characterized by excessive immature white blood cells in the blood and bone marrow.

ADA-SCID – Severe combined immunodeficiency due to a defective gene for adenosine deaminase.

Adeno-associated virus (AAV) – A single-stranded, linear DNA virus of the parvovirus family that is unable to reproduce without a co-infecting helper virus such as adenovirus or herpesvirus.

Adenoids – Patches of lymphatic tissue between the back of the nose and the throat.

Adenosine deaminase (ADA) – A cellular enzyme that converts deoxyadenosine to deoxyinosine. Defects in the enzyme cause SCID.

Ammonia – A bodily waste product that results from the digestion of protein.

Anemia – A condition characterized by a lack of red blood cells or a lack of functional red blood cells.

Atherosclerosis – A disease involving the accumulation of fats and cholesterols on the inner walls of the arteries.

Cas9 – The DNA-cleaving enzyme of the CRISPR/Cas9 complex.

CFTR – The gene that is defective in cystic fibrosis. This gene encodes the cystic fibrosis transmembrane conductance regulator protein.

CRISPR/Cas9 – A DNA editing system composed of two RNAs (crRNA and tracrRNA) and the DNA-cleaving Cas9 protein.

(continued)

(continued)

crRNA – One of the two bacterial RNAs that bind with the Cas9 protein to generate the CRISPR complex with DNA-cleaving activity.

Deoxyadenosine – A toxic biomolecule that accumulates in the absence of adenosine deaminase.

Deoxyinosine – A nontoxic derivative of deoxyadenosine produced by the enzymatic activity of adenosine deaminase.

Diploid genome – A genome that has two copies of each chromosome and thus 2 copies of each gene, one maternal and one paternal.

Dominant-negative genetic disease – An illness produced by having one copy of a defective gene. In these illnesses, the protein product of the defective gene overcomes the function of the protein from the wild-type gene.

E1A – An adenoviral protein that binds the host pRB protein.

E1B-55K protein – An adenoviral protein that degrades cellular p53 protein.

E2F – A cellular transcription factor critical for activating genes in the S phase of the cell cycle.

Ex vivo – An event taking place outside of the living organism.

Gamma c – A protein that serves as a subunit of several different interleukin receptors.

Gammaretrovirus – A type of retrovirus with a propensity for integrating the proviral DNA adjacent to the transcriptional start sites of chromosomal genes.

Gene-editing – Any technique that can alter the nucleotide sequence of a gene.

Glybera – A commercial therapeutic using the AAV1 vector to deliver the lipoprotein lipase gene.

Granulocyte – A white blood cell possessing granules in its cytoplasm; includes neutrophils, basophils, and eosinophils.

Granulocyte-macrophage colony-stimulating factor (GM-CSF) – A protein secreted by several cell types that stimulates the proliferation and differentiation of granulocyte cells and macrophages.

Gutless adenovirus – Adenovirus vectors with all the viral genes removed.

Hematopoietic stem cell – The cell type in the bone marrow that gives rise to all the blood cells.

Hemoglobin – The protein that carries oxygen in our red blood cells.

Hemophilia – A bleeding disorder caused by a defective gene for the blood-clotting protein known as factor VIII.

Hepatic artery – A blood vessel that distributes blood to the liver, pancreas, and gallbladder as well as to the stomach and duodenal portion of the small intestine.

Homozygous – Having two identical copies of a gene.

Human telomerase reverse transcriptase (hTERT) – A cellular enzyme that is important for maintaining the integrity of the ends of chromosomes known as telomeres.

Hutchinson-Gilford syndrome – A genetic disorder characterized by rapid aging; also known as progeria.

IL2RG – The gene that encodes the common gamma chain protein that is a subunit for multiple interleukin receptors.

In situ – In the normal location.

In vivo – Taking place in a living organism.

(continued)

(continued)

Interleukin – A group of proteins, produced by leukocytes, that regulate immune responses.

ITR elements – Inverted terminal repeat sequences located at the ends of adeno-virus or adeno-associated virus genomes that are important for genome replication.

Lamin A – A cellular protein that provides a structural and regulatory function for the cell nucleus.

Leber's congenital amaurosis – A genetic disease that causes severe vision loss at birth.

Lipoprotein lipase (LPL) – A cellular enzyme that helps break down certain fats.

Luxturna – An AAV2-based therapy for the treatment of progressive blindness caused by Leber's congenital amaurosis.

Macular degeneration – A disease where the center of the retina deteriorates leading to distortion or loss of vision.

Metastatic – Relating to tumor cells that have separated from the primary tumor and moved to other parts of the body.

Microenvironment – The immediate surrounding cells and structures located in proximity to a particular cell or tissue, including tumors.

Moloney murine leukemia virus (MoMLV) – A type of retrovirus causing leukemia in mice.

Muscular dystrophies – A group of genetic diseases that cause progressive weak-ness and/or loss of muscles.

Off-target effects – Any unwanted or adverse effect that occurs when a thera-peutic acts on a molecule, cell, or tissue other than the intended target.

Oncolytic – Having the capacity to lyse or kill a tumor cell.

ONYX-015 – A mutant form of adenovirus that cannot produce the viral protein known as E1B-55K.

Ornithine transcarboxylase (OTC) – A liver enzyme that detoxifies the ammonia produced by the digestion of protein.

Packaging cells – Cells that have been engineered to support the replication and virion assembly of viral vectors.

Packaging sequence – A viral genomic sequence that is required for the packing of newly made viral genomes into newly made virions.

Parkinson's disease – A progressive neurological disorder that impairs movement and causes tremors.

pRB – A cellular protein that regulates the activity of the cellular E2F protein.

Progeria – A genetic disorder characterized by rapid aging; also known as Hutchinson-Gilford syndrome.

Recessive genetic disease – An illness that requires defects in both copies of a gene for the disease to manifest.

S phase – The stage of the cell cycle where chromosomal DNA replication occurs.

SCID X1 – A type of severe combined immunodeficiency disease caused by a gene defect on the X chromosome.

Scleroderma – hardening of the skin.

(continued)

(continued)

Self-inactivating vectors (SIN) – Retroviral vectors that lack a functional promoter and enhancer in their LTR regions.

Serotypes – Different strains of an organism that are distinguishable by serological methods.

Severe combined immunodeficiency (SCID) – Any genetic defect that causes a life-threatening dysfunction of the immune system.

sgRNA – A laboratory-produced RNA that combines the crRNA and tracrRNA into one molecule.

Sickle cell anemia – A genetic disease characterized by a defect in the hemoglobin gene.

Spinal muscular atrophy (SMA) – A genetic disease characterized by the loss of motor neurons due to a defect in survival motor neuron 1 protein.

Stem cells – Specialized cells in the body that can differentiate into many types of cells and from which all other cell types are replenished.

Survival motor neuron 1 (SMN1) – A cellular protein required for the health and survival of motor neurons.

Talimogene laherparepvec (T-vec) – A herpes simplex virus carrying a transgene that encodes a cytokine called granulocyte-macrophage colony-stimulating factor (GM-CSF).

T-cell lymphoblastic leukemia – A leukemia characterized by cancerous immature T cells.

tracrRNA – One of the two bacterial RNAs that bind with the Cas9 protein to generate the CRISPR complex with DNA-cleaving activity.

Transgene – A foreign gene that is introduced into another organism. Transgenes are often carried by viral vectors for delivery to the target organism.

Virotherapy – Any use of a virus in the treatment of disease.

Wild-type – A gene or protein that has normal function.

X chromosome – The sex chromosome that is found in both females (XX) and males (XY).

X-linked myotubular myopathy (XLMTM) – A genetic disease causing muscle weakness due to mutations in the gene for the protein myotubularin.

Zolgensma – An AAV9 vector used to deliver a transgene called survival motor neuron one (SMN1) that is used to treat spinal muscular atrophy (SMA).

Ψ – A sequence in a viral genome that is necessary for incorporating newly made viral genomes into virions.

Abbreviations

AAV – adeno-associated virus
ADA – adenosine deaminase
ADA-SCID – adenosine deaminase-severe combined immunodeficiency disease
AdV – adenovirus
AML – acute myeloid leukemia
CFTR – cystic fibrosis transmembrane conductance regulator
CRISPR/Cas9 – clustered regularly interspaced short palindromic repeats/CRISPR associated protein 9
crRNA – CRISPR RNA
GM-CSF – granulocyte-macrophage colony-stimulating factor
hTERT – human telomerase reverse transcriptase
IL2RG – interleukin 2 Receptor Subunit Gamma
IRT – inverted terminal repeat
LPL – lipoprotein lipase
MoMLV – Moloney murine leukemia virus
OTC – ornithine transcarboxylase
pRB – retinoblastoma protein
RBC – red blood cell
SCID – severe combined immunodeficiency
sgRNA – single guide RNA
SIN – self-inactivating
SMA – spinal muscular atrophy
SMN1 – survival motor neuron 1
tracrRNA – trans-activating CRISPR RNA
T-vec – Talimogene laherparepvec
XLMTM – X-linked myotubular myopathy
Ψ – packaging sequence

Epilogue

In the last 150 years, we've progressed from profound ignorance to considerable knowledge about our viral world. Thousands of viruses have been identified and the genomes of many of these known viruses have been sequenced to reveal their genetic secrets. Examining these sequences reveals how viral genomes are organized, what genes they have, and how their genomes change with mutation and genetic exchanges. In addition to genomes, we've probed virion morphology and examined the structures and functions of numerous viral proteins. Comparing proteins and genomes from distantly related viruses divulges evolutionary relationships and even gives us glimpses into their ancient origins. For decades we've studied infected cells, plants, and animals to define the viral life cycle and the impact of infection on cell biology. We now have a fundamental appreciation for how viruses subvert our cells to favor viral reproduction, how these viral-induced changes adversely affect cell function, and how this cellular damage leads to systemic disease. Importantly, these viral studies have informed us about the workings of our immune system and the multitude of host defenses that work ceaselessly to keep us healthy and alive. Intrinsic, innate, and acquired immunity collectively battle viral infections and provide both immediate and long-lasting protection against these intimate invaders. We've also learned that viruses can be oncogenic agents that contribute significantly to the human cancer burden. Using oncogenic viruses as tools, we've discovered elegant and diverse growth signaling cascades and identified critical anti-oncogenes, such as the p53 protein, that protect us from many oncogenic events. In particular, the retroviruses and the small DNA viruses were instrumental in elucidating the basic molecular mechanisms for cellular transformation and the resultant cancers. Yet for all our accumulated knowledge and wisdom about how viruses work, we still stand on the precipice of the viral unknown. Even for the most well-studied viruses, our understanding is incomplete with more questions than answers. We slowly chip away at the knowledge void with every experiment and each scientific study conducted, but the void is huge considering that we've only encountered a small fraction of the millions of different viruses that co-inhabit our planet. The enormity of the undiscovered viral world presages that there is still much to learn and that there will inevitably be both surprises and dangers as humans and unknown viruses intersect.

The SARS coronavirus 2 (SARS-CoV-2) pandemic that emerged in 2020 is just the latest example of a dangerous viral surprise. This novel zoonotic virus crossed over into human populations in late 2019 and turned out to be an insidious pathogen with some unexpected properties. The scientific and medical communities had a general knowledge of common coronaviruses and the original SARS virus, but SARS-CoV-2 defied expectations. With complete ignorance about this new virus, initial responses to the viral spread were mostly guesswork and were often wrong. Like the original SARS, SARS-CoV-2 spreads via the respiratory tract, but unlike SARS, SARS-CoV-2 transmits efficiently from human to human and causes mostly asymptomatic infections, a fact that went unnoticed for too long. National and international shutdowns, universal mask-wearing, social distancing, and other containment practices simply weren't implemented rapidly enough to curtail the viral spread. Consequently, this propensity for symptomless infections allowed the virus to spread quickly, continuously, and silently as it bypassed borders and exploded from a local epidemic in Wuhan, China to a global pandemic in just a few months. Sadly, in addition to asymptomatic or mild infections, the virus is also capable of causing severe morbidity and even death in a significant subset of infected individuals. With a sudden surge in cases and no specific knowledge about best care practices, the first waves of the pandemic saw a high death rate in the elderly and people with comorbidities. Over the next year as medical knowledge slowly accumulated the death to case ratio gradually improved although not before nearly three million people died worldwide.

While our inexperience with this new virus caused a global disaster, our scientific knowledge eventually saved us. Decades of experience with viruses, molecular biology, and vaccines were marshaled for the quickest vaccine development ever. From previous coronavirus research, we knew the likely vaccine target, the viral spike protein that mediates attachment to the host cell receptor. If a vaccine could evoke effective neutralizing antibodies against the spike protein this should prevent viral infection and disease. Vaccine science has several platforms that potentially are capable of delivering the spike protein, including live attenuated SARS-CoV-2, inactivated SARS-CoV-2, vector viruses, purified protein, DNA vaccines, and RNA vaccines. Dozens of labs in academia and the biotech industry explored these options leading to multiple, highly protective vaccines available to the public by late 2020 and early 2021. The development, testing, and deployment of these vaccines in under a year was a remarkable testament to the scientific acumen and manufacturing infrastructure of many countries. Mass vaccination of the world's population should eventually curtail the spread of SARS-CoV-2 by producing herd immunity. Once we reach herd immunity the pandemic will wane although SARS-CoV-2 will likely become an endemic virus that is always circulating at a low level. While the world will survive this COVID pandemic, SARS-CoV-2 is a grim reminder that we share our planet with a vast legion of diverse and constantly changing viruses, many of which are hostile to humans. Our best defense is to continually seek more knowledge about our viral co-inhabitants of Earth as we try to truly become the masters of our planet.

Lastly, it's important not to categorize viruses solely as a curse to humans just because they've caused countless illnesses and deaths. In addition to their disease-causing properties, viruses have made positive contributions to humanity as well. Viral invasions of our genome, primarily by retroviruses, provided genetic diversity that was a critical driver of our evolution. By seeding our genomes with new genetic information, viruses provided the raw materials for genetic experimentation that lead to new genes and new transcriptional regulatory pathways. From promoting placental development to enhancing our neuroanatomy, viral invasions shaped the very essence of our humanness. Furthermore, by exerting selective pressure to survive ceaseless viral attacks, viruses impacted our biology by molding our immunity and honing it into the intricate and robust system that effectively protects us from continuous assaults by the microbial world. In recent times we've even begun to harness viruses for our own benefits. Not only are viruses exceptional scientific tools for probing the molecular biology and biochemistry of our cells, but their use in clinical applications is burgeoning. Initially, medical practice primarily used viruses in vaccines, either as attenuated or inactivated whole viruses. More recently, we are developing bacteriophages to fight antibiotic-resistant bacterial infections and viral vectors to kill cancer cells. Retargeting the intrinsic killing properties of viruses to destroy rogue bacteria and cancer cells is an elegant and powerful approach that should someday be a useful complement to traditional medicinal therapies. Similarly, repurposing the ability of viruses to invade our cells into a delivery mechanism for exogenous genes or gene-editing machinery has enormous potential to treat previously intractable genetic diseases. With these novel approaches, viruses are poised to become important weapons in our clinical arsenal. While we must always be vigilant for the next lethal virus, with imagination, creativity, and knowledge science can and will exploit viruses increasingly for the benefit of humankind.

Index

Printed in the United States
by Baker & Taylor Publisher Services